ORGANIC POLYMER CHEMISTRY

ORGANIC
POLYMER CHEMISTRY

AN INTRODUCTION TO THE ORGANIC CHEMISTRY OF ADHESIVES, FIBRES, PAINTS, PLASTICS AND RUBBERS

Second edition

K. J. SAUNDERS

Department of Applied Chemical and Biological Sciences
Ryerson Polytechnical Institute, Toronto

LONDON   NEW YORK
CHAPMAN AND HALL

First published in 1973 by
Chapman and Hall Ltd
11 New Fetter Lane, London EC4P 4EE
Second edition 1988
Published in the USA by
Chapman and Hall
29 West 35th Street, New York, NY 10001

Printed in Great Britain by
J. W. Arrowsmith Ltd, Bristol

ISBN 0 412 27570 8

British Library Cataloguing in Publication Data

Saunders, K. J. (Keith John)
    Organic polymer chemistry: an introduction
    to the organic chemistry of adhesives, fibres,
    paints, plastics and rubbers.—2nd ed.
    1. Polymers and polymerization
    I. Title
    547.7   QD381

    ISBN 0–412–27570–8

Library of Congress Cataloging in Publication Data

Saunders, K. J. (Keith J.), 1931–
    Organic polymer chemistry.

    Includes index.
    1. Polymers and polymerization.   2. Plastics.
I. Title.
TP1140.S32   1988   668.9   87–31974
ISBN 0–412–27570–8

*This book is dedicated with gratitude to
my parents, Leonard and Marjorie Saunders,
for their sacrifices in earlier years and to
my wife, Jeannette, for her steadfast encouragement
in recent times.*

CONTENTS

PREFACE

This book deals with the organic chemistry of polymers which find technological use as adhesives, fibres, paints, plastics and rubbers. For the most part, only polymers which are of commercial significance are considered and the primary aim of the book is to relate theoretical aspects to industrial practice. The book is mainly intended for use by students in technical institutions and universities who are specializing in polymer science and by graduates who require an introduction to this field. There are available several books dealing with the physical chemistry of polymers but the organic chemistry of polymers has not received so much attention. In recognition of this situation and because the two aspects of polymer chemistry are often taught separately, this book deals specifically with organic chemistry and topics of physical chemistry have been omitted. Also, in this way the book has been kept to a reasonable size. This is not to say that integration of the two areas of polymer science is undesirable; on the contrary, it is important that the inter-relationship should be appreciated.

I was gratified by the favourable comments prompted by the first edition of the book and I have therefore retained the same organization in this second edition. Nevertheless, the book has been extensively revised to reflect the developments which have taken place. The most noticeable features of the period since the publication of the first edition have been the continued dominance of the same bulk commodity polymers and the appearance of several new speciality engineering thermoplastics. I have aimed to be comprehensive in scope and so both the well-established and the newer polymers are dealt with in this edition.

K.J.S.
Department of Applied Chemical and Biological Sciences,
Ryerson Polytechnical Institute, Toronto, Ontario, Canada.

1

BASIC CONCEPTS

1.1 DEFINITIONS

A *polymer* may be defined as a large molecule comprised of repeating structural units joined by covalent bonds. (The word is derived from the Greek: poly – many, meros – part.) In this context, a large molecule is commonly arbitrarily regarded either as one having a molecular weight of at least 1000 or as one containing 100 structural units or more. By a structural unit is meant a relatively simple group of atoms joined by covalent bonds in a specific spatial arrangement. Since covalent bonds also connect the structural units to one another, polymers are distinguished from those solids and liquids wherein repeating units (ions, atoms or molecules) are held together by ionic bonds, metallic bonds, hydrogen bonds, dipole interactions or dispersion forces.

The term *macromolecule* simply means a large molecule (Greek: macros – large) and is often used synonymously with 'polymer'. Strictly speaking, the terms are not equivalent since macromolecules, in principle, need not be composed of repeating structural units though, in practice, they generally are.

It may be noted that 'polymer' is often also used to refer to the massive state. Then the term refers to a material whose molecules are polymers, i.e. a polymeric material. Likewise, the term *resin* is sometimes used to refer to any material whose molecules are polymers. Originally this term was restricted to natural secretions, usually from coniferous trees, used mainly in surface coatings; later, similar synthetic substances were included. Now the term is generally used to indicate a precursor of a cross-linked polymeric material, e.g. epoxy resin and novolak resin. (See later.)

1.2 SCOPE

A great variety of polymeric materials of many different types is to be found throughout countless technological applications. For the purposes of this book it is convenient to divide these materials according to whether they are inorganic or organic and whether they are naturally occurring or synthetic. Using this classification, the diverse nature and widespread application of polymeric materials is illustrated by Fig. 1.1. This book is concerned solely with technologically useful organic polymeric materials. These materials are commonly classified as adhesives, fibres, paints, plastics and rubbers according to their use. Although these are all polymeric materials, they clearly possess a great diversity of properties about which few generalizations can be made. It is significant, however, that no low molecular weight organic compounds are useful in the above applications. The physical properties of an individual polymeric material are largely determined by molecular weight, strength of intermolecular forces, regularity of polymer structure and flexibility of the polymer molecule.

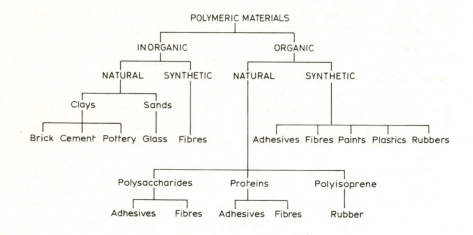

Fig. 1.1   Some applications of polymeric materials.

1.3   RISE OF THE CONCEPT OF POLYMERS

Nowadays the concept of polymers (or, simply, big molecules) is easy to accept but this has not always been the case. The rise of the concept is of interest and a brief historical review is given below.

By the 1850s the existence of atoms and molecules was accepted but mainly in respect of simple inorganic compounds. The application of these ideas to more complex organic materials was not understood so clearly. At least, by this time the notion that organic compounds all contained a mysterious 'vital force' derived from living things had been finally abandoned as more and more organic compounds were synthesized in the laboratory.

In 1858 Kekulé suggested that organic molecules were somewhat larger than the simple inorganic molecules and consisted of atoms linked in chains by bonds. This led to the realization that the order in which atoms are arranged in the molecule is significant, i.e. the meaning of 'structure' was appreciated. These ideas, aided by improving methods of elemental analysis, resulted in the elucidation of the structure of many simple organic compounds such as acetic acid and alcohol. However, virtually nothing was deduced about the structure of more complex organic materials such as rubber, cellulose and silk. All that was clear was that these materials had elemental analyses which were quite similar to those of the simple compounds whose structures were known.

The first significant information came in 1861 when Graham found that solutions of such natural materials as albumin, gelatin and glue diffused through a parchment membrane at a very slow rate. Materials of this kind were called *colloids* (Greek: kolla – glue). In contrast, solutions of materials like sugar and salts diffused readily; these substances were called *crystalloids* since they were generally crystalline. The reason for this difference was not clear, but it was generally supposed that the colloid solute particles were rather large and therefore their passage through the membrane was hindered. There were a few tentative suggestions that the colloids had high molecular weights so that a solute particle was large simply because it comprised one large molecule. However, this view was not at all acceptable to most scientists of the day. At this time the current practices of organic chemistry demanded the preparation of crystalline compounds of great purity with exact elemental analyses and sharp melting points. It was generally felt that if the colloids were 'cleaned up' they would crystallize and reveal themselves as 'normal' low molecular weight compounds. This view was apparently reinforced by the fact that many inorganic materials of low molecular weight can be prepared so that they behave as colloids, e.g. colloidal arsenious sulphide, gold and silver chloride. It was held, quite rightly, that in these cases the colloidal particles are aggregates of smaller particles held together by secondary valency forces of some kind. (Incidentally, the physically associated groups were frequently called 'polymers' in the literature.) This view was very much in keeping with the great emphasis which was being placed on van der Waals forces in the 1890s and early 1900s. By analogy, the organic colloids were assumed to be molecular aggregates or micelles, a concept still to be found in the literature of the 1940s.

The first worker to take a clearly opposite view was Staudinger in Germany in a paper of 1920. He maintained that the colloidal properties of organic materials are due simply to the large size of the individual molecules and that such macromolecules contain only primary valency bonds. Staudinger's initial evidence was mainly negative. Firstly, he demonstrated that the organic materials retain their colloidal properties in all solvents in which they dissolve. This is in contrast to the inorganic association colloids which often lose their colloidal characteristics on change of solvent. Secondly, he showed that, contrary to then current expectations, chemical modification does not destroy the colloidal properties of the organic materials. At that time it was commonly held that natural rubber was a cyclic material composed of isoprene residues linked in rings of various sizes as, for example, in the following structure:

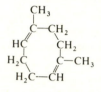

Such molecules were then supposed to aggregate by virtue of secondary valency forces arising from the presence of double bonds. (Incidentally, the cyclic structure was held to account for the fact that no end-groups could be detected.) However, Staudinger showed that the hydrogenation of natural rubber produces a saturated material which still exhibits colloidal properties. Thus he demonstrated that secondary forces originating from the unsaturation of natural rubber are unlikely to be responsible for colloidal behaviour. Staudinger proposed the long chain structures, which are accepted today, for several polymers. Additional support for the existence of macromolecules came with the development of methods of molecular weight determination. Until this time, only cryoscopic methods were available and these were inadequate for the very high molecular weights involved; also, it was commonly held that the laws which hold for ordinary solutions were not applicable to colloidal solutions.

Thus by about 1930 the concept of polymers was firmly established even if not universally accepted. The macromolecular viewpoint was finally secured largely by the work of Carothers in the USA. This work was begun in 1929 and had as its objectives, clearly stated at the outset, the preparation of polymers of definite structure through the use of established reactions of organic chemistry and the elucidation of the relationship between structure and properties of polymers. These researches were brilliantly successful and finally dispelled the mysticism surrounding this field of chemistry.

One outstanding result of Carothers' work was the commercial development of nylon. Nylon stockings came on the market in 1940, when polymers, in terms of popular acceptance, might be said to have arrived. The theoretical, practical and economic foundations had been laid and since this date progress has been phenomenal.

1.4　GENERAL METHODS OF PREPARATION OF POLYMERS

There are three general methods by which polymers may be prepared from relatively simple starting materials (*monomers*). Each of these methods is briefly described in this section; more detailed considerations are to be found throughout later chapters.

1.4.1　Polymerization through functional groups

In this type of polymerization, reaction proceeds between pairs of functional groups associated with two different molecules. Provided all the reacting molecules have at least two reactive groups, a sequence of reactions occurs and a polymer is formed. It will be apparent that the structural units of the polymer will contain a group whose arrangement of atoms is not to be found in the starting material. A polymerization of this kind occurs when ω-aminoundecanoic acid is heated:

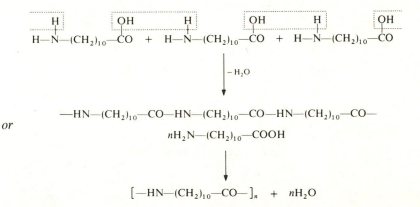

or

The resulting polymer (nylon 11) consists essentially of a long chain of repeating $-HN-(CH_2)_{10}-CO-$ groups (the structural units); in a typical commercial material n has a value of the order of 100. The polymer contains amide groups ($-CO-NH-$) (which are not present in the starting material) at regular intervals along the chain and is, therefore, a polyamide. It may be noted that in the representation of polymers it is common to leave unspecified the nature of the end-groups (as above). This practice is justified since the number of end-groups is very small compared to the number of repeating units.

In the above example of polymerization through functional groups, the single monomer (ω-aminoundecanoic acid) contains two types of functional group, namely amino and carboxyl. In practice it is more usual to use a mixture of two monomers, each having only one type of functional group; such monomers are generally more readily obtainable. This technique is illustrated by the formation of a polyamide (nylon 6,6) from hexamethylene-diamine and adipic acid:

$$n\text{H}_2\text{N}\text{---}(\text{CH}_2)_6\text{---NH}_2 \;+\; n\text{HOOC}\text{---}(\text{CH}_2)_4\text{---COOH}$$

$$\left[\text{---HN---}(\text{CH}_2)_6\text{---NH---OC---}(\text{CH}_2)_4\text{---CO---}\right]_n \;+\; 2n\text{H}_2\text{O}$$

Further examples of polymerizations of this kind are as follows:

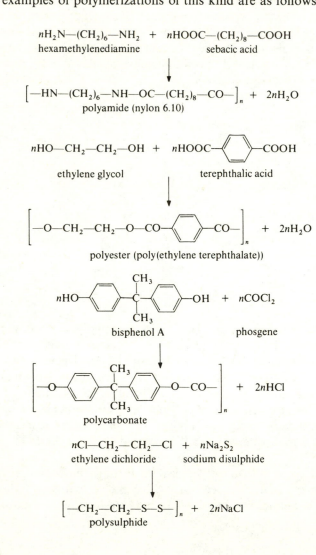

$$\underset{\text{hexamethylenediamine}}{n\text{H}_2\text{N}\text{---}(\text{CH}_2)_6\text{---NH}_2} \;+\; \underset{\text{sebacic acid}}{n\text{HOOC}\text{---}(\text{CH}_2)_8\text{---COOH}}$$

$$\underset{\text{polyamide (nylon 6.10)}}{\left[\text{---HN---}(\text{CH}_2)_6\text{---NH---OC---}(\text{CH}_2)_8\text{---CO---}\right]_n} \;+\; 2n\text{H}_2\text{O}$$

$$\underset{\text{ethylene glycol}}{n\text{HO---CH}_2\text{---CH}_2\text{---OH}} \;+\; \underset{\text{terephthalic acid}}{n\text{HOOC---}\langle\text{---}\rangle\text{---COOH}}$$

$$\underset{\text{polyester (poly(ethylene terephthalate))}}{\left[\text{---O---CH}_2\text{---CH}_2\text{---O---CO---}\langle\text{---}\rangle\text{---CO---}\right]_n} \;+\; 2n\text{H}_2\text{O}$$

bisphenol A   phosgene

polycarbonate

$$\underset{\text{ethylene dichloride}}{n\text{Cl---CH}_2\text{---CH}_2\text{---Cl}} \;+\; \underset{\text{sodium disulphide}}{n\text{Na}_2\text{S}_2}$$

$$\underset{\text{polysulphide}}{\left[\text{---CH}_2\text{---CH}_2\text{---S---S---}\right]_n} \;+\; 2n\text{NaCl}$$

It will be noticed that in all the above examples the production of polymer is accompanied by the formation of a secondary product; in each case there is elimination of some small molecule as a by-product, e.g. water or hydrogen chloride. However, polymerization through functional groups does not always result in such a by-product. For example, in the reaction between an isocyanate, such as 1,6-hexamethylene diisocyanate, and a glycol, such as 1,4-butanediol, polymer is the sole product:

$$n\text{OCN}-(\text{CH}_2)_6-\text{NCO} \quad + \quad n\text{HO}-(\text{CH}_2)_4-\text{OH} \longrightarrow$$

$$[-\text{OC}-\text{HN}-(\text{CH}_2)_6-\text{NH}-\text{CO}-\text{O}-(\text{CH}_2)_4-\text{O}-]_n$$

The polymer contains urethane groups along the chain and is, therefore, a polyurethane. The significance of the absence of by-products in reactions of this kind is referred to later in this chapter (section 1.4.5).

The chemistry involved in polymerizations through functional groups is essentially the chemistry of simple organic reactions wherein the corresponding functional groups are present in small molecules. These polymerizations nearly always proceed in a *stepwise* manner and the polymer chain is therefore built up relatively slowly by a sequence of discreet interactions between pairs of functional groups. Each interaction is chemically identical since each involves the same kind of functional group. It may also be noted that the reactivity of a functional group at the end of a polymer chain is similar to that of the corresponding functional group in a monomer molecule. Thus a functional group at the end of a polymer chain can react with a functional group which is either in a monomer molecule or at the end of another polymer chain. The growth of a polymer chain is therefore somewhat fitful. Since there is no inherent termination reaction, the molecular weight of the polymer continues to increase with time until, ideally, no more functional groups remain available for reaction. By far the greatest majority of polymerizations through functional groups proceed in a stepwise manner but a few have been found to involve a chain reaction (see section 1.4.5).

1.4.2 Polymerization through multiple bonds

This method of polymer preparation may be simply regarded as the joining together of unsaturated molecules through the multiple bonds. There is essentially no difference in the relative positions of the atoms in the unsaturated molecules and in the structural units of the polymer and there is no change in composition. Polymerizations of this kind may be divided into the various categories which follow.

(a) Vinyl polymerization

The most common unsaturated compounds which undergo polymerization through their multiple bonds contain carbon–carbon double bonds and are ethylene derivatives. The simplest monomer is ethylene itself, the polymerization of which may be written:

$$nH_2C=CH_2 \longrightarrow [-CH_2-CH_2-]_n$$

The resulting polymer, polyethylene, consists essentially of a long chain of repeating $-CH_2-CH_2-$ groups (the structural units); in a typical commercial material n has a value of the order of 1000. Further examples of polymerizations of ethylene derivatives are as follows:

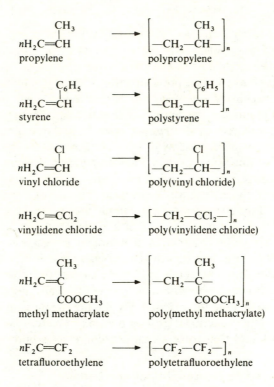

Polymerizations of the above type are often referred to as vinyl polymerizations although, strictly speaking, vinyl compounds are only those containing the $H_2C=CH-$ group. It is convenient, however, to consider all these polymerizations under one heading.

Vinyl polymerization, as illustrated by the above reactions, involves a three-part process, namely *initiation*, in which is formed an active species capable of starting polymerization of the otherwise unreactive vinyl compound; *propagation*, in which high molecular weight polymer is formed; and *termination*, in which deactivation occurs to produce the final stable polymer. The active species in vinyl polymerizations may be of three different types, namely free radicals, anions and cations and these possibilities give rise to three distinct methods of accomplishing polymerization.

Free radical polymerization
In free radical polymerization, initiation may be brought about by light or heat; most commonly, however, it is achieved by the addition of a material which, on heating, decomposes into free radicals (which may be defined as molecules containing an unpaired electron). Examples of frequently used initiators are benzoyl peroxide and azobisisobutyronitrile which give rise to free radicals as follows:

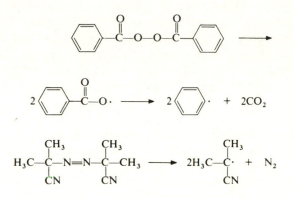

The initiation of polymerization is therefore a two-step sequence. The first step is the dissociation of the initiator, as illustrated above. This may be represented as:

$$I—I \longrightarrow 2I\cdot$$

The second step is the addition of the initiator fragment radical ($I\cdot$) to a vinyl monomer molecule ($H_2C=CHR$) to give an initiated monomer radical:

$$I\cdot \quad + \quad H_2C=CHR \longrightarrow I—CH_2—\dot{C}HR$$

This new radical then adds further monomer molecules in rapid succession to form a polymer chain. In this propagation the active centre remains, being continuously relocated at the end of the chain:

$$I—CH_2—\dot{C}HR \xrightarrow{H_2C=CHR} I—CH_2—CHR—CH_2—\dot{C}HR$$

$$\downarrow H_2C=CHR$$

$$I—CH_2—CHR—CH_2—CHR—CH_2—\dot{C}HR$$

$$\downarrow nH_2C=CHR$$

$$I\left[CH_2—CHR\right]_{n+2}—CH_2—\dot{C}HR$$

Propagation continues until the growing long chain radical becomes de-activated. Such termination is commonly by reaction with another long chain radical in one of two ways:

(i) Combination:

$$\sim CH_2-\dot{C}HR \quad + \quad RH\dot{C}-CH_2\sim \quad \longrightarrow \quad \sim CH_2-CHR-CHR-CH_2\sim$$

(ii) Disproportionation:

$$\sim CH_2-\dot{C}HR \quad + \quad RH\dot{C}-CH_2\sim \quad \longrightarrow \quad \sim CH_2-CH_2R \quad + \quad CHR=CH\sim$$

where $\sim CH_2-\dot{C}HR$ represents the last unit in a growing polymer chain.

During a polymerization reaction both of these mechanisms may operate together or one only may occur to the exclusion of the other. The actual mode of termination depends on the experimental conditions and the monomer involved. Termination of a growing polymer chain may also occur by *transfer*. In this case, however, deactivation of the chain radical results in the formation of a new free radical. Transfer reactions have the general form:

$$\sim CH_2-\dot{C}HR \quad + \quad AB \quad \longrightarrow \quad \sim CH_2-CHRA \quad + \quad B\cdot$$

where AB may be monomer, polymer, solvent or added modifier. Depending on its reactivity, the new free radical (B•) may or may not initiate the growth of another polymer chain. The transfer of hydrogen from monomer to growing polymer may be given as an example of termination by transfer:

$$\sim CH_2-\dot{C}HR \quad + \quad H_2C=CHR \quad \longrightarrow \quad \sim CH_2-CH_2R \quad + \quad H_2C=\dot{C}R$$

Anionic polymerization

In anionic polymerization of vinyl monomers the active centre is a carbanion. Substances which initiate this type of polymerization are of two kinds:

(i) *Ionic compounds* of the type X^+Y^- or *ionogenic compounds* of the type $X^{\delta+}Y^{\delta-}$ where the actual or potential anion Y^- is able to add to the carbon–carbon double bond to form a carbanion which can then propagate. Initiators of this kind include alkali metal alkyls, aryls, alkoxides and amides. The initiation step which occurs with potassium amide will serve to illustrate the mode of reaction of such initiators:

$$K^+H_2\underset{\cdot\cdot}{N}^- \quad + \quad H_2C=CHR \quad \longrightarrow \quad H_2N-CH_2-\bar{C}HR \; K^+$$

(ii) *Free metals* which are able to transfer an electron to the monomer with the consequent formation of an anion-radical. Alkali metals (M•) are the most common initiators of this type and may initiate polymerization by two different kinds of electron-transfer reactions:

(a) Direct transfer of an electron to the monomer to form the initiating species:

$$M\cdot \quad + \quad H_2C=CHR \quad \longrightarrow \quad H_2\dot{C}-\bar{C}HR \; M^+$$

In most solvents the resultant anion-radical rapidly dimerizes to give a dicarbanion which functions as the actual initiating species:

$$2H_2\dot{C}\!-\!\bar{\ddot{C}}HR\;M^+ \longrightarrow M^+R H\bar{\ddot{C}}\!-\!CH_2\!-\!CH_2\!-\!\bar{\ddot{C}}HR\;M^+$$

(b) Transfer of an electron to an intermediate compound (A) to form an ion radical which subsequently transfers the electron to the monomer to form the initiating species:

$$M\cdot \;+\; A \longrightarrow M^+A^{\bar{\cdot}}$$

$$M^+A^{\bar{\cdot}} \;+\; H_2C\!=\!CHR \longrightarrow H_2\dot{C}\!-\!\bar{\ddot{C}}HR\;M^+ \;+\; A$$

The resultant anion-radical generally dimerizes to give a dicarbanion which functions as the actual initiating species (see above). An example of this type of initiator system is the solution obtained by adding sodium to naphthalene in an inert solvent such as tetrahydrofuran. The reactions involved are:

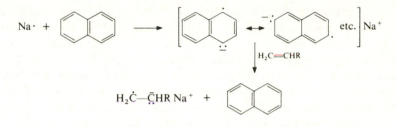

$$2\cdot CH_2\!-\!\bar{\ddot{C}}HR\;Na^+ \longrightarrow Na^+ R H\bar{\ddot{C}}\!-\!CH_2\!-\!CH_2\!-\!\bar{\ddot{C}}HR\;Na^+$$

It might be supposed that whatever the nature of the initiation of anionic polymerization, the propagation step could be regarded as analogous to the free radical propagation step, being represented as:

$$\sim CH_2\!-\!\bar{\ddot{C}}HR \;+\; H_2C\!=\!CHR \longrightarrow \sim CH_2\!-\!CHR\!-\!CH_2\!-\!\bar{\ddot{C}}HR$$

However, in general the presence of a counter (or gegen) ion in close proximity to the active centre has a profound effect. Thus, whilst in free radical polymerization the growth of the propagating chain is independent of the initiator used, the same cannot be said of anionic polymerization. In particular, the separation between the carbanion end-group and the counter ion is the primary factor determining the stereochemistry of the propagation reaction. Also in contrast to free radical polymerization, true termination reactions are absent from anionic polymerizations. Under vigorous reaction conditions the active centre may be destroyed by hydride elimination:

$$\sim CH_2\!-\!\bar{\ddot{C}}HR\;M^+ \longrightarrow \sim CH\!=\!CHR \;+\; MH$$

A similar reaction may also result in transfer of activity to monomer:

$$\sim CH_2\!-\!\bar{\ddot{C}}HR\;M^+ \;+\; H_2C\!=\!CHR \longrightarrow \sim CH\!=\!CHR \;+\; H_3C\!-\!\bar{\ddot{C}}HR\;M^+$$

The presence of a solvent may provide another mode of transfer. The mechanism of such a reaction clearly depends on the nature of the solvent; one possibility is proton transfer from the solvent (S–H):

$$\sim CH_2—\bar{C}HR\ M^+\ +\ S—H\ \longrightarrow\ \sim CH_2—CH_2R\ +\ S^-M^+$$

Impurities containing active hydrogen also participate in transfers of the above type. Carbon dioxide inhibits polymerization by forming a carboxylate anion which is not sufficiently reactive to initiate further polymerization:

$$\sim CH_2—\bar{C}HR\ M^+\ +\ \overset{\overset{O}{\parallel}}{C}{=}O\ \longrightarrow\ \sim CH_2—CHR—\overset{\overset{O}{\parallel}}{C}—O^-M^+$$

Provided inert solvents and pure reactants are used, most monomers under appropriate conditions will give rise to systems in which active carbanion end-groups are always present. The indefinite activity of the growing chains has led to the rather inappropriate term, 'living polymers' for these materials.

Anionic polymerizations of a rather different kind to those discussed above may be effected by the use of *co-ordinqtion catalysts*. This general type of catalysis was discovered by Ziegler who used it to polymerize ethylene (1953); Natta extended its use to the polymerization of other unsaturated monomers (1954–60). (See later.) Hence these catalysts are often referred to as 'Ziegler-Natta catalysts'. The catalysts are typically obtained by mixing an alkyl or aryl of a metal from Groups I–III of the Periodic Table with a compound (commonly a halide) of a transition metal of Groups IV–VIII. Several thousand permutations are possible and a great number of combinations are cited in the literature. Examples of organometallic compounds which have been used are phenyllithium, diethylberyllium, diethylzinc, and triethylaluminium. Examples of transition metal compounds which have been used are titanium tetrachloride, vanadium oxychloride (VOCl), molybdenum pentachloride and tungsten hexachloride. Probably the best-known example of this class of catalyst is that obtained from triethylaluminium and titanium tetrachloride. The structure of these catalysts and their mode of operation have been the subject of extensive investigation and much information has resulted. However, the matter is by no means settled and this information has not yet been integrated into a generally accepted theoretical framework. By way of illustration of proposals which have gained wide acceptance the system based on triethylaluminium and titanium tetrachloride will be discussed specifically. When triethylaluminium and titanium tetrachloride are mixed together in a hydrocarbon solvent (e.g. n-heptane) at about room temperature, hydrocarbon gases are evolved and there is precipitated a black-brown solid which is the active polymerization catalyst. The gases comprise mainly ethane with small amounts of ethylene and n-butane; some polyethylene is also formed. The initial reaction between the two components is generally regarded as involving a series of alkylations and reductive dealkylations of the following kind:

$$(C_2H_5)_3Al \; + \; TiCl_4 \longrightarrow (C_2H_5)_2AlCl \; + \; C_2H_5TiCl_3$$
$$C_2H_5TiCl_3 \longrightarrow \cdot C_2H_5 \; + \; TiCl_3$$
$$(C_2H_5)_3Al \; + \; TiCl_3 \longrightarrow (C_2H_5)_2AlCl \; + \; C_2H_5TiCl_2$$
$$C_2H_5TiCl_2 \longrightarrow \cdot C_2H_5 \; + \; TiCl_2$$
$$(C_2H_5)_3Al \; + \; C_2H_5TiCl_3 \longrightarrow (C_2H_5)_2AlCl \; + \; (C_2H_5)_2TiCl_2$$
$$(C_2H_5)_2TiCl_2 \longrightarrow \cdot C_2H_5 \; + \; C_2H_5TiCl_2$$

Several other alkylation reactions may be written. The formation of ethyl free radicals by decomposition of unstable alkyltitanium compounds accounts for the evolution of the hydrocarbon gases. Thus the final product is a complex mixture of organo-aluminium and -titanium compounds, lower titanium chlorides and some organic fragments. The actual composition of the product is dependent on the relative proportions of the starting materials and the time and temperature of reaction.

It is generally supposed that the active component of the catalyst mixture is a complex formed between titanium trichloride and triethylaluminium. This complex is envisaged as forming at the surfaces of titanium trichloride crystals; its formation may be regarded as the strong adsorption of the aluminium compound on to the crystal surfaces. By analogy with the bridged structures known to be present in dimeric aluminium alkyls, the complex is assumed to have the following structure:

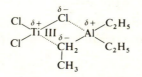

In such a complex the titanium has unfilled $3d$-orbitals to which may be co-ordinated π-electrons from the double bond of the vinyl monomer. Thus a possible representation of initiation and propagation is as follows:

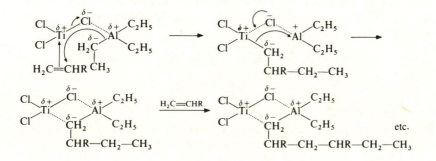

The essential feature of this mechanism is that monomers are inserted, one after the other, into a polarized titanium–carbon bond. The polymer therefore grows out of the active centre, rather as a hair grows from the root. It will be noticed that the propagating end of the polymer chain is negatively charged and therefore the reaction may be regarded as an anionic polymerization. Chain growth may be terminated by several types of transfer, e.g.

Internal hydride transfer:

$$Cat-CH_2-CHR-[CH_2-CHR]_n-CH_2-CH_3 \longrightarrow$$
$$Cat-H + H_2C=CR-[CH_2-CHR]_n-CH_2-CH_3$$

Transfer to monomer:

$$Cat-CH_2-CHR-[CH_2-CHR]_n-CH_2-CH_3 + H_2C=CHR \longrightarrow$$
$$Cat-CH_2-CH_2R + H_2C=CR-[CH_2-CHR]_n-CH_2-CH_3$$

True (kinetic) termination may be brought about by the addition of an active hydrogen compound such as an alcohol:

$$Cat-CH_2-CHR-[CH_2-CHR]_n-CH_2-CH_3 + R'-OH \longrightarrow$$
$$Cat-OR' + H_3C-CHR-[CH_2-CHR]_n-CH_2-CH_3$$

Cationic polymerization

In cationic (or more specifically, carbocationic) polymerization of vinyl monomers the active centre is a carbenium ion. Substances which initiate this type of polymerization are of three kinds:

(i) *Protic acids,* e.g. sulphuric, perchloric and trifluoroacetic acids. The initiation step consists of the transfer of a proton to the monomer:

$$HA + H_2C=CHR \longrightarrow H_3C-\overset{+}{C}HR\ A^-$$

Propagation then proceeds via the carbenium ion produced:

$$H_3C-\overset{+}{C}HR\ A^- + H_2C=CHR \longrightarrow H_3C-CHR-CH_2-\overset{+}{C}HR\ A^-\ \text{etc.}$$

Chain growth continues until either chain transfer or termination occurs. In chain transfer, there is no loss of active centres and the kinetic chain remains operational. Various types of chain transfer are possible, e.g.

Transfer to counter ion:

$$\sim CH_2-\overset{+}{C}HR\ A^- \longrightarrow \sim CH=CHR + HA$$

Transfer to monomer:

$$\sim CH_2-\overset{+}{C}HR\ A^- + H_2C=CHR \longrightarrow \sim CH=CHR + H_3C-\overset{+}{C}HR\ A^-$$

In true (kinetic) termination there is irreversible loss of propagating ability. This may be brought about in several ways, the most important of which is neutralization. In this process, the propagating carbenium ion and counter-ion interact to give an electrically neutral species. With trifluoracetic acid, for example, termination occurs by ester formation:

$$\sim CH_2 \overset{+}{-} CHR \ CF_3COO^- \longrightarrow \ \sim CH_2 - CHR - OCO - CF_3$$

(ii) *Carbenium ion salts*, e.g. trityl hexachloroantimonate $(Ph_3C^+SbCl_6^-)$. Here, initiation involves direct cationation of the monomer, e.g.

$$Ph_3C^+SbCl_6^- \ + \ H_2C=CHR \longrightarrow Ph_3C-CH_2 \overset{+}{-} CHR \ SbCl_6^-$$

Propagation and chain transfer then occur as discussed above. The nature of termination reactions is obscure. The use of carbenium ion salts as polymerization initiators appears to be restricted to aromatic and vinyl ether monomers. The salts do not initiate polymerization of aliphatic olefins.

(iii) *Metal halides* of the type which catalyses the Friedel-Crafts reaction, e.g. aluminium trichloride, boron trifluoride and ferric chloride. The pure, anhydrous metal halides do not initiate polymerization; they are active only in the presence of co-initiators. Co-initiators are commonly compounds containing active hydrogen, e.g. alcohols, protic acids and water. The co-initiator (QH) co-ordinates with the metal halide (MX_n) to form a complex protic acid which transfers a proton to the monomer:

$$MX_n \ + \ QH \longrightarrow [QMX_n]^- H^+$$

$$[QMX_n]^- H^+ \ + \ H_2C=CHR \longrightarrow H_3C \overset{+}{-} CHR \ [QMX_n]^-$$

Propagation then proceeds as follows:

$$H_3C \overset{+}{-} CHR \ [QMX_n]^- \ + \ H_2C=CHR \longrightarrow H_3C-CHR-CH_2 \overset{+}{-} CHR \ [QMX_n]^- \ etc.$$

Chain transfer takes place as discussed above. The nature of termination reactions has not been fully established. The advantage of the metal halide initiators over protic acid initiators is their ability to extend the lifetime of the kinetic chain and thus give polymers of higher molecular weight. For commercial purposes, the metal halides are by far the most important type of initiators used for cationic polymerization.

By way of conclusion to this short discussion of vinyl polymerization, it may be noted that not all vinyl monomers can be polymerized to high molecular weight polymers by all three of the general methods described, namely free radical, anionic and cationic polymerization. Table 1.1 indicates the general applicability of the three methods in homogeneous systems to some vinyl monomers. Often the effectiveness, or otherwise, of the methods can be related to the polarity of the monomer double bond. Electron-releasing substituents favour the formation of carbenium ions and render the monomer susceptible to cationic polymerization. Thus isobutene (with two electron-releasing methyl groups) and vinyl ethers (in which the resonance

Table 1.1 General applicability of polymerization methods in homogeneous systems to some vinyl monomers

Monomer	Structure	Polymerization method		
		Cationic	Free radical	Anionic
Isobutene	$H_2C=C(CH_3)_2$	Yes	No	No
Vinyl ethers	$H_2C=CHOR$	Yes	No	No
Ethylene	$H_2C=CH_2$	Yes	Yes	No
Vinyl esters	$H_2C=CHOCOR$	No	Yes	No
Vinyl halides	$H_2C=CHX$	No	Yes	No
Acrylic esters	$H_2C=CHCOOR$	No	Yes	Yes
Acrylonitrile	$H_2C=CHCN$	No	Yes	Yes
Vinylidene halides	$H_2C=CX_2$	No	Yes	Yes
1-Nitro-1-alkenes	$H_2C=CRNO_2$	No	No	Yes
Vinylidene cyanide	$H_2C=C(CN)_2$	No	No	Yes
Styrene	$H_2C=CHC_6H_5$	Yes	Yes	Yes

Monomers are arranged in approximate order of increasing susceptibility to anionic polymerization.

effect, due to delocalization of an unshared electron pair on the oxygen atom, outweighs the inductive effect exerted by the ether group to give an overall electron-releasing effect) undergo cationic polymerization exclusively. On the other hand, electron-withdrawing substituents favour the formation of carbanions and render the monomer susceptible to anionic polymerization. Thus vinylidene cyanide (with two electron-withdrawing cyano-groups) and 1-nitro-1-alkenes (in which the resonance and inductive effects exerted by the nitro-group reinforce one another and outweigh the inductive effect exerted by the alkyl group to give an overall electron-withdrawing effect) undergo anionic polymerization exclusively. Between these two extremes lie those monomers wherein the electron-releasing and -withdrawing effects are less pronounced; these monomers undergo free-radical polymerization and show a tendency to undergo either cationic or anionic polymerization, depending on the nature of the substituent. A few monomers such as styrene (in which the phenyl group can function as either an electron source or electron sink) can be polymerized by all three methods. It may be noted here that the fact that a monomer has a high tendency to polymerize does not necessarily mean it forms high molecular weight polymer. The molecular weight is also governed by factors such as transfer and termination reactions.

(b) Diene polymerization

Conjugated dienes comprise the second group of unsaturated compounds which undergo polymerization through their multiple bonds. The most

common dienes used for the preparation of commercially important polymers are butadiene, chloroprene and isoprene. These are normally represented by the following structures:

$$H_2C{=}CH{-}CH{=}CH_2$$
butadiene

$$\overset{\displaystyle CH_3}{\overset{|}{H_2C{=}C{-}CH{=}CH_2}}$$
isoprene

$$\overset{\displaystyle Cl}{\overset{|}{H_2C{=}C{-}CH{=}CH_2}}$$
chloroprene

Such monomers can give rise to polymers which contain various isomeric
structural units. Each of the above structures contains a 1,2- and a 3,4-double
bond and there is thus the possibility that either double bond may participate
independently in polymerization, giving rise to 1,2-units and 3,4-units respectively:

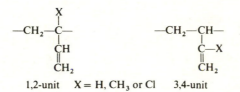

1,2-unit   X = H, CH₃ or Cl   3,4-unit

With symmetrical dienes such as butadiene, these two units become identical.
A further possibility is that both bonds are involved in polymerization
through conjugate reactions, giving rise to 1,4-units. A 1,4-unit may occur as
either the *cis*- or the *trans*-isomer:

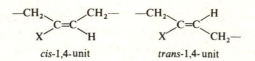

cis-1,4-unit   *trans*-1,4-unit

Generally speaking, the polymer obtained from a conjugated diene contains more than one of the above structural units. The relative frequency of
each type of unit is governed by the nature of the initiator and the experimental conditions as well as the structure of the diene. Each of the three
general methods of accomplishing vinyl polymerization, described above,
may be used for the polymerization of conjugated dienes.

In free radical polymerization, the various structural units may be envisaged as arising from the addition of one or other end of a monomer molecule
to one of the resonant forms of the growing polymeric radical, e.g.

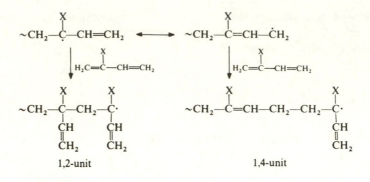

1,2-unit   1,4-unit

and

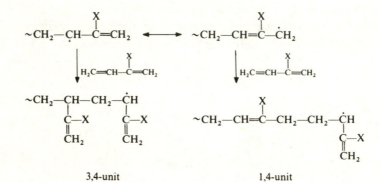

3,4-unit   1,4-unit

In practice, a preponderance of 1,4-units is found; this is attributable to the greater accessibility of the primary radical compared to that of the secondary radical. The *trans*-form of the growing radical is slightly more favoured energetically than the *cis*-form and hence 1,4-*trans*-units predominate in the polymer, particularly when polymerization is carried out at low temperatures.

With regard to anionic polymerization of conjugated dienes, alkali metal alkyls and free alkali metals are the most commonly used initiators. The resultant polymers generally have much higher contents of 3,4-units compared to the polymers prepared by free.radical polymerization. When an alkali metal alkyl (M^+R^-) is the initiator, the initiation reaction may be represented as:

$$M^+R^- \; + \; H_2C{=}CH{-}\underset{\underset{X}{|}}{C}{=}CH_2 \longrightarrow R{-}CH_2{-}\overset{..}{C}H{-}\underset{\underset{X}{|}}{C}{=}CH_2 \; M^+$$

$$R{-}CH_2{-}CH{=}\underset{\underset{X}{|}}{C}{-}\overset{..}{C}H_2 \; M^+$$

Addition of a monomer molecule to one of the resonant forms of the anion then results in the formation of the corresponding structural unit. When an alkali metal (M·) is the initiator, the initiation reaction may be represented as:

$$M\cdot \; + \; H_2C{=}\underset{X}{\overset{|}{C}}{-}CH{=}CH_2 \quad \longrightarrow \quad M^+H_2\bar{\underset{..}{C}}{-}\underset{X}{\overset{|}{C}}{=}CH{-}\dot{C}H_2$$

The anion-radical produced may either dimerize or add on another alkali metal atom; in either event a dicarbanion is formed and this functions as the actual initiating species:

$$2M^+H_2\bar{\underset{..}{C}}{-}\underset{X}{\overset{|}{C}}{=}CH{-}\dot{C}H_2 \quad \longrightarrow \quad M^+H_2\bar{\underset{..}{C}}{-}\underset{X}{\overset{|}{C}}{=}CH{-}CH_2{-}CH_2{-}CH{=}\underset{X}{\overset{|}{C}}{-}\bar{C}H_2 \; M^+$$

$$M^+H_2\bar{\underset{..}{C}}{-}\underset{X}{\overset{|}{C}}{=}CH{-}\dot{C}H_2 \; + \; M\cdot \quad \longrightarrow \quad M^+H_2\bar{\underset{..}{C}}{-}\underset{X}{\overset{|}{C}}{=}CH{-}\bar{C}H_2 \; M^+$$

Polymerization then proceeds in the manner of the alkyl-initiated reaction described above. It may be noted here that lithium (both as the free metal and as alkyls) stands in marked contrast to the other alkali metals in giving rise to polymers with high proportions of 1,4-units. This is attributable to the covalent character of the lithium–carbon bond, as is discussed later (Chapter 20).

The use of Ziegler-Natta catalysts in the polymerization of conjugated dienes has been widely investigated. It is characteristic of these catalysts that the resulting polymers often contain a very high proportion of one type of structural unit. By appropriate choice of catalyst, polydienes comprised almost exclusively of cis-1,4-, trans-1,4-, 1,2-, and 3,4-units have been obtained. The mechanisms of such reactions are, at present, somewhat obscure but presumably the diene molecule co-ordinates with the metal–carbon bond of the catalyst in a manner similar to that involving the lithium–carbon bond mentioned above.

In contrast to free radical and anionic polymerization, the cationic polymerization of conjugated dienes has received little study. Generally, rather low molecular weight polymers are produced and these have not attained commercial significance. A feature of the products obtained by the polymerization of conjugated dienes using metal halide catalysts is that they contain an appreciable proportion of cyclized structures. The linear portions of the polymers consist mainly of trans-1,4-units.

The profound effect of the initiator on the microstructure of the products obtained by polymerization of conjugated dienes is illustrated by Table 1.2.

Table 1.2 Effect of initiator on microstructure of polyisoprene [1]

Initiator	Structural units (%)			
	cis-1,4-	trans-1,4-	1,2-	3,4-
Free radical	22	65	6	7
Na	0	43	6	51
Na n-Bu	4	35	7	54
Li	94	0	0	6
Li n-Bu	93	0	0	7
$TiCl_4$-$AlEt_3$	96	0	0	4
α-$TiCl_3$-$AlEt_3$	0	91	0	9
V(acetylacetonate)$_3$-$AlEt_3$	–	–	–	90
$SnCl_4$*	0	89	6	5

* The figures given for this initiator refer to the composition of the linear portion of the polymer.

(c) Hetero-multiple bond polymerization

In both the above types of polymerization through multiple bonds, namely vinyl and diene polymerization, a carbon–carbon double bond is the active site. However, multiple bonds involving elements besides carbon may also be utilized in the preparation of polymers, which then contain hetero-atoms in the main chain. The most common unsaturated monomers used for the preparation of polymers of this kind are carbonyl compounds. Formaldehyde has been the most widely studied in this respect and its polymers are of commercial importance. The product of polymerization of formaldehyde may be regarded as a polyether:

$$n\mathrm{H_2C{=}O} \longrightarrow [\mathrm{-CH_2-O-}]_n$$

and is discussed in Chapter 9. As will be shown, polymerization may be accomplished by the use of both anionic and cationic initiators, but free radical initiators are ineffective.

A further example of this category of polymerization is the polymerization of monoisocyanates through the carbon–nitrogen double bond. This polymerization is carried out at low temperatures (to limit cyclization) with an anionic initiator, e.g. sodium cyanide:

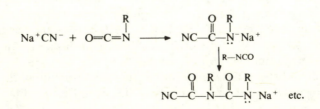

1.4.3   Polymerization through ring-opening

Many cyclic compounds undergo ring-opening reactions which lead to polymer formation. Usually, the structural units of polymers prepared in this way have the same composition as the monomer and there is essentially no change in the relative positions of the atoms. Examples of cyclic compounds which have been found to undergo polymerization are as follows: N-carboxy-α-aminoacid anhydrides (Leuchs' anhydrides), cyclic ethers, cyclic imines, cyclic sulphides, cyclopropanes, lactams and lactones. It will be apparent from this list that cyclic compounds which polymerize usually contain at least one hetero-atom. Normally, polymerizations through ring-opening are accomplished by use of either anionic or cationic initiators. Despite the large numbers of cyclic compounds which have been investigated as monomers, commercial importance has been achieved by only two types, namely cyclic ethers and lactams. The polymerization of each of these classes of monomer is illustrated by the following examples:

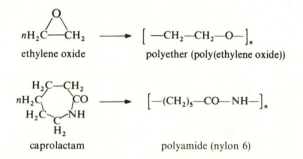

ethylene oxide   polyether (poly(ethylene oxide))

caprolactam   polyamide (nylon 6)

These types of polymerization are considered in detail in later discussions of polyethers (Chapter 9) and polyamides (Chapter 10).

1.4.4   Polymer modification

For completeness, this fourth general method of preparing polymers is included at this point although it cannot be regarded as a method of polymerization. This technique consists simply of subjecting an existing polymer to such chemical reaction that a different polymer is obtained. For example, poly(vinyl acetate) may be subjected to alcoholysis by treatment with methanol to give poly(vinyl alcohol):

In this case, the modification sequence can be extended by treating the new polymer with an aldehyde to give a poly(vinyl acetal):

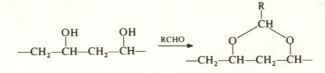

Similarly, cellulose may be modified through the hydroxyl groups to give esters and ethers, e.g. cellulose acetate, cellulose nitrate and methyl cellulose, which are all important commercial materials.

1.4.5   Classification of polymerization reactions

In 1929 Carothers made the proposition that all polymers could be divided into two types, namely *condensation polymers* and *addition polymers*. A condensation polymer was defined as a polymer in which the structural unit contains fewer atoms than the monomer (or monomers) from which the polymer is derived. An addition polymer was defined as a polymer in which the structural unit has the same molecular formula as the monomer. A limitation of this classification is that some polymers may be included in both categories. For example, polyethylene is usually prepared by the polymerization of ethylene:

$$n\mathrm{H_2C{=}CH_2} \longrightarrow [\mathrm{-CH_2-CH_2-}]_n$$

The polymer may thus be counted as an addition polymer. A polymer which may be considered to be the same (if any differences in molecular weight are disregarded) can be formed from decamethylene bromide and sodium in the Wurtz reaction:

$$n\mathrm{Br(CH_2)_{10}Br} \;+\; 2n\mathrm{Na} \longrightarrow [\mathrm{-CH_2-CH_2-}]_{5n} \;+\; 2n\mathrm{NaBr}$$

On the basis of this reaction, polyethylene may be counted as a condensation polymer.

The foregoing proposals by Carothers are also commonly used as the basis of a scheme of classification of polymerization reactions. In this scheme, the processes by which polymers are formed are divided into *condensation polymerizations* and *addition polymerizations*. Thus a condensation polymerization leads to a polymer in which the structural unit contains fewer atoms than the monomer whilst an addition polymerization results in a polymer having a structural unit with the same molecular formula as the monomer. A limitation of this classification is that a somewhat anomalous situation arises when polymerization through functional groups (see section 1.4.1) is considered. If there is elimination of a by-product then clearly the reaction counts as a condensation polymerization but if no by-product is evolved then the reaction must be regarded as an addition polymerization. Since whether or not the structural unit differs in composition from the monomer is of no particular significance it is not very desirable so to separate reactions which are similar in all other respects. Many authors therefore apply the term condensation polymerization to polymerization through functional groups irrespective of whether or not a by-product is formed. Some authors, however, prefer to restrict the term to only those cases where there is elimination of a small molecule: they then apply the term *rearrangement polymerization* (or, sometimes, *poly-addition*) to polymerization which proceeds through the interaction of functional groups without elimination of a small molecule. Addition polymerization has been defined above as leading to a polymer with a structural unit having the same molecular formula as the monomer. On this basis, all polymerizations through multiple bonds are to be classified as addition polymerizations. Generally polymerizations through ring-opening also come into this category, but the polymerization of N-carboxy-α-aminoacid anhydrides may be cited as an exception since a by-product is formed:

$$n\begin{array}{c} R_2C{\overset{CO}{\underset{|}{\diagup}}}O \\ HN{\diagdown}CO \end{array} \longrightarrow [-NH-CR_2-CO-]_n + nCO_2$$

An alternative method of classifying polymerization reactions is according to mechanism. In this scheme, processes are divided into *stepwise polymerizations* and *chain polymerizations*. In a stepwise polymerization the polymer is built up relatively slowly by a sequence of discrete reactions; the initiation, propagation and termination reactions are essentially similar, each having the same rate and mechanism. A feature of this type of polymerization is that a monomer molecule is capable of reacting with another monomer molecule or with a polymer molecule with equal facility. As a result there is rapid disappearance of monomer at an early stage in the reaction but a high reaction conversion is required for the attainment of high molecular weight. It has been shown previously that polymerizations through functional groups generally proceed in a stepwise manner. As an exception to this rule may be cited the polymerization of benzyl chlorides, e.g. 2,5-dimethylbenzyl chloride:

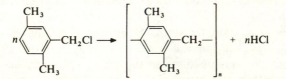

Although polymerization involves functional groups (phenyl and chloromethyl) and a by-product is eliminated, a stepwise reaction does not occur. In fact, polymerization proceeds through a chain reaction involving cationic species. (Catalytic quantities of metal halides initiate the polymerization.)

In contrast to stepwise polymerization, chain polymerization involves a chain reaction in the kinetic as well as the structural sense and a polymer molecule grows extremely rapidly once initiation has occurred. Typically, several thousand structural units are added in the space of approximately one second. The initiation, propagation and termination reactions are significantly different in rate and mechanism. In this case a monomer molecule cannot react with another monomer molecule but only with an active end-group on a polymer radical or ion. Thus high molecular weight polymer and monomer are present throughout the reaction. As has been shown previously, most polymerizations through multiple bonds and through ring-opening involve a chain reaction. By way of exception to this rule may be cited polymerization based on the Diels-Alder reaction; polymerization is through multiple bonds but proceeds in a stepwise manner, e.g.

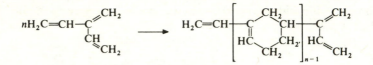

In this example 2-vinylbutadiene acts as both the diene and the dienophile to give a polycyclohexene. (For clarity, only one of the possible isomeric structural units is shown.)

1.5   POLYMERIZATION TECHNIQUES

In principle, a polymerization reaction may be carried out in the solid phase, the liquid phase or the gas phase. In practice, commercial scale polymerizations are almost always conducted in the liquid phase. It may be noted, however, that gas phase processes for polyolefins are of importance. (See Chapter 2.) Liquid phase polymerizations may be subdivided into four types according to the nature of the physical system employed. All of these variations find widespread application throughout the polymerization methods discussed in the previous section, as is illustrated by Table 1.3 which gives the techniques commonly used in the production of some commercial polymers.

In *bulk polymerization*, the system is essentially composed of only monomer/polymer. This technique is most commonly used for polymerizations which proceed through functional groups in a stepwise manner and then the method merely involves heating the straight monomer or monomer mixture (sometimes with addition of a small amount of a catalyst to increase the reaction rate). The system is maintained in a fluid state by keeping the

Table 1.3 Polymerization techniques used in the production of some commercial polymers

Polymer	Polymerization technique
Polyamides	Bulk
Polycarbonate	Bulk
Poly(ethylene terephthalate)	Bulk
Polysulphides	Suspension
Polyethylene (low density)	Bulk*
Polyethylene (high density)	Solution
Poly(methyl methacrylate)	Bulk, Suspension
Polypropylene	Solution
Polystyrene	Solution
Poly(vinyl acetate)	Emulsion
Poly(vinyl chloride)	Suspension
Polyisoprene	Solution
Styrene-butadiene copolymer	Emulsion
Polyformaldehyde	Solution
Polycaproamide (nylon 6)	Bulk

* See page 50.

temperature sufficiently high. As has been mentioned previously, in this type of reaction there is a progressive increase in molecular weight and the high viscosity of the resultant polymer melt can lead to handling difficulties. When the technique of bulk polymerization is applied to polymerizations which involve chain reactions, the straight monomer is heated with a small amount of appropriate initiator. Again there is a substantial rise in viscosity as the concentration of polymer (which is soluble in the monomer) increases and this can lead to difficulty in dissipating the high exothermic heat of reaction which is usually a feature of such polymerizations. Because of the possibility of localized overheating leading to degradation and discoloration of the polymer, bulk polymerization is seldom practised with large batches. It may be noted that bulk polymerization results in relatively pure polymer.

In *solution polymerization*, the monomer is dissolved in a solvent prior to polymerization. This technique is commonly employed for the ionic polymerization of gaseous vinyl monomers. The solvent facilitates contact of monomer and initiator (which may or may not be soluble in the solvent) and assists dissipation of exothermic heat of reaction. A limitation of this technique is the possibility of chain transfer to the solvent with consequent formation of low molecular weight polymer. An added disadvantage is the need to remove the solvent in order to isolate the solid polymer. In this respect, it is common practice to use a solvent in which the monomer but not the polymer is soluble; the polymer is then obtained directly as a slurry and little further purification is necessary.

In *suspension polymerization*, the monomer is dispersed in water in small droplets (generally about 10^{-2}–10^{-1} cm in diameter) maintained by vigorous stirring. This technique is extensively used for the free radical polymerization of vinyl monomers. A monomer-soluble initiator is added and polymerization occurs within each droplet. Generally a material such as poly(vinyl alcohol) or gelatin is added to provide a protective coating for the droplets; this prevents the droplets from cohering when they are at the stage of being composed of a sticky mixture of monomer and polymer. Besides facilitating the removal of exothermic heat of reaction, suspension polymerization has the advantage that the polymer is obtained in the form of small beads which are easily collected and dried. The polymer is relatively free from contaminants and there are no solvent recovery considerations.

In *emulsion polymerization*, the monomer is dispersed in water containing a soap (usually about 5%) to form an emulsion; such a dispersion is stable and its existence is not dependent on continued agitation. This technique is extensively used for the free radical polymerization of diene monomers in the preparation of synthetic rubbers. In this case a water-soluble initiator is used and the course of the polymerization is considerably different from that followed in the systems described previously. At the start of an emulsion polymerization three components are present:

(i) Relatively large droplets of monomer, about 10^{-4} cm in diameter, stabilized by soap molecules around the periphery.
(ii) Aggregates (micelles) of 50–100 soap molecules swollen with monomer to a diameter of about 10^{-6} cm.
(iii) The aqueous phase containing a few monomer molecules and the initiator which gives rise to free radicals.

The monomer droplets and the micelles swollen with monomer compete for the free radicals generated in the aqueous phase, but since there are many more micelles than droplets in the system most of the free radicals enter micelles. Polymerization is initiated within individual micelles. The monomer consumed during the resulting polymerization is replenished by diffusion of new monomer molecules from the aqueous phase, which in turn, is kept saturated with monomer from the droplets of monomer. Polymerization continues within a given micelle until a second free radical enters the micelle, in which case termination quickly occurs because of the small volume of the reaction locus. The micelle then remains inactive until a third free radical enters, and so on. As reaction proceeds the micelles become larger and are disrupted to form particles of polymer swollen with monomer which are stabilized by soap molecules around the periphery. Monomer continues to diffuse into these particles and polymerization is maintained therein until the monomer supply is exhausted. The final product is a stable dispersion (*latex*) of polymer particles with diameter of about 10^{-5}–10^{-4} cm. The polymer is isolated by 'breaking' the latex, usually by the addition of acid which converts the soap to fatty acid. In some instances, the latex is used directly without coagulation; such is the case in, for example, the preparation of poly(vinyl acetate) latex paints. An attractive feature of emulsion polymerization is that it is possible to prepare very high molecular weight material at high rates of conversion. A limitation of the method is the difficulty of washing the product free of soap residues which impair the electrical insulation properties and optical clarity.

1.6   POLYMER STRUCTURE

Up to this point, polymers have been considered simply as large molecules composed of repeating structural units and there has been little consideration of *micro-structure*, i.e. the arrangement of the structural units relative to one another. Several variations are possible and these can have pronounced effects on the properties of a polymeric material. The more important of these structural arrangements may be conveniently considered under the four headings which now follow.

1.6.1   Polymer geometry

In most of the examples given so far, the polymer molecules have been *linear*, that is to say, the molecules have a thread-like shape. This description may be misleading since it may be taken to mean that these molecules resemble a thread in its most commonly encountered form, i.e. in a more or less fully extended state. In fact, linear polymer molecules may have various conformations. In amorphous materials the molecules are highly kinked in an irregular fashion (randomly coiled) (see Fig. 1.2a) whilst in crystalline materials the molecules are generally spiral or zig-zag. The point is that a thread could be made to resemble these conformations.

Not all polymers, however, are linear; various *non-linear* forms are also possible. In *branched* polymers, linearity is destroyed by the presence of side-chains. Branched polymers commonly arise in two types of polymerization process:

(i) *Free radical polymerization of vinyl and diene monomers.* In the previous discussion of free radical vinyl polymerization it was mentioned that various transfer reactions may occur. One possibility is the transfer of hydrogen from within a polymer to a growing polymer, e.g.

$$\sim\!CH_2\!-\!CHR\!\sim \quad + \quad \sim\!CH_2\!-\!\dot{C}HR \quad \longrightarrow \quad \sim\!CH_2\!-\!\dot{C}R\!\sim \quad + \quad \sim\!CH_2\!-\!CH_2R$$

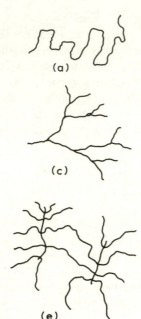

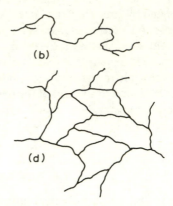

Fig. 1.2 Planar representation of polymer molecules. (a) Randomly coiled linear polymer. (b) Slightly branched polymer. (c) Highly branched polymer. (d) Cross-linked polymer with 3-functional junctions as might arise by continued reaction of (c). (e) Cross-linked polymer with 4-functional junctions, as might arise by cross-linking an unsaturated linear polymer.

The resultant free radical may initiate polymerization of monomer, in which case a branch is produced:

$$\sim CH_2-\overset{\cdot}{C}R\sim \;\;+\;\; H_2C{=}CHR \longrightarrow \begin{array}{l} \sim CH_2-CR\sim \\ \qquad\quad | \\ CH_2-\overset{\cdot}{C}HR \text{ etc.} \end{array}$$

In general, this type of branching arises when the propagating radical is highly energetic (i.e. lacking in resonance stabilization) and the polymer contains readily replaceable hydrogen atoms (i.e. the resultant free radical is resonance stabilized). Thus branching occurs in, for example, polyacrylates and poly(vinyl acetate) (see Chapters 6 and 5). Similarly, branching also occurs in free radical diene polymerization. As was mentioned previously, the polydienes contain 1,2- and 3,4- units. These units contain pendant vinyl groups which are susceptible to free radical attack and consequent branching; these units also contain allylic hydrogen atoms which are readily abstracted to give sites for branching. The branched polymers obtained in vinyl and diene polymerization may generally be regarded essentially as linear polymers with relatively few side-chains (see Fig. 1.2b).

(ii) *Polymerization of polyfunctional monomers.* In the previous discussion of polymerization through functional groups, the reaction between a diisocyanate and a glycol was seen to result in a linear polyurethane. If, however, a triisocyanate and a glycol are used the resultant polyurethane is branched; the reaction may be represented as follows:

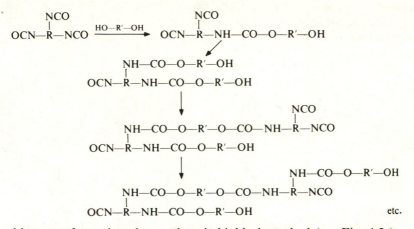

In this type of reaction the product is highly branched (see Fig. 1.2c) and cross-linking (see later) occurs before a high molecular weight can be reached. It will be apparent from this example that in polymerization through functional groups, the *functionality* of the reactants determines whether a linear or branched polymer is obtained. The functionality of a compound is the number of groups per molecule which are capable of undergoing the polymerization reaction. If all the reacting molecules are difunctional a linear polymer results; if polyfunctional (i.e. trifunctional or greater) molecules are present branching occurs. It may be noted that these comments regarding functionality are, in essence, applicable to polymerization through multiple bonds if each multiple bond is regarded as being difunctional. Thus vinyl compounds, conjugated dienes (where normally only one double bond is involved) and carbonyl compounds are designated as difunctional. As has been seen, these monomers give rise to essentially linear polymers (although subsequent transfer reactions may lead to some branching). Likewise, cyclic monomers are regarded as difunctional.

The final structural form to be described in this section relates to *cross-linked* or *network* polymers. Such polymers are commonly formed in two ways:

(i) *Polymerization of branched materials.* In the polyurethane reaction described above which involves a triisocyanate and a glycol, the branched structures initially formed contain reactive groups, i.e. isocyanate and hydroxyl groups. Thus if reaction is allowed to proceed, the branched structures link up with one another to form large continuous three-dimensional network (see Fig. 1.2d). Materials such as epoxy, phenol-formaldehyde and silicone resins are utilized commercially in this way.

(ii) *Cross-linking of linear polymers.* It was seen above that some branching occurs during the polymerization of conjugated dienes and some vinyl monomers. At sufficiently high degrees of conversion, continued reaction may result in the formation of cross-links. For example, the combination of two growing branches on two different polymer chains would result in a cross-link, e.g.

$$
\begin{array}{ccc}
\sim CH_2\!-\!CR \sim & & \sim CH_2\!-\!CR \sim \\
| & & | \\
CH_2 & & CH_2 \\
| & & | \\
CHR & & CHR \\
\cdot & \longrightarrow & | \\
CHR & & CHR \\
| & & | \\
CH_2 & & CH_2 \\
| & & | \\
\sim CH_2\!-\!CR \sim & & \sim CH_2\!-\!CR \sim
\end{array}
$$

In the preparation of commercial polymers it is usual to stop polymerization before the extent of such reactions becomes appreciable. However, the deliberate cross-linking of linear polymers in subsequent operations is common technological practice. For example, diene rubbers are cross-linked (vulcanized) in virtually all applications; likewise, linear polyesters used in the preparation of glass fibre laminates and castings become cross-linked during the process. In both these examples, cross-linking is accomplished by the incorporation of a reagent which forms bridges between polymer chains by interaction at unsaturated sites along the chains (see Chapters 20 and 11). In the rigid product obtained from the polyester the degree of cross-linking is large compared to that in a rubber vulcanizate where the flexible molecular chains are constrained by network junctions at only infrequent intervals (see Fig. 1.2e).

1.6.2 Structural unit variety

All the polymers considered so far have been *homopolymers*, that is, each polymer has consisted of repeating structural units of the same kind (disregarding terminal units). It is, however, possible to prepare polymers which contain more than one type of structural unit; such polymers are distinguished by the term *copolymer*. Most important commercial copolymers contain two kinds of structural unit whilst a few have three units. (The latter products are often called *terpolymers*.) Copolymers may be prepared by the same general methods described above for homopolymers but, of course, more than one monomer must be used. Depending on the monomers chosen and the experimental techniques used, various distributions of structural units within the polymer chain may be achieved. Various possible arrangements of the two structural units A and B are shown in Fig. 1.3.

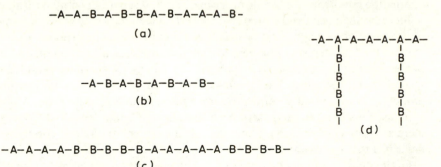

—A–A–B–A–B–B–A–B–A–A–A–B–

(a)

—A–B–A–B–A–B–A–B—

(b)

—A–A–A–A–A–A–A–A—
 | |
 B   B
 | |
 B   B
 | |
 B   B
 | |
 B   B
 | |

(d)

—A–A–A–A–B–B–B–B–B–A–A–A–A–A–B–B–B–B—

(c)

Fig. 1.3 Copolymer types: (a) Random copolymer (b) Alternating copolymer (c) Block copolymer (d) Graft copolymer.

Random copolymers are prepared by the polymerization of an appropriate mixture of monomers. Many commercial products of this type are available, being mostly based on vinyl monomers and/or conjugated dienes, e.g. vinyl chloride-vinyl acetate, vinylidene chloride-vinyl chloride, styrene-butadiene, ethylene-propylene-isoprene copolymers.

Few truly alternating copolymers have been prepared; the best known example is the product obtained by the free radical copolymerization of an approximately equimolar mixture of maleic anhydride and styrene. There is virtually no tendency for the radical end of a growing chain to react with its own type of monomer. The structure of the product is therefore essentially as follows:

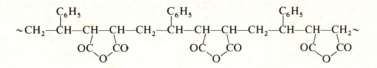

It may be noted that this structure could be regarded as a homopolymer having only the following structural unit:

$$-CH_2-\underset{\underset{C_6H_5}{|}}{CH}-CH-CH- \\ OC \quad CO \\ \diagdown O \diagup$$

However, it is usual to consider as copolymers products obtained from a mixture of monomers when each of the monomers is separately capable of forming a homopolymer (under appropriate conditions). Thus such polymers as polyamides derived from diamines and dibasic acids are not counted as alternating copolymers since the monomers are not separately polymerizable.

Block copolymers can be prepared by several techniques, of which anionic polymerization offers the best possibilities for controlling the product. In this method the first step is to polymerize a single monomer, allowing reaction to proceed until the monomer is exhausted. To the 'living polymer' is added a second monomer which then forms the second block. When the second monomer is exhausted a third monomer may be added, and so on. Many combinations of monomers have been investigated and a few block copolymers are now commercially available, e.g. the styrene-butadiene copolymer described in Chapter 20.

Graft copolymers may be prepared in three general ways, namely transfer grafting, irradiation grafting and chemical grafting. Transfer grafting is most commonly free radical initiated. Typically, a vinyl or diene polymer is treated with a peroxide in the presence of vinyl monomer. Transfer occurs between the polymer chain and radicals derived from the initiator; the resultant polymer chain radical then initiates polymerization of the monomer, e.g.

$$\sim CH_2\!-\!CHR \sim \;+\; R'O\cdot \longrightarrow \;\sim CH_2\!-\!\dot{C}R\sim \;+\; R'OH$$

$$\sim CH_2\!-\!\dot{C}R \sim \;+\; nH_2C\!=\!CHR'' \longrightarrow \;\sim CH_2\!-\!\overset{\displaystyle [CH_2-CHR''-]_n}{\underset{\displaystyle |}{C}}R \sim$$

Grafting is invariably accompanied by formation of homopolymer of the monomer to be grafted. In irradiation grafting, an essentially similar process is involved except that the reactive sites on the polymeric substrate are created by irradiation (commonly with ultraviolet light). In chemical grafting, reactive groups present along the polymer chain are used as sites for grafting. Both free radical and ionic reactions have been utilized in this technique. One method involves irradiation of the polymeric substrate in the presence of oxygen to produce peroxide groups which can be subsequently decomposed thermally in the presence of monomer to initiate free radical grafting, e.g.

$$\sim CH_2\!-\!CHR \sim \;+\; O_2 \longrightarrow \;\sim CH_2\!-\!\overset{\displaystyle OOH}{\underset{\displaystyle |}{C}}R \sim \longrightarrow \;\sim CH_2\!-\!\overset{\displaystyle O\cdot}{\underset{\displaystyle |}{C}}R \sim \;+\; HO\cdot$$

$$\sim CH_2\!-\!\overset{\displaystyle O\cdot}{\underset{\displaystyle |}{C}}R \sim \;+\; nH_2C\!=\!CHR' \longrightarrow \;\sim CH_2\!-\!\overset{\displaystyle O\!\!-\!\![CH_2-CHR'-]_n}{\underset{\displaystyle |}{C}}R \sim$$

An example of ionic grafting is provided by the formation of poly(styrene-g-acrylonitrile) from poly(p-lithiostyrene) (prepared by treatment of poly(p-iodostyrene) with n-butyllithium):

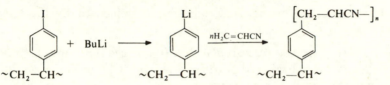

At the present time, graft copolymers have achieved limited commercial importance.

1.6.3   Structural unit orientation

A further possible variable in polymer structure arises when vinyl and diene monomers are polymerized.

In vinyl polymerization, three different types of linkage may be formed since each monomer molecule can assume either of two orientations as it adds to the preceding unit:

$$\sim CH_2-\overset{*}{C}HR \; + \; H_2C{=}CHR \longrightarrow$$

> $\sim CH_2-CHR-CH_2-\overset{*}{C}HR$
> head-to-tail linkage
>
> $\sim CH_2-CHR-CHR-\overset{*}{C}H_2$
> head-to-head linkage

$$\sim CHR-\overset{*}{C}H_2 \; + \; H_2C{=}CHR \longrightarrow$$

> $\sim CHR-CH_2-CH_2-\overset{*}{C}HR$
> tail-to-tail linkage
>
> $\sim CHR-CH_2-CHR-\overset{*}{C}H_2$
> tail-to-head linkage

(*signifies a radical, anion or cation)

The linkages are identified by referring to the substituted end of the monomer molecule as the 'head' and to the unsubstituted end as the 'tail'. In the above scheme, the tail-to-head linkage is equivalent to the head-to-tail linkage. Thus, in theory, three arrangements are possible in the polymer; the linkages may be all head-to-tail, they may be alternately head-to-head and tail-to-tail, or they may be mixed:

$$\sim CH_2-CHR-CH_2-CHR-CH_2-CHR-CH_2-CHR-CH_2-CHR \sim$$
head-to-tail polymer

$$\sim CH_2-CHR-CHR-CH_2-CH_2-CHR-CHR-CH_2-CH_2-CHR \sim$$
head-to-head/tail-to-tail polymer

$$\sim CH_2-CHR-CH_2-CHR-CHR-CH_2-CHR-CH_2-CH_2-CHR \sim$$
random polymer

It follows that a polymer which contains head-to-head linkages must also contain tail-to-tail linkages and vice versa. All the vinyl polymers which have been examined so far are very predominantly head-to-tail in their orientation. The methods which have been used to investigate the structures of specific polymers are described later in the chapters relating to these polymers.

The orientation of monomer addition is influenced by the following factors:

(i) *Stability of the new free radical or ion.* It has been seen previously that the first step in free radical vinyl polymerization is the homolytic dissociation of the initiator, which may be represented as:

$$I\!\!-\!\!I \longrightarrow 2I\cdot$$

The second step involves the addition of an initiator fragment to a monomer molecule; this reaction may occur in either of the following ways:

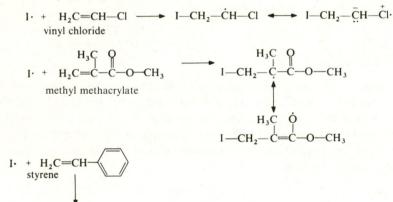

$$I\!\!-\!\!CH_2\!\!-\!\!\overset{\cdot}{C}HR \qquad (1)$$

$$I\cdot\ +\ H_2C\!\!=\!\!CHR$$

$$H_2\overset{\cdot}{C}\!\!-\!\!CHR\!\!-\!\!I \qquad (2)$$

In general, the radical formed in reaction (1) is more stable than that formed in reaction (2) because in this radical the substituent R is able to make the more effective contribution to resonance stabilization. The extent of such stabilization depends on the facility with which R can delocalize the odd electron; the effect is particularly marked when R contains suitably placed multiple bonds. The resonance stabilization of radicals derived from some common vinyl monomers is illustrated below:

$$I\cdot\ +\ H_2C\!\!=\!\!CH\!\!-\!\!Cl \longrightarrow I\!\!-\!\!CH_2\!\!-\!\!\overset{\cdot}{C}H\!\!-\!\!Cl \longleftrightarrow I\!\!-\!\!CH_2\!\!-\!\!\overset{-}{C}H\!\!-\!\!\overset{+}{C}l\cdot$$
vinyl chloride

methyl methacrylate

styrene

This type of stabilization is not possible in radicals formed in reaction (2). The new free radical may participate in the first step of the propagation reaction in either of the following ways:

$$I\text{—}CH_2\text{—}\overset{\centerdot}{C}HR + H_2C\text{=}CHR \longrightarrow \begin{cases} I\text{—}CH_2\text{—}CHR\text{—}CH_2\text{—}\overset{\centerdot}{C}HR \quad (3) \\ \text{head-to-tail linkage} \\ \\ I\text{—}CH_2\text{—}CHR\text{—}CHR\text{—}\overset{\centerdot}{C}H_2 \quad (4) \\ \text{head-to-head linkage} \end{cases}$$

In reaction (3) the stabilizing influence of R is preserved whilst in reaction (4) it is not; hence reaction (3) is energetically more probable. The same situation applies to the succeeding steps in the propagation process. Thus during propagation, head-to-tail addition is energetically preferred to head-to-head addition. It may be noted that a tail-to-tail addition is energetically favourable since it results in the more stable radical ending:

$$\sim CHR\text{—}\overset{\centerdot}{C}H_2 + H_2C\text{=}CHR \longrightarrow \sim CHR\text{—}CH_2\text{—}CH_2\text{—}\overset{\centerdot}{C}HR$$

However, this reaction requires the prior formation of the less probable radical ending, $\sim CHR\text{—}\overset{\centerdot}{C}H_2$. On these grounds, therefore, it is to be predicted that head-to-tail linkages will predominate over head-to-head and tail-to-tail linkages in the final polymer. Since the stabilization of carbanions and carbocations follows similar principles to those discussed above, it is to be expected that the head-to-tail structure will also be favoured in ionic vinyl polymerizations.

(ii) *Steric hindrance*. In the free radical polymerization of a vinyl monomer $H_2C\text{=}CHR$, the substituent group R will offer steric hindrance to the approaching initiator fragment in the initiation step. This will result in a preference for reaction (1) above rather than reaction (2); similarly reaction (3) is more probable than reaction (4). Thus the steric effects reinforce the stabilization effects. Similar considerations apply to ionic vinyl polymerizations and the head-to-tail structure is again favoured.

(iii) *Electrostatic forces*. In ionic vinyl polymerizations it is clear that electrostatic interaction between the polymeric ion pair and a permanent or induced dipole in the approaching monomer molecule will also have an important directing influence on the mode of addition. This influence will tend to favour arrangements wherein all the structural units are oriented in the same way relative to one another, i.e. head-to-tail arrangements.

It has been shown previously that the polymerization of unsymmetrical conjugated dienes may give rise to 1,2-, 3,4- and 1,4-units. The 1,2- and 3,4-units may be regarded as arising when the two double bonds in the diene react independently. The diene may then be considered to be a vinyl monomer and the previous observations relating to the orientation of structural units in vinyl polymers may be applied to the 1,2- and 3,4-units in polydienes. In the case of the 1,4-units, however, a somewhat different situation arises. Consecutive 1,4-units may be arranged as follows (for isoprene):

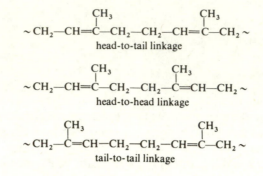

head-to-tail linkage

head-to-head linkage

tail-to-tail linkage

In free radical diene polymerization, the head-to-tail orientation predominates but some head-to-head and tail-to-tail arrangements also occur. The two radical endings which would be involved in the formation of the three linkages shown above are:

The stabilities of these radicals are similar and there are no steric restrictions to dictate whether the next monomer molecule is linked in the 1- or 4-position. Presumably, therefore, the preponderance of head-to-tail orientation is due to the electrostatic forces arising from the dipole of the diene. It is to be expected that this effect is still more marked in the ionic polymerization of conjugated dienes.

1.6.4 Polymer tacticity

A further possible variable in polymer structure arises when the backbone of the polymer molecule contains a carbon atom attached to two different side-groups. Such polymers may have various configurational arrangements or tacticity. Polymers with a regular arrangement are known as *tactic* polymers whilst those with a random arrangement are *atactic*.

The simplest type of tacticity relates to vinyl polymers of structure $[-CH_2-CHR-]_n$ and $[-CH_2-CRR'-]_n$. In such polymers every substituted carbon atom is asymmetric, i.e. is attached to four different atoms or groups and may thus have two possible configurations. This situation arises because every carbon atom is joined to two residual portions of the polymer chain and these two residues will be different. If the two residues are denoted by A and B then the group $-CHR-$ (and similarly $-CRR'-$) may have either of the tetrahedral configurations shown in Fig. 1.4*. It is thus apparent that an enormous number of different configurational arrangements (optical isomers) are possible for the vinyl polymer, but only a few of these are of concern here. If the two forms shown in Fig. 1.4 are designated arbitrarily as right-handed (*d*-) and left-handed (*l*-) configurations (with no implication as to direction of any optical rotation), the various stereoisomers can be represented in the manner illustrated in Fig. 1.5. When there is a repetition of the same configuration at each progressive asymmetric carbon atom along the polymer chain the polymer is *isotactic* whilst if there is an alternation of configuration the polymer is *syndiotactic*. These are the two simplest types of tacticity and several vinyl polymers have been obtained in these forms. The nature of tacticity is further illustrated in Fig. 1.6 wherein an isotactic,

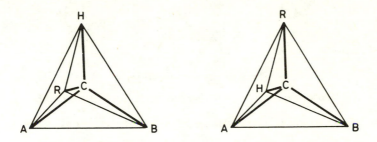

Fig. 1.4   Configurations at asymmetric carbon atoms.

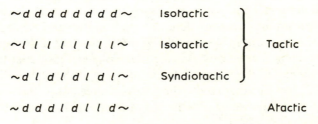

Fig. 1.5   Representation of polymer tacticity.

---

* It is conceivable that the centre carbon atom of a polymer backbone containing an odd number of carbon atoms may not be asymmetric. This can occur only when A and B (Fig. 1.4) have identical structures, i.e. the same lengths and the same terminal groups. If A and B also have the same overall configuration, the centre carbon atom is not asymmetric since it is joined to two identical groups. (A carbon atom whose asymmetry depends on the configuration of substituent groups is said to be pseudo-asymmetric.) This is a very unlikely set of circumstances in an actual polymer.

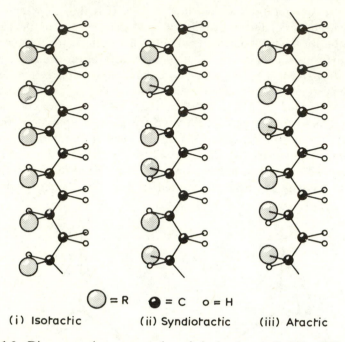

$\bigcirc$ = R   ⬤ = C   o = H

(i) Isotactic   (ii) Syndiotactic   (iii) Atactic

Fig. 1.6 Diagrammatic representation of vinyl polymers $[-CH_2-CHR-]_n$.

syndiotactic and atactic vinyl polymer are diagrammatically represented. For clarity, the main chain is shown as having a planar zig-zag form though this may not be the actual conformation of the chain; for instance, in an isotactic polymer the R group would interfere with one another and the chain adopts a helical conformation. Inspection of Fig. 1.6 shows that if the backbone of the polymer chain is visualized as being in the plane of the paper then in the isotactic polymer all the R groups lie on one side of the plane whilst in the syndiotactic they lie alternately above and below. A further method of depicting these relationships is by means of the Fischer projection formulae which are shown in Fig. 1.7.

The possibility of tacticity is not, of course, restricted to vinyl polymers. Other types of polymer which contain asymmetric carbon atoms and which have been obtained in tactic forms include poly(propylene oxide) and natural rubber hydrochloride:

$$\left[-CH_2-\underset{\displaystyle CH}{\overset{\displaystyle CH_3}{|}}-O-\right]_n \qquad \left[-CH_2-\underset{\displaystyle Cl}{\overset{\displaystyle CH_3}{\underset{|}{\overset{|}{C}}}}-CH_2-CH_2-\right]_n$$

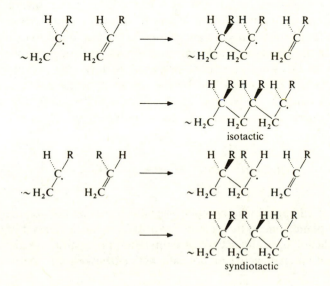

Fig. 1.7 Projection formulae of polymers depicted in Fig. 1.6.

It may also be noted that a polymer may have more than one asymmetric carbon atom in its structural unit, in which case the number of possible configurations is greatly increased; a few stereoregular polymers of this type have been prepared.

The configuration about an asymmetric carbon atom in a polymer is determined at the stage of monomer addition. This is illustrated in the following representation of the propagation reaction in a free radical vinyl polymerization, wherein tacticity is determined by the mode of presentation of monomer units:

Ionic vinyl polymerizations may be represented in a similar manner.

Normally, free radical polymerization results in essentially atactic polymers although steric and electrostatic effects generally favour one configuration or the other and truly random polymers are probably rare. In some cases, free radical polymerization results in polymers which are largely syndiotactic; examples of monomers which behave in this way are chlorotrifluoroethylene, methyl methacrylate and vinyl acetate. More commonly, however, highly tactic polymers are prepared by means of ionic polymerizations. Most of the common vinyl polymers have been obtained in isotactic and syndiotactic forms by the use of ionic initiators, particularly lithium alkyls and Ziegler-Natta catalysts.

Since the factors underlying tacticity in polymers are the same as those which give rise to optical activity in low molecular weight compounds, it is apparent that there exists the possibility that some polymers may be optically active. In practice, common vinyl polymers are not optically active. This observation is readily explained in the cases of atactic and syndiotactic polymers where there are equal numbers of d- and l-configurations in each polymer molecule and internal compensation results. In the case of an isotactic polymer there are normally equal numbers of molecules with all d- and all l-configurations, i.e. the material is a racemate. Even when an isotactic polymer can be prepared so that the numbers of molecules with all d- and all l-configurations are not equal, significant optical activity does not arise. In a vinyl polymer molecule with either all d- or all l-configurations each asymmetric carbon atom has a mirror image on the opposite side of the mid-point of the chain and the molecule as a whole is optically inactive. The presence of dissimilar end-groups or occasional anomalous modes of addition would produce a negligible optical activity. On the other hand, several polymers with all d- or all l-configurations in which the asymmetric carbon atoms are joined to four completely different groups have been found to exhibit optical activity. In these cases internal compensation does not occur. Optically active polymers of this kind have been prepared both by the polymerization of optically active monomers and by the use of optically active catalysts. L-Propylene oxide is an example of an optically active monomer which has been polymerized to an optically active polymer (using ferric chloride as catalyst):

$$n\,H_2C{-}\overset{*}{H}C\underset{}{\overset{\displaystyle O}{\triangle}}CH_2 \longrightarrow \left[-CH_2{-}\overset{\displaystyle \overset{CH_3}{|}}{\underset{\underset{H}{|}}{\overset{*}{C}}}{-}O{-} \right]_n$$

The use of an optically active catalyst is illustrated by the polymerization of *trans*-1,3-pentadiene in the presence of a Ziegler-Natta catalyst of vanadium trichloride and optically active triisoamylaluminium; although the monomer is not itself optically active an optically active *trans*-1,4-polymer is produced:

$$n\text{H}_3\text{C}-\text{CH}=\text{CH}-\text{CH}=\text{CH}_2 \quad \longrightarrow \quad \left[\begin{array}{c} \text{CH}_3 \\ | \\ \overset{*}{\text{C}}-\text{CH}=\text{CH}-\text{CH}_2- \\ | \\ \text{H} \end{array} \right]_n$$

which may be written as

$$\left[\begin{array}{c} \text{CH}_3 \\ | \\ -\text{CH}_2-\overset{*}{\text{C}}-\text{CH}=\text{CH}- \\ | \\ \text{H} \end{array} \right]_n$$

Finally, it may be noted that some polymers have been obtained in which optical activity is ascribed mainly to conformational asymmetry. In these cases there is a predominance of either right-handed or left-handed enantiomorphs of helical polymer molecules, in contrast to the more usual situation wherein equal amounts of the two enantiomorphs are produced and there is no resultant optical activity. Optically active polymers of this type have been obtained from α-olefins possessing optically active side chains, e.g. 3-methyl-1-pentene, 4-methyl-1-hexene and 5-methyl-1-heptene. Isotactic polymers from these monomers have greatly enhanced optical activity compared to the monomer. Since these polymers are vinyl polymers this optical activity cannot be associated with the asymmetry of the carbon atom in the polymer backbone (for the reasons given above). Thus it is supposed that the presence of optically active side groups favours a particular screw sense of the helix so that the resultant polymer shows a large optical rotation. Optical activity of this type has not been observed when the side groups are not asymmetric.

1.7 POLYMER UTILIZATION

Finally, in this chapter, some aspects of the relation between polymer structure and the practical use of polymeric materials are briefly discussed.

In order to be useful as a solid material of construction a material must be capable of being manipulated into some desired shape, be it electrical insulation, fibre, film, mechanical component or whatever. There are basically two means by which a material may be shaped, namely mechanical methods and moulding methods. Mechanical methods are either reduction processes such as carving, drilling and grinding or enlargement processes such as bonding, nailing or twisting into fibres. In contrast, moulding methods involve making the material undergo liquid flow into the required shape and then setting the material so that the shape is retained. Moulding methods are generally attractive in that they normally involve few operations and result in little wastage and they are nearly always used for the manipulation of synthetic organic polymeric materials.

A material may undergo liquid flow either as a melt (at a temperature high enough to give a sufficiently low viscosity) or as a solution (or related form such as a latex or paste). In the latter case it is usually necessary to remove the solvent so as to leave the solute in the required shape. It has been seen previously that polymeric materials such as polyethylene, polystyrene, and poly(vinyl chloride) are comprised of long chain molecules which are not linked one to another by covalent bonds. Thus when a sample of such a material is heated under pressure the individual polymer molecules slide past each other and the sample takes up a new shape. When the material is cooled it again becomes solid and the new shape is retained. This process of heating, re-shaping and cooling may be repeated indefinitely (provided, of course, thermal decomposition does not occur). Polymers which may be treated in this way, which are reversibly fusible, are said to be *thermoplastic*. Such polymers are commonly processed (shaped) in the melt by calendering, extrusion and injection moulding. In addition, these polymers are often soluble and so may also be manipulated as solutions; for example, cellulose acetate film is manufactured by casting a solution and then evaporating the solvent. In the case of cross-linked polymers a very different situation exists. Here the molecular chains are joined together by covalent bonds and so the chains cannot slide past one another on the application of heat and pressure (unless degradation occurs). Polymers of this kind, which are infusible, are said to be *thermoset*. Since such polymers are also insoluble they cannot be processed by the methods which are commonly applied to thermoplastic materials; mechanical methods only may be used. This difficulty may be circumvented, however, and it is possible to obtain a cross-linked polymer in a desired shape by moulding methods. This may be accomplished by carrying out the shaping operations before cross-linking occurs. Thus an intermediate which can be processed either as a melt or as a solution is transformed into the required shape and then chemical reaction is initiated so that the shape becomes composed of cross-linked polymeric material. In practice, two types of intermediates are used:

(i) *Low molecular weight polyfunctional materials* (commonly called *resins*) which are initially fusible but which on sustained heating form branched structures and, eventually, cross-linked polymers. Materials of this kind, which are irreversibly fusible, are said to be *thermosetting*. This type of system is commonly used for such products as alkyd, epoxy, phenol-formaldehyde and urea-formaldehyde polymers in such applications as adhesives, compression moulding powders and paints.

(ii) *Linear polymers containing reactive sites* which are initially fusible but which on sustained heating become cross-linked. This type of system is used for rubbers which are shaped by such processes as calendering, compression moulding and extrusion and then subsequently vulcanized. A system of this kind is also used in the preparation of glass fibre laminates and castings from unsaturated linear polyesters (see section 1.6.1).

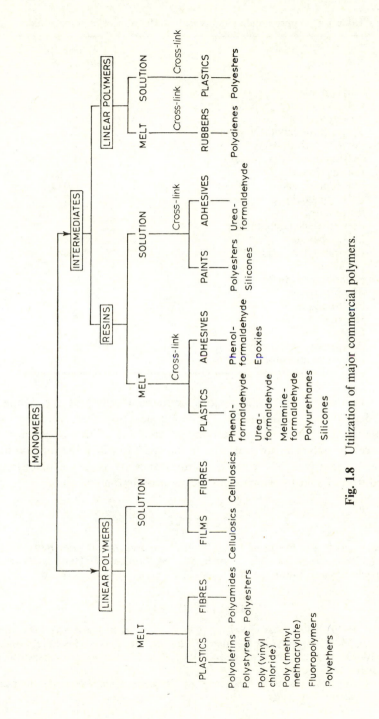

Fig. 1.8 Utilization of major commercial polymers.

Table 1.4 Approximate 1986 world production of major polymers

Polymer	Production (million tonnes)
Polyethylene	24
Polypropylene	10
Polystyrene	6
Acrylonitrile-butadiene-styrene	2
Poly(vinyl chloride)	14
Poly(methyl methacrylate)	1
Polyacrylonitrile	2
Fluoropolymers	0.03
Polyformaldehyde	0.2
Polyamides	4
Poly(ethylene terephthalate)	7
Polycarbonate	0.3
Poly(phenylene oxide)	0.2
Cellulosics	3
Phenol-formaldehyde	2
Aminopolymers	6
Polyurethanes	3
Silicones	0.3
Epoxies	0.4
Polydienes	13
Polyesters (unsaturated)	1

The means by which the major commercial polymers are commonly utilized are illustrated in Fig. 1.8. These polymers (together with a few others of interest) are considered in subsequent chapters of this book. For convenience, the polymers are dealt with in the order in which they appear in this diagram. Some idea of the relative importance of these polymers may be gained from Table 1.4 which gives global production statistics.

REFERENCES

1. Stearns, R. S. and Foreman, L. E. (1959) *J. Polymer Sci.*, **41**, 381; Bawn, C. E. H. (1962) *Proc. Chem. Soc.*, 165; Cooper, W. and Vaughan, G. (1967) in *Progress in Polymer Science*, Vol. I, (ed. A. D. Jenkins), Pergamon Press, Oxford.

BIBLIOGRAPHY

Flory, P. J. (1953) *Principles of Polymer Chemistry*, Cornell University Press, Ithaca.
Kaufman, M. (1963) *The First Century of Plastics*, The Plastics Institute, London.
Battaerd, H. A. J. and Tregear, G. W. (1967) *Graft Copolymers*, Interscience Publishers, New York.
Lenz, R. W. (1967) *Organic Chemistry of Synthetic High Polymers*, Interscience Publishers, New York.

Smith, D. A. (ed.) (1968) *Addition Polymers: Formation and Characterization*, Butterworths, London.

Szwarc, M. (1968) *Carbanions, Living Polymers and Electron Transfer Processes*, Interscience Publishers, New York.

Cowie, J. M. G. (1973) *Polymers: Chemistry and Physics of Modern Materials*, Intertext, London.

Boor, J. (1979) *Ziegler-Natta Catalysts and Polymerizations*, Academic Press, Inc., New York.

Bovey, F. A. and Winslow, F. H. (eds.) (1979) *Macromolecules—An Introduction to Polymer Science*, Academic Press, New York.

Brydson, J. A. (1982) *Plastics Materials*, 4th edn, Butterworth Scientific, London.

Kennedy, J. P. and Marechal, E. (1982) *Carbocationic Polymerization*, John Wiley and Sons, New York.

Billmeyer, F. W. (1984) *Textbook of Polymer Science*, 3rd edn, John Wiley and Sons, New York.

Elias, H-G. (1984) *Macromolecules I and II*, (Translated by J. W. Stafford), 2nd edn, Plenum Press, New York.

Cowie, J. M. G. (ed.) (1985) *Alternating Copolymers*, Plenum Press, New York.

Goodman, I. (ed.) (1982, 1985) *Developments in Block Copolymers*, Vol. 1, Vol. 2, Elsevier Applied Science Publishers, London.

2

POLYOLEFINS

2.1 SCOPE

For the purposes of this chapter, polyolefins are defined as polymers based on unsaturated aliphatic hydrocarbons containing one double bond per molecule. Polymers derived from unsaturated aromatic hydrocarbons and dienes are considered in later chapters. At the present time the principal commercial polyolefins are polyethylene (polythene), polypropylene, polyisobutene, polybutene and poly(4-methyl-1-pentene) together with related copolymers. These polymers are considered individually in subsequent sections after a brief account of relevant raw materials.

2.2 RAW MATERIALS

The present world commercial production of the lower olefins is based almost entirely on petroleum and natural gas.

The precise composition of crude petroleum varies widely from one source to another, but the principal components are always hydrocarbon paraffins (40–75%), cycloparaffins or naphthenes (mainly cyclopentane and cyclohexane derivatives) (20–50%) and aromatic hydrocarbons (5–20%). The first step in petroleum refining is distillation into broad fractions, which typically have the boiling ranges and compositions given in Table 2.1. None of these distillates contain significant amounts of olefins. The lower olefins, ethylene, propylene and butenes are produced principally by subsequent 'cracking' operations.

Many complex reactions occur during cracking but the main ones are dehydrogenation and chain-scission:

$$C_nH_{2n+2} \longrightarrow C_nH_{2n} + H_2$$
$$C_{m+n}H_{2(m+n)+2} \longrightarrow C_mH_{2m} + C_nH_{2n+2}$$

The extent of these reactions, relative to one another, depends on the nature of the starting material and the conditions used. Thus two types of cracking process are commonly carried out, namely catalytic cracking and thermal cracking.

Table 2.1 Typical petroleum fractions

Fraction	Distillation range (°C)	Approximate number of carbon atoms in component molecules
Gas	up to 25	C_{1-4}
Naphtha	20–100	C_{4-7}
Gasoline (petrol)	70–200	C_{6-12}
Kerosene (paraffin)	175–275	C_{9-16}
Gas oil (diesel oil)	200–400	C_{15-25}
Paraffin wax	230–300 (6–10 kPa)	C_{18-35}
Lubricating oil	300–365 (6–10 kPa)	C_{25-40}
Asphalt (pitch)	Residue	C_{30-70}

(i) *Catalytic cracking.* In this process the primary objective is the production of gasoline. The gasoline obtained by the direct distillation of crude petroleum (straight-run gasoline) is insufficient to meet demand and additional quantities of gasoline are obtained by the cracking of the higher boiling petroleum fractions. Typically cracking is carried out by heating the higher fraction, such as gas oil, at 450–550°C in the presence of an alumina/silica catalyst. In addition to gasoline, there is obtained a gaseous fraction comprising C_{1-4} paraffins and olefins. The olefin content is generally low and catalytic cracking represents a relatively small direct source of olefins; the paraffin content, however, may be converted to olefins by subsequent thermal cracking (see below). A related process is *reforming*, in which the quality of straight-run gasoline is improved by heating at 450–530°C with hydrogen in the presence of a platinum catalyst. (Naphtha may be treated similarly.) Branched paraffins and aromatic compounds are formed and raise the 'octane' (anti-knock) rating of the gasoline. In addition, a gaseous fraction is obtained; this is composed principally of C_{1-4} paraffins which may be converted to olefins by thermal cracking.

(ii) *Thermal cracking (pyrolysis).* In this process, feed-stocks are cracked by heating in the absence of a catalyst. The primary objective of the process is generally to obtain high yields of olefins and thermal cracking is the main commercial source of olefins. Suitable feed-stocks include LPG (liquefied petroleum gas, consisting mainly of propane and butane, obtained from the catalytic cracking processes described above and from the gas fractions of petroleum distillation) and petroleum fractions in the naphtha to gas oil range. Cracking is carried out in tubular coils at 700–850°C with residence times of under 1 second. To avoid coke formation, the hydrocarbon vapour is mixed with steam and the cracking units are commonly known as steam crackers. The reaction mixture is liquefied and separated by fractional distillation. In Western Europe and Japan, naphtha and gas oil represent the main source of olefins.

A further source of olefins is provided by natural gas. The composition of natural gas varies widely according to source, but the principal component is invariably methane (60–99%), the remainder being C_{2-5} paraffins, carbon dioxide, nitrogen and, in some cases, sulphur compounds. Only the so-called 'wet' gases which contain substantial proportions of NGL (natural gas liquids, consisting mainly of ethane, propane, butane and natural gasoline) are suitable for the production of olefins. The conversion is accomplished by thermal cracking as described above. In the United States, NGL represents the main source of olefins. Since NGL is associated with North Sea oilfields, this source of olefins is also utilized in the United Kingdom.

Although olefins higher than the butenes may be formed in the various operations described above, the main current commercial route to 4-methyl-1-pentene is the dimerization of propylene in the presence of an alkali metal or alkali metal alkyl. The dimerization of propylene can lead to several isomeric hexenes, but catalysts which particularly favour the formation of 4-methyl-1-pentene have been found. An especially effective catalyst is potassium on a potassium carbonate support; with a reaction temperature of 160°C the yield of required isomer is approximately 85% [1]. The following mechanism has been suggested for this reaction:

$$H_2C{=}CH{-}CH_3 \xrightarrow{\ K\ } H_2\bar{C}{=}CH{-}\overset{..}{C}H_2\,K^+$$

$$H_2C{=}CH{-}\overset{..}{C}H_2\,K^+ + \overset{\overset{\textstyle CH_3}{|}}{CH}{=}CH_2 \longrightarrow H_2C{=}CH{-}CH_2{-}\overset{\overset{\textstyle CH_3}{|}}{CH}{-}\overset{..}{C}H_2\,K^+$$

$$H_2C{=}CH{-}CH_2{-}\overset{\overset{\textstyle \cdot CH_3}{|}}{CH}{-}\overset{..}{C}H_2\,K^+ + H_2C{=}CH{-}CH_3 \rightleftharpoons$$

$$H_2C{=}CH{-}CH_2{-}\overset{\overset{\textstyle CH_3}{|}}{CH}{-}CH_3 + H_2C{=}CH{-}\overset{..}{C}H_2\,K^+$$

2.3 POLYETHYLENE

2.3.1 Development

The conversion of ethylene to high molecular weight polymer was first accomplished in 1933 by Fawcett and Gibson of Imperial Chemical Industries Ltd (UK) during an investigation of the reaction between ethylene and benzaldehyde at high pressure. The first commercial high pressure polyethylene plant began production in 1939 and until 1955 all commercial polyethylene was produced by high pressure processes. In 1953 Ziegler in Germany, Phillips Petroleum Co. (USA) and Standard Oil Co. (Indiana) (USA) almost simultaneously discovered methods whereby high pressures could be avoided. The first plant using the Zeigler process came on stream in 1955 (Farbwerke Hoechst A.G.) whilst the first Phillips and Standard Oil plants began production in 1957 and 1961 respectively. The polymers obtained were different from those produced under high pressure, being more linear in structure and with densities of 0.95–0.96 g/cm^3 compared to about 0.92 g/cm^3 for the high pressure polymer. This new material became known as high density polyethylene (HDPE) and the older material as low density polyethylene (LDPE). All of the new low pressure methods were solution or slurry processes. In 1968, Union Carbide Corp. (USA) began commercial production of high density polyethylene using a low pressure gas-phase process. Then in 1977, a major development occurred. Union Carbide announced that technology had been developed whereby a polyethylene with low density could be obtained using a low pressure gas-phase process. This material became known as linear low density polyethylene (LLDPE). Since 1979, the material has made significant inroads into the low density polyethylene market because of its favourable production economics and product performance characteristics. At the present time, about 65% of the global production of polyethylene is LDPE/LLDPE and 35% is HDPE.

Polyethylene is now one of the major commercial polymers and is used in such diverse applications as chemical plant, domestic goods, electrical insulation, packaging film and toys. Film applications account for about 70% of LDPE/LLDPE production whilst injection moulding and blow moulding take about 75% of HDPE production.

2.3.2 Preparation

(a) Low density polyethylene (LDPE)

Low density polyethylene is prepared by methods which involve the use of high pressures. Commercial processes are generally operated at 150–350 MPa (1500–3500 atmospheres) and temperatures of 80–300°C. Polymerization is initiated by free radical producing agents such as azodiisobutyronitrile, benzoyl peroxide or oxygen (which probably leads to in situ peroxide formation), and the choice of initiator determines the appropriate reaction conditions. For the oxygen-initiated reaction, for example, the optimum conditions are about 150 MPa (1500 atmospheres) and 200°C with 0.03–0.1% oxygen. Generally, a continuous process is operated and polymerization is carried out either in narrow-bore tubular reactors or in autoclaves fitted with stirrers. The reaction is highly exothermic and efficient heat dissipation is essential. Insufficient control of reaction variables is liable to lead to explosive formation of carbon, hydrogen and methane. In some processes a diluent such as benzene or water is added, mainly to serve as a heat-exchange medium but also to assist in the removal of polymer from the reactor. In a typical process, 10–30% of the monomer is converted to polymer per cycle. After pressure let-down, the polymer is separated from unreacted ethylene (which is recycled) and diluent, if any, and then the polymer is extruded as ribbon and granulated.

The most noteworthy feature of the above type of process is, of course, the very high pressure involved. It is necessary to use high pressure in order to obtain high molecular weight polymer. It seems that growing polyethylene radicals have a very limited life available for reaction with monomer and unless they have reacted within a certain interval termination occurs. Thus high molecular weights are favoured by the increased monomer concentration which is obtained at high pressures. It may be noted that whilst commercial processes involve what is strictly the gas phase (since temperatures are above the critical temperature of 9.7°C), at the high pressures used the gas resembles a liquid and contains polymer in solution. Thus these processes resemble conventional bulk polymerization.

In the previous chapter the polymerization of ethylene is shown as leading to a linear polymer as follows:

$$n\mathrm{H_2C{=}CH_2} \longrightarrow [\mathrm{-CH_2-CH_2-}]_n$$

Whilst this is an essentially correct representation of the high pressure processes, the polymer obtained by these methods is, in fact, somewhat branched. Examination of the infrared spectrum of the polymer shows the presence of 20–30 methyl groups per 1000 carbon atoms in the chain. Methyl groups must be terminal groups, but this number is larger than can be accounted for by methyl chain ends in a linear polymer; hence branches must be present. Further infrared studies (see Reference 2 for an account of this work) indicate that practically all the branches are short, being either ethyl or butyl groups. Production of these branches has been attributed to intramolecular transfer or 'backbite'. Butyl group formation may be represented as follows, being favoured by the stability of the six-membered ring transition state:

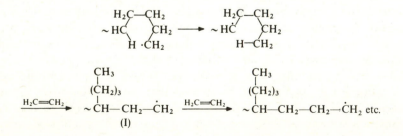

Besides propagating the chain as shown, radical I may also backbite to form ethyl groups:

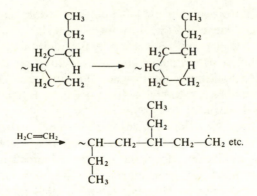

In addition to these short branches, there are also a few long branches (probably about 3 per 1000 carbon atoms) which arise through intermolecular transfer:

$$\sim CH_2 - CH_2 \sim \xrightarrow{\sim CH_2 - \dot{C}H} \sim CH_2 - \dot{C}H \sim \xrightarrow{H_2C=CH_2}$$

$$\sim CH_2 - CH - CH_2 - \dot{C}H_2 \xrightarrow{H_2C=CH_2} \sim CH_2 - CH - CH_2 - CH_2 - CH_2 - \dot{C}H_2 \text{ etc.}$$

(See Fig. 2.1.)

(b) High density polyethylene (HDPE)

High density polyethylene is produced by four methods. The first three involve solution or slurry processes which differ mainly in the nature of the catalyst used. The fourth method involves the gas phase.

(i) Ziegler processes

Commercial Ziegler polyethylene processes are generally operated at pressures only slightly above atmospheric, namely 0.2–0.4 MPa (2–4 atmospheres) and at temperatures of 50–75°C. Polymerization is conducted in the presence of Ziegler-Natta catalysts, the nature of which is discussed in Chapter 1. For the polymerization of ethylene the catalyst is usually based on titanium tetrachloride/aluminium alkyl (e.g. diethylaluminium chloride). The catalyst may be prepared *in situ* by adding the components separately to the reactor as solutions in diluents such as diesel oil, heptane or toluene or the components may be pre-reacted and the catalyst added as a slurry in a liquid diluent. These operations must be conducted in an inert atmosphere (usually of nitrogen) since oxygen and water reduce the effectiveness of the catalyst and may even cause explosive decomposition. In a typical process, ethylene and the catalyst and diluent are fed continuously into the reactor. At the reaction temperatures generally used, the polymer is only sparingly soluble in the hydrocarbon diluent and therefore forms as a slurry, which is continuously removed. The reaction is quenched by the addition of alcohols such as methanol, ethanol or isopropanol and the resulting metallic residues are extracted with alcoholic hydrochloric acid. This purification step is unnecessary if supported catalysts with high activity are used. Finally, the polymer is centrifuged, dried, extruded and granulated.

The mechanism of vinyl polymerization effected by Ziegler-Natta catalysts is discussed in the previous chapter. The polyethylene obtained by Ziegler processes differs significantly from the polymer obtained by high pressure processes in that much less branching is present. Such branching as does occur is thought to arise by oligomerization of ethylene to low molecular weight α-olefins which then copolymerize with the ethylene. Spectroscopic evidence indicates the presence of 5–7 ethyl groups per 1000 carbon atoms but butyl groups appear to be absent.

(ii)   *Phillips processes*

Phillips processes are usually operated under conditions which are intermediate between those used in high pressure processes and Ziegler processes. Pressures are generally 3–4 MPa (30–40 atmospheres) and temperatures are 90–160°C. Polymerization is effected by means of a chromium oxide catalyst. Typically, the catalyst is prepared by impregnating silica or silica-alumina (which is the support) with an aqueous solution of a chromium salt and heating the product in air at 400–800°C. The final product contains about 5% of chromium oxides, mainly chromium trioxide. In a typical process, ethylene is passed into a suspension of the catalyst in a hydrocarbon such as cyclohexane. Depending mainly on the temperature at which polymerization is carried out, the polymer is obtained either in solution or as a slurry. Solution processes are normally run at 120–160°C, at which temperatures the polymer is soluble in the diluent. Hot polymer solution is continuously drawn from the reactor; unreacted ethylene is flashed off and suspended catalyst is removed by filtration or centrifuging. The solution is then cooled to precipitate the polymer which is separated by centrifuging. Slurry processes, on the other hand, are usually run at 90–100°C, at which temperatures the polymer has low solubility in the diluent. The slurry, which consists of polymer granules each formed around separate catalyst particles, is drawn off continuously. Unreacted ethylene is flashed off and then the polymer is separated by centrifuging. The polymer so obtained is contaminated with a small amount of catalyst which is not usually removed.

At the present time the nature of polymerization with Phillips catalysts is incompletely understood. A possible mechanism is discussed briefly under Standard Oil processes. Whatever their mode of reaction may be, these catalysts lead to almost completely linear polyethylene. No ethyl or butyl branches have been definitely detected although the number of methyl groups (up to about 3 per 1000 carbon atoms) may be somewhat greater than the number of chain ends.

(iii) *Standard Oil processes*

Standard Oil processes are similar in many ways to the Phillips processes described above. Pressures of 4–10 MPa (40–100 atmospheres) are employed whilst temperatures are in the region of 200–300°C. Polymerization is effected by means of a catalyst consisting of a metal oxide, such as molybdenum trioxide on a support which may be alumina, titanium dioxide or zirconium dioxide. Typically, the catalyst is prepared by impregnating alumina with ammonium molybdate and heating the product in air at 500–600°C. Before use, the catalyst is activated by heating in a reducing atmosphere such as hydrogen at 300–650°C; alternatively, activation may be by addition of a reducing promoter such as a Group IA or IIA metal or hydride. An essential difference between the Phillips and Standard Oil catalysts appears to be that the latter require a reducing agent to render them active whilst the former are reduced to the active state in the presence of ethylene alone. The Standard Oil catalysts seem to be rather more effective in solution processes than in slurry processes. Solution processes are carried out in a manner similar to that described in the preceding account of Phillips processes.

The nature of polymerization with Standard Oil catalysts is not at present completely understood. Nor is it certain whether the Phillips and Standard Oil catalysts operate by the same or different mechanisms. However, the polyethylene obtained by the Standard Oil processes resembles that produced in the Phillips processes in that it, too, is almost completely linear.

One suggested mode of polymerization with metal oxide catalysts envisages reaction in a chemisorbed monomer layer. In this mechanism, ethylene is chemisorbed on the catalyst surface by interaction of the monomer π-electrons and unpaired electrons of the transition metal. Initiation involves the supply of an unpaired electron from a metal atom to an adjacent adsorbed monomer molecule; propagation then ensues by reaction of the product with a neighbouring adsorbed monomer molecule. The process may be depicted as follows:

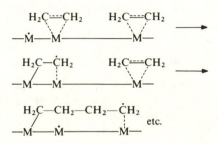

The process resembles free radical polymerization, but propagation is not by free radicals but by 'bound ion-radicals' which are fixed to the catalyst surface. A feature of this proposal is that the metal-carbon bond which is the active polymerization site moves through the adsorbed monomer layer. (See References 3 and 4 for a fuller account of the mechanism of polymerization with metal oxides.)

(iv)   *Union Carbide processes* [5]
In Union Carbide processes, ethylene is polymerized in the gas phase. Pressures of 0.7–2 MPa (7–20 atmospheres) and temperatures of about 100°C are used. Polymerization is effected by proprietary catalysts, which consist of supported organochromium compounds such as chromacene ((C_5H_5)$_2$Cr). The processes use a fluidized bed and are inherently simple in that ethylene serves as the fluidizing gas (as well as reactant) and polyethylene is the bed material. The polymer is produced as granules (from which catalyst is not removed), which can be used directly. Since no solvent is involved, gas phase processes are simpler to operate and have lower energy consumption than other processes for high density polyethylene.

(c)   *Linear low density polyethylene (LLDPE)*

Linear low density polyethylene is actually a copolymer of ethylene and 5–12% by weight of an α-olefin such as 1-butene, 1-hexene or 1-octene. The copolymer is produced by both solution and gas phase processes.

(i) *Solution processes.* Few details of solution processes for linear low density polyethylene have been published. Polymerization is carried out at 2.5–10 MPa (25–100 atmospheres) and 25–300°C. Transition metal catalysts are used.

(ii) *Gas phase processes.* The Union Carbide processes for high density polyethylene described above are also used for linear low density polyethylene. Indeed, many plants have swing capacity.

Linear low density polyethylene contains short branches which arise from the inclusion of a higher α-olefin in the polymerization system. Long branches, however, are absent and this is the meaning of 'linear' in this context. (See Fig. 2.1.)

(d)   *Very low density polyethylene (VLDPE)*

Polyethylene with density lower than 0.915 g/cm^3 is generally regarded as a distinct type of polyethylene. Polymers with densities ranging between 0.890 and 0.915 g/cm^3 are commercially available. These polymers are copolymers of ethylene and α-olefins and are made by processes similar to those described

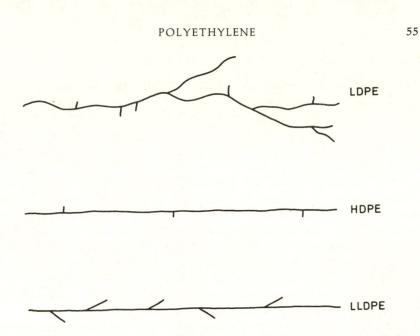

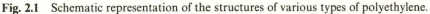

Fig. 2.1 Schematic representation of the structures of various types of polyethylene.

above for the gas phase preparation of linear low density polyethylene. The lower density is achieved by using higher levels of comonomer. Compared to other types of polyethylene, very low density polyethylene has greater flexibility. Applications include film, tubing and squeeze bottles.

(e) High molecular weight polyethylenes

Two types of polyethylenes with high molecular weight are commercially available, namely high molecular weight high density polyethylene (HMW-HDPE) and ultrahigh molecular weight polyethylene (UHMWPE).

High molecular weight high density polyethylene is produced by processes which are similar to those used for conventional high density polyethylene. (See section 2.3.2(b)). Commercial polymers typically have molecular weight (weight average) in the range 200 000–500 000.

Ultrahigh molecular weight polyethylene is prepared by a special Ziegler process. In this case, the final product is not granulated because extrusion causes chain scission, with a corresponding reduction in molecular weight. Commercial polymers have molecular weight (weight average) of at least 3 million.

Compared to other types of polyethylenes, high molecular weight polyethylenes show improved impact strength, abrasion resistance and environmental stress cracking resistance. HMW-HDPE finds application in film, piping and blow-moulded containers. UHMWPE is used for components of material handling equipment and for container linings.

2.3.3   Properties

As has been noted in the preceding section, the polyethylenes obtained in the various commercial processes differ in the amount of branching which is present. The presence of branches reduces the ability of polymer chains to pack together closely and regularly, i.e. to crystallize. As a result the various types of polyethylene have somewhat different properties. Thus the more highly branched polymers have the lower density, crystalline melting point, stiffness, surface hardness, and softening temperature and greater permeability to gases and vapours. To a first approximation, these properties are dependent only on the degree of branching. Other physical properties are more difficult to correlate since they are also affected by variations in average molecular weight and molecular weight distribution. Comparative values for some properties of typical commercial grades of polyethylenes are given in Table 2.2. The electrical insulating properties of polyethylene are outstanding. Since it is a non-polar material, properties such as dielectric constant and power factor are almost independent of the frequency and temperature. Dielectric constant increases slighltly with increasing density.

Chemically, polyethylene can be regarded as a high molecular weight paraffin and as may be expected it is a rather inert material. Since there is no specific interaction, such as hydrogen bonding, with any solvent and since the polymer is crystalline, there is no solvent for polyethylene at room temperature. At elevated temperatures, however, there is appreciable solubility in hydrocarbons and halogenated hydrocarbons, e.g. toluene, xylene and dichloroethylene. The temperature necessary to dissolve polyethylene in a given solvent increases as the crystallinity of the polymer increases and ranges from about 60°C to 80°C for commercial materials. It may be noted here that some liquids cause *environmental stress cracking* of polyethylene. If the polymer is stressed in the presence of liquids such as alcohols, esters and ketones, fracture occurs at stresses much lower than those required to cause failure in the absence of the liquid.

Table 2.2   Typical values for various properties of polyethylene and polypropylene

	LDPE	LLDPE	HDPE	Polypropylene (homopolymer)
Density (g/cm^3)	0.92	0.94	0.96	0.90
Crystalline melting point (°C)	108	123	133	176
Tensile strength (MPa)	10	10	28	28
(lbf/in^2)	1500	1500	4000	4000
Elongation at break (%)	450	700	500	200
Hardness (Shore D)	45	55	65	75
Softening point, Vicat (°C)	95	–	120	150

Polyethylene is unaffected by most acids, alkalis and aqueous solutions. Strong oxidizing agents such as concentrated nitric acid and concentrated solutions of hydrogen peroxide and potassium permanganate oxidize the polymer, resulting in an increase in power factor and a deterioration of mechanical properties. Resistance to these reagents increases with increase in density because of diminished permeability. Oxidation of polyethylene also occurs in air on exposure to ultraviolet light and/or elevated temperature. The reaction of polyethylene, and other hydrocarbon polymers with oxygen proceeds by a free radical mechanism and may be represented by the following scheme:

Initiation

$$Polymer \longrightarrow R\cdot$$

Propagation

$$R\cdot \; + \; O_2 \longrightarrow ROO\cdot$$
$$ROO\cdot \; + \; RH \longrightarrow ROOH \; + \; R\cdot$$

Termination

$$2R\cdot \longrightarrow R{-}R$$
$$R \; + \; ROO\cdot \longrightarrow ROOR$$
$$2ROO\cdot \longrightarrow ROOR \; + \; O_2$$

The initiation step consists of the formation of polymeric free radicals ($R\cdot$) by rather ill-defined reactions. Such reactions probably occur at sensitive irregularities in the polymer. The propagation reaction is a two-step sequence which involves, firstly, combination with oxygen and, secondly, abstraction of hydrogen from the polymer (RH) to form a hydroperoxide (ROOH). Such abstraction will involve the most labile hydrogen atoms which, in polyethylene, are those attached to the tertiary carbon atoms at points of branching. The hydroperoxide formed in each propagation sequence can undergo decomposition in the following ways:

$$ROOH \longrightarrow RO\cdot \; + \; HO\cdot$$
$$2ROOH \longrightarrow RO\cdot \; + \; ROO\cdot \; + \; H_2O$$

These reactions represent a further source of initiating free radicals and there is thus an acceleration in the rate of uptake of oxygen. (Oxidation processes which show autoacceleration are often termed *autoxidation*.) Hydroperoxides are also particularly susceptible to decomposition by ions of such metals as chromium, copper and iron, e.g.

$$ROOH \; + \; Co^{2+} \longrightarrow RO\cdot \; + \; OH^- \; + \; Co^{3+}$$
$$ROOH \; + \; Co^{3+} \longrightarrow ROO\cdot \; + \; H^+ \; + \; Co^{2+}$$

Thus the presence of traces of metals in the polymer results in enhanced response to attack by oxygen. Hydroperoxides may also undergo decomposition of a different kind, as a result of which polymer chain scission occurs.

This process is illustrated by the following sequence which may be envisaged for polyethylene:

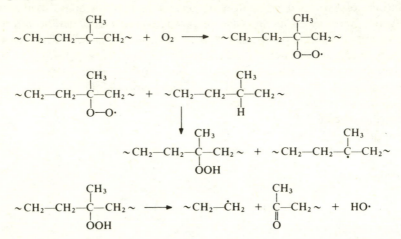

Chain scission, of course, results in a reduction in the average molecular weight of the polymer and this is manifested in the deterioration in mechanical properties which occurs when polyethylene is subjected to oxidation. It will be noticed that in the previously given representation of the reaction between hydrocarbon polymers and oxygen, termination is shown as occurring by combination. Combination reactions lead to an increase in molecular weight and to cross-linking; such reactions generally become more pronounced at higher levels of oxidation and then lead to brittle, insoluble products.

Since oxidation of polyethylene can occur during processing and on exposure to sunlight with resulting loss of mechanical and electrical strength, antioxidants are commonly added to the polymer. Antioxidants are generally hindered phenols such as 4-methyl-2,6-*tert*-butylphenol (II) and octadecyl 3-(3′,5′-di-*tert*-butyl-4′-hydroxyphenyl)propionate (III).

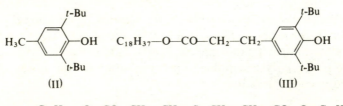

Extraction of hydrogen from the antioxidant (AH) by a peroxy radical interrupts the propagation process:

$$AH \ + \ ROO\cdot \longrightarrow ROOH \ + \ A\cdot$$

The new free radical A· is unreactive because of aromatic stability and unable to propagate an oxidative chain. Commonly, such chainbreaking antioxidants are used in conjunction with peroxide-decomposing antioxidants such as dilauryl β, β'-thiodipropionate (DLTP) (IV). These antioxidants decompose hydroperoxides in a complex series of non-radical reactions (see Reference 6 for an account of this mechanism) so that initiation of the oxidative process does not occur.

Polyethylene may also be cross-linked (intentionally) by exposure to high energy radiation such as X-rays, γ-rays and fast electrons. In addition to cross-linking, irradiation leads to the formation of small quantities of hydrogen, methane, ethane and propane. The yield of paraffins is greatest with low density polyethylene and it seems that the short branches are removed intact from the main chain and appear as saturated hydrocarbons in the gaseous products. Irradiation also results in some unsaturation in the polymer chains. These various reactions are illustrated in the following sequence:

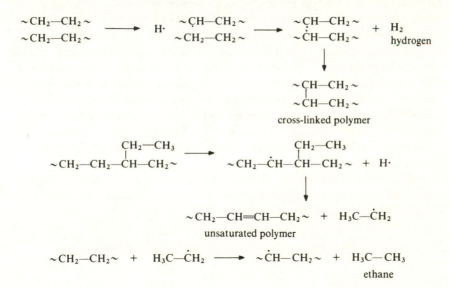

The presence of cross-links in irradiated polyethylene gives the product enhanced temperature resistance. For example, irradiated articles retain their shape at 140°C if no strain is put on them, whereas untreated articles have negligible strength at this temperature. Polyethylene may also be cross-linked by heating with peroxides such as dicumyl peroxide. The peroxide decomposes to give free radicals which abstract hydrogen from the polymer; the resulting polymeric radicals then combine to form a network. Cross-linked polyethylene finds some application as cable insulation which is subjected to elevated temperatures.

As is the case with simple paraffins, polyethylene may be halogenated. Chlorinated polyethylene is available commercially but has achieved little importance. Polyethylene may be chlorinated in solution, using solvents such as hot carbon tetrachloride, chloroform and chlorobenzene. Alternatively, a suspension of the polymer may be chlorinated; suspensions are prepared either by cooling a solution of polymer in carbon tetrachloride or by polymerizing ethylene in the presence of water. The chlorination of polyethylene is usually conducted at temperatures between 45°C and 75°C and is initiated by light or peroxides. Initiation involves production of chlorine atoms which then propagate the following chain reaction:

$$\sim CH_2 - CH_2 \sim \ + \ Cl \cdot \ \longrightarrow \ \sim CH_2 - \overset{\cdot}{C}H \sim \ + \ HCl$$
$$\sim CH_2 - \overset{\cdot}{C}H \sim \ + \ Cl_2 \ \longrightarrow \ \sim CH_2 - CHCl \sim \ + \ Cl \cdot$$

Infrared analyses indicate that chlorination proceeds initially mainly by substitution of secondary hydrogen atoms, giving –CHCl– groups; only at later stages do $-CCl_2-$ groups become appreciable. The replacement of hydrogen atoms by chlorine reduces the ability of the polyethylene to crystallize. At about 25–40% chlorine by weight most of the original crystallinity has been lost and the material is rubbery; at chlorine contents above about 50%, intermolecular forces increase and the product becomes hard. Chlorinated polyethylene containing 25–30% chlorine finds limited use as an additive to rigid poly(vinyl chloride) to give improved impact strength. Chlorinated polyethylene containing 36–48% chlorine is available for use as an elastomer. The elastomer shows good oil, heat and ageing resistance but difficulty of achieving satisfactory cross-linking (with peroxides) has restricted development.

When polyethylene is chlorinated in the presence of sulphur dioxide, chlorosulphonated polyethylene is formed. This product is of some commercial importance and is described below.

2.4   CHLOROSULPHONATED POLYETHYLENE

2.4.1   Preparation

In the preparation of chlorosulphonated polyethylene the polymer (commonly, a low density polyethylene with molecular weight (M_n) of about 20 000) is treated with chlorine in the presence of a small amount of sulphur dioxide. Typically, the reaction is carried out in solution in hot carbon tetrachloride. Both chloride and sulphonyl chloride groups are introduced into the polymer, the degree of substitution and the ratio of the two types of groups depending on the reaction conditions. Commercial products generally contain about 30% chlorine and 1.5% sulphur, which corresponds to approximately one chlorine group per 7 carbon atoms and one sulphonyl chloride group per 85 carbon atoms. Such a product may be represented as follows:

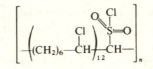

The reaction probably proceeds by the following radical mechanism:

$$\sim CH_2{-}CH_2 \sim \; + \; Cl\cdot \longrightarrow \; \sim CH_2{-}\overset{\cdot}{C}H \sim \; + \; HCl$$

$$\sim CH_2{-}\overset{\cdot}{C}H \sim$$

$$\xrightarrow{Cl_2} \; \sim CH_2{-}CHCl\sim \; + \; Cl\cdot$$

$$\xrightarrow{SO_2} \; \sim CH_2{-}\underset{\underset{SO_2\cdot}{|}}{C}H\sim \; \xrightarrow{Cl_2} \; \sim CH_2{-}\underset{\underset{SO_2Cl}{|}}{C}H\sim \; + \; Cl\cdot$$

2.4.2 Properties

Before vulcanization, chlorosulphonated polyethylene is a tacky, rubbery material of low tensile strength; it is soluble in chlorinated hydrocarbons. The sulphonyl chloride groups are reactive and may be used for cross-linking the polymer. Vulcanization is generally carried out by heating with metal oxides such as litharge or magnesium oxide in the presence of a little water. Reaction probably proceeds as follows:

$$\underset{\underset{\underset{\sim CH\sim}{|}}{\overset{\overset{Cl}{|}}{SO_2}}}{} \; + \; H_2O \; \longrightarrow \; \underset{\underset{\underset{\sim CH\sim}{|}}{\overset{\overset{OH}{|}}{SO_2}}}{} \; + \; HCl$$

$$MgO \; + \; 2HCl \; \longrightarrow \; MgCl_2 \; + \; H_2O$$

$$\begin{matrix} \sim CH\sim \\ | \\ SO_2 \\ | \\ OH \\ MgO \\ OH \\ | \\ SO_2 \\ | \\ \sim CH\sim \end{matrix} \; \longrightarrow \; \begin{matrix} \sim CH\sim \\ | \\ SO_2 \\ | \\ O \\ Mg \; + \; H_2O \\ O \\ | \\ SO_2 \\ | \\ \sim CH\sim \end{matrix}$$

The cross-linked product has good resistance to chemical attack, especially attack by ozone, oxygen and other oxidizing agents. As a rubber, the material has excellent mechanical properties and these are maintained over long periods of use at elevated temperatures. The rather high cost of the material limits its use to such applications as sheeting and wire and cable coating intended for service in demanding conditions.

2.5   ETHYLENE COPOLYMERS

Although ethylene may be copolymerized with numerous monomers, relatively few ethylene copolymers are manufactured. The more important comonomers are of three kinds, namely propylene, higher olefins, and carboxylic acids and esters.

Ethylene-propylene copolymers are important materials and are discussed in sections 2.7 and 2.8. Copolymers containing small amounts of α-olefins such as 1-butene, 1-hexene and 1-octene are normally referred to as linear low density polyethylene and are described in section 2.3.

Various ethylene copolymers are available in which the second comonomer is a carboxylic acid or ester such as vinyl acetate, acrylic acid, methacrylic acid, methyl acrylate or ethyl acrylate. These copolymers are produced by free radical high pressure polymerization using processes similar to those described in section 2.3.2(a). The random introduction of rather bulky side chains into the polymer leads to a progressive reduction in crystallinity and stiffness which is directly proportional to the molar content of the comonomer, until at about 20 mole % the copolymer is completely amorphous. At the same time, the presence of polar pendant groups leads to chain interaction and increased toughness.

Ethylene-vinyl acetate (EVA) copolymers with vinyl acetate content in the range 5–20% by weight are widely used. Compared to low density polyethylene, these copolymers show improved clarity, low temperature flexibility, impact strength, flexibility and stress crack resistance. They have greater oil and grease resistance but are more permeable to gases and water vapour; they are also less heat stable. Copolymers with about 5% vinyl acetate are used for film extrusion; such film finds application in food packaging. Copolymers with about 10% vinyl acetate are used for extrusion and injection moulding of such items as medical tubing, squeeze bulbs and gaskets. Copolymers with about 20% vinyl acetate find use as coatings and hot melt adhesives.

Ethylene-acrylic acid (EAA) copolymers containing up to 6.5% by weight acrylic acid and ethylene-methacrylic acid (EMAA) copolymers containing up to 15% by weight of methacrylic acid are used for extrusion coating on to aluminium foil and for laminations with metals and glass fibre. These copolymers are comparable to ethylene-vinyl acetate copolymers but have improved adhesion to polar substrates in a variety of aggressive environments. There is a tendency for the copolymers to cross-link at elevated temperatures through acid anhydride formation.

Ethylene-methyl acrylate (EMA) copolymers containing about 20% by weight methyl acrylate and ethylene-ethyl acrylate (EEA) copolymers containing up to about 20% by weight ethyl acrylate are available [7]. These copolymers are quite similar to ethylene-vinyl acetate copolymers but show greater thermal stability and can be processed over a wider temperature range. On the other hand, the acrylate copolymers are less satisfactory as packaging film, having poorer optical properties. Ethylene-methyl acrylate copolymers find use in extrusion coating and lamination, imparting heat-sealability and improved adhesion. Ethylene-ethyl acrylate copolymers are used for extrusion and injection moulding of such items as hose, gaskets, hospital sheeting and handle grips.

Copolymers of ethylene and methacrylic acid are also the basis of interesting materials which have been called *ionomers*. These are prepared by copolymerizing ethylene with 1–10% methacrylic acid using a high pressure process. The copolymer is then treated with a derivative of a metal (generally sodium or zinc) and some of the carboxylic groups are ionized. The sodium salt, for example, may be represented as follows:

$$\sim CH_2-\underset{\underset{\displaystyle COO^-}{|}}{\overset{\overset{\displaystyle CH_3}{|}}{C}}-CH_2-CH_2-CH_2-CH_2-CH_2-\underset{\underset{\displaystyle COO^-}{|}}{\overset{\overset{\displaystyle CH_3}{|}}{C}}-CH_2-CH_2 \sim$$

$$Na^+ \quad Na^+ \qquad\qquad Na^+ \quad Na^+$$

$$\sim CH_2-\underset{\underset{\displaystyle CH_3}{|}}{\overset{\overset{\displaystyle COO^-}{|}}{C}}-CH_2-CH_2-CH_2-CH_2-CH_2-\underset{\underset{\displaystyle CH_3}{|}}{\overset{\overset{\displaystyle COO^-}{|}}{C}}-CH_2-CH_2 \sim$$

Although the presence of carboxyl groups reduces crystallinity in the polymer, the ionic cross-links give enhanced stiffness and toughness and as a result the ionomers have many physical properties which are similar to those of low density polyethylene. They are, however, much clearer in thick sections and have better abrasion resistance and oil resistance. On the other hand, the ionomers are somewhat moisture sensitive and have low softening points. The ionic cross-links are stable at normal ambient temperatures but reversibly break down on heating so that ionomers can be melt processed by the normal techniques used for thermoplastics. These copolymers find application as extrusion coating materials and as skin packaging film.

2.6 POLYPROPYLENE

2.6.1 Development

It has been previously noted that co-ordination catalysts were originally employed by Ziegler in 1953 to effect the polymerization of ethylene. Natta extended the use of the catalysts to higher olefins and obtained isotactic polypropylene in 1954. The first commercial production of the polymer was by Montecatini in 1957. Polypropylene is now an important commercial polymer, being used principally in the injection moulding of diverse articles such as luggage, sterilizable hospital equipment, storage battery cases and washing machine agitators. Film finds application for food wrapping and fibre is widely used as the tufting material in carpets.

2.6.2   Preparation

High pressure, free radical processes of the type used to prepare polyethylene are not satisfactory when applied to propylene and other α-olefins bearing a hydrogen atom on the carbon atom adjacent to the double bond. This is attributed to extensive transfer of this hydrogen to propagating centres ($R\cdot$):

$$R\cdot \; + \; H-CH_2-CH=CH_2 \longrightarrow RH \; + \; H_2C-CH=CH_2$$
$$\updownarrow$$
$$H_2C=CH-\dot{C}H_2$$

The resulting allyl radical is resonance stabilized and has a reduced tendency to react with another monomer molecule.

Although the Phillips and Standard Oil processes can be used to prepare polypropylene, the polymer yields tend to be low and it appears that these processes have not been used for commercial production of polypropylene. Until about 1980, polypropylene has been produced commercially only by the use of Ziegler-Natta catalysts. Commonly a slurry process is used and is carried out in much the same manner as described previously for the preparation of polyethylene (see section 2.3.2(b)). In the case of polypropylene, some atactic polymer is formed besides the required isotactic polymer; but much of this atactic material is soluble in the diluent (commonly heptane) so that the product isolated is largely isotactic polymer. Recently, there has been a marked shift towards processes involving gas phase polymerization and liquid phase polymerization. Few details of these newer processes have been published. Gas phase processes resemble those described previously for the preparation of polyethylene (see section (2.3.2(b)) and swing plants are now feasible. In liquid phase processes polymerization is conducted in liquid propylene, typically at 2 MPa (20 atmospheres) and 55°C. Concurrently with these developments, new catalyst systems have been introduced. These materials have very high activity and the reduced levels that are required make it unnecessary to remove catalyst from the final polymer. Also, the new catalyst systems lead to polypropylene with higher proportions of isotactic polymer and removal of atactic polymer is not necessary.

2.6.3   Properties

As a first approximation, the physical properties of polypropylene resemble those of high density polyethylene but there are some significant differences.

For comparison, typical values for some properties of a standard commercial grade of polypropylene are given in Table 2.2. The most noticeable attribute of polypropylene is the increased softening point and consequent higher maximum service temperature. Properties such as tensile strength and stiffness, whilst comparable to high density polyethylene at room temperature, are maintained to a greater extent at elevated temperatures. At 140°C polypropylene is still sufficiently stiff for a strain-free article to retain its shape, whereas polyethylene is molten at this temperature. At room temperature the impact strength of polypropylene is comparable to that of high density polyethylene; however, the impact strength of polypropylene falls markedly as the temperature is reduced whereas polyethylene shows little change.

The solubility characteristics of polypropylene are similar to those of polyethylene. Thus polypropylene is insoluble at room temperature but is soluble in hydrocarbons and chlorinated hydrocarbons at temperatures above about 80°C. Polypropylene does not suffer environmental stress cracking of the kind shown by polyethylene.

In polypropylene, each alternate carbon atom in the polymer chain bears a tertiary hydrogen atom which is relatively labile. Thus, compared to polyethylene, polypropylene is more prone to attack by oxidizing agents. Of particular significance is the susceptibility of the polymer to oxidation by air at elevated temperatures, and antioxidants are generally added. As has been mentioned earlier, when polyethylene is exposed to high energy radiation or is heated with a peroxide, cross-linking occurs to give a useful material. When polypropylene is similarly treated, the cross-linking reactions are accompanied by an approximately equal amount of chain scission and useful properties are not developed.

Polypropylene can be chlorinated and chlorosulphonated in much the same way as polyethylene, but the reactions are accompanied by severe degradation and some cross-linking. These processes have, therefore, found little commercial utilization.

2.7 PROPYLENE COPOLYMERS

Four different types of propylene copolymers are of commercial significance; · in each case, ethylene is the comonomer.

The first materials are generally referred to as propylene random copolymers and are made by introducing small amounts of ethylene (normally 2–5%) into the polymerization reactor. The resulting random copolymers show greatly improved clarity and somewhat improved toughness compared to the homopolymer. Stiffness, however, is reduced.

The second type of materials are commonly called impact polypropylene or propylene block copolymers. The latter term is misleading since the materials are actually blends of polypropylene with up to 25% ethylene-propylene copolymer. These products are manufactured by using two polymerization reactors in sequence. Polypropylene homopolymer is formed in the first reactor and then passed to the second. Here ethylene and propylene are copolymerized to give a rubbery polymer which becomes intimately mixed with the homopolymer. Such reactor-made blends show significantly improved impact strength compared to the homopolymer.

The third type of materials are random copolymers of propylene and ethylene and involve more nearly equal amounts of each olefin. These materials are known as ethylene-propylene rubbers and are considered in the next section.

The fourth type of materials are blends of propylene with up to 65% ethylene-propylene rubber and are commonly referred to as polyolefin thermoplastic elastomers. These materials are considered in section 2.9

2.8 ETHYLENE-PROPYLENE RUBBERS

2.8.1 Preparation

Commercial ethylene-propylene rubbers (EPR or EPM*) generally contain about 35 mole % propylene although rubbery properties are shown by copolymers with a propylene content ranging from 30–60 mole %. At the present time, these materials are prepared exclusively by Ziegler-type processes. Generally, true solution processes are preferred in which a soluble catalyst system is used and the polymer remains in solution rather than form a slurry. A common soluble catalyst system is based on vanadium oxychloride/aluminium trihexyl. Catalysts of this type favour the formation of amorphous atactic polymers and lead to narrower molecular weight distributions than solid catalysts. Typically, polymerization is carried out at about 40°C in a solvent such as chlorobenzene or pentane and the polymer is isolated by precipitation with an alcohol.

The ethylene-propylene copolymers prepared by these means are random copolymers and do not contain long blocks of either ethylene or propylene units. Thus crystallization cannot occur and these products are completely amorphous.

In the ethylene-propylene copolymers described above there are virtually no double bonds; consequently vulcanization by conventional techniques using sulphur is not possible and peroxides have to be used. This limitation may be overcome by introducing unsaturation into the polymer by use of a third monomer in the copolymerization process. The third monomer is a non-conjugated diene; one of its double bonds enters into the polymerization

---

* M in this nomenclature refers to the methylene groups in the saturated polymer chains.

process becoming incorporated in the main polymer chain whilst the other double bond does not react and is left in a side chain and is available for subsequent vulcanization. Ethylene-propylene-diene terpolymers (EPTR or EPDM) which are sulphur-curable are important commercial rubbers. Various dienes have been found to be suitable for these terpolymers; the following are examples of commercially used compounds:

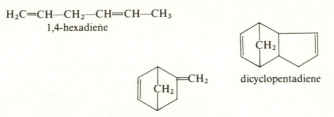

$$H_2C{=}CH{-}CH_2{-}CH{=}CH{-}CH_3$$
1,4-hexadiene

dicyclopentadiene

2-methylene-5-norbornene

The amount of unsaturation introduced is generally within the range 2–15 double bonds per 1000 carbon atoms, distributed randomly throughout the polymer. The structure of an ethylene-propylene-1,4-hexadiene terpolymer might therefore be represented as follows:

$$\sim CH_2{-}CH_2{-}CH_2{-}\underset{\underset{CH_3}{|}}{CH}{-}CH_2{-}CH_2{-}CH_2{-}\underset{\underset{CH_3}{|}}{CH}{-}CH_2{-}\underset{\underset{CH_2{-}CH{=}CH{-}CH_3}{|}}{CH}{-}CH_2{-}CH_2\sim$$

2.8.2 Properties

Uncured ethylene-propylene copolymers are soluble in hydrocarbons and have rather poor physical properties; useful technological properties are developed only on vulcanization. As mentioned above, the saturated copolymers are vulcanized by heating with peroxides whilst the terpolymers are vulcanized by conventional sulphur systems. The peroxide-cured rubbers have somewhat better heat ageing characteristics and resistance to compression set but sulphur-cured rubbers are more convenient to process and allow greater compounding freedom.

Ethylene-propylene rubbers are made from low cost monomers and have a low specific gravity (about 0.87); they therefore have considerable commercial potential as general purpose rubbers. Since the vulcanizates are saturated they have outstanding resistance to degradation by heat, oxygen and ozone. The limitations of these rubbers are low tear strength, difficulty in bonding to other elastomers and poor tack properties. Typical applications include weather strip for automobile windscreens, radiator and heater hose, and sheeting for outside use.

2.9 POLYOLEFIN THERMOPLASTIC ELASTOMERS

Thermoplastic elastomers are materials which possess, at normal temperatures, the characteristic resilience and recovery from extension of cross-linked elastomers but which exhibit plastic flow at elevated temperatures and can be fabricated by the usual techniques applied to thermoplastics, e.g. injection moulding and extrusion. These effects are associated with linear polymers containing segments which give rise to inter-chain attraction. At normal temperatures, these interactions have the effects of conventional covalent cross-links and confer elasticity but at elevated temperatures the secondary forces are inoperative and the material exhibits thermoplastic behaviour. Thus a thermoplastic elastomer has two intermingled polymeric systems, each with its own phase and softening temperature. In the useful temperature range one phase (the soft phase) is above its softening temperature whilst the other (the hard phase) is below its softening temperature. The hard phase thus acts to anchor the chains of the soft phase and the material is elastomeric. At temperatures above the softening temperature of the hard phase, the restrictions imposed by the hard phase disappear and the material behaves as a viscous liquid. On cooling, the hard phase re-solidifies and the material regains its elastomeric characteristics.

Several types of thermoplastic elastomers have been developed, including the following:

		Market share [8]
(i)	Polyolefin thermoplastic elastomers	18%
(ii)	Polyamide thermoplastic elastomers (section 10.3)	small
(iii)	Polyester thermoplastic elastomers (section 11.8)	6%
(iv)	Polyurethane thermoplastic elastomers (section 16.6.3)	19%
(v)	Styrenic thermoplastic elastomers (section 20.5.5)	52%

Polyolefin thermoplastic elastomers are generally blends of polypropylene with up to 65% ethylene-propylene rubber and it is supposed that short propylene blocks in the latter co-crystallize with segments of the polypropylene chains to give microcrystalline regions which act as cross-links. A recent development in this field has been the use of highly cross-linked ethylene-propylene rubbers (and other rubbers) in the blends to give so-called thermoplastic vulcanizates (TPVs). In these blends the rubber is present as finely dispersed particles in a polypropylene matrix. Compared to the simple blends, these materials have generally enhanced properties.

Polyolefin thermoplastic elastomers are characterized by low cost, good mechanical properties and excellent resistance to oxygen, ozone and polar solvents; they are attacked by non-polar and chlorinated solvents, especially

at elevated temperatures. The major use of polyolefin thermoplastic elastomers is in the automotive field for such parts as radiator grilles, headlight
surrounds and rub strips.

2.10 POLYISOBUTENE

2.10.1 Preparation

It was noted in the previous chapter (section 1.4.2 (a)) that the polymerization
of isobutene could be accomplished only with cationic initiators. Aluminium
chloride and boron trifluoride are the preferred initiators for commercial
processes; the separate addition of a co-initiator is not generally necessary and
adventitious substances possibly fulfil this role. At ordinary temperatures
polymerization is extremely rapid and leads to low molecular weight polymers which are viscous oils or sticky solids. However, at low temperature
(-80 to $-100°C$) high molecular weight material is produced. Even at these
low temperatures the reaction is complete in a few seconds and it is necessary
to have particularly efficient means of dissipating the heat evolved. Conventional batch processes are unsuitable and continuous processes are used in
which only small quantities of reactants are involved at any one moment. In
one process (Badische Anilin- & Soda-Fabrik A. G.), solutions of isobutene
and boron trifluoride in liquid ethylene are mixed on a moving belt so that
the polymerizing system is in the form of a thin film and heat is removed by
the vaporization of the solvent. The polymer is then mixed with an alkali or
ethanol to deactivate the initiator and treated with steam to remove water-
soluble contaminants. Another process for polyisobutene (Standard Oil Co.
(N.J.) (USA)) follows closely the procedure outlined in Section 2.11.2 for the
manufacture of butyl rubber.

It may be noted that polymers of isobutene have two methyl groups on
alternate carbon atoms along the chain and two hydrogen atoms on the other
chain carbon atoms:

The possibilities of atactic, isotactic and syndiotactic structures therefore do
not arise. Polyisobutene appears to be completely linear; no evidence for
branching has been found.

2.10.2 Properties

The physical properties of polyisobutene are very dependent on molecular
weight. Polymers with average molecular weight (M_w) of about 15 000 are

sticky viscous liquids whilst those with molecular weight of 100 000–200 000 are rubber-like, resembling unmilled crepe rubber.

Polyisobutene is non-crystalline when unstretched and is therefore soluble at room temperature in hydrocarbons and halogenated hydrocarbons. The material is resistant to most acids, alkalis and aqueous solutions, as would be expected from its saturated hydrocarbon structure and absence of tertiary hydrogen atoms. The lack of tertiary hydrogen atoms renders polyisobutene more resistant to oxidation than polypropylene; also, the less numerous and partially shielded methylene groups in polyisobutene are less reactive than those in polyethylene. However, polyisobutene is rather susceptible to thermal degradation since chain scission is favoured by the greater stability of the resultant tertiary free radical:

$$\sim CH_2-\underset{\underset{CH_3}{|}}{\overset{\overset{CH_3}{|}}{C}}\sim \longrightarrow \sim \dot{C}H_2 \ + \ \cdot\underset{\underset{CH_3}{|}}{\overset{\overset{CH_3}{|}}{C}}\sim$$

Polyisobutene may be chlorinated but the reaction is accompanied by severe degradation.

A limitation of polyisobutene is its tendency to cold flow and, as a result, the polymer finds little use in self-supporting form. Applications are restricted mainly to adhesives, fabric and paper coatings, and blends with other polymers. Low molecular weight polyisobutene is also used in caulking compounds.

2.11 BUTYL RUBBER

2.11.1 Development

As mentioned above, a major limitation of polyisobutene is its tendency to cold flow. It is to be expected that this limitation would be substantially overcome by cross-linking the polymer. However, vulcanization by conventional techniques using sulphur is not possible since the polymer is saturated and heating with peroxides leads to extensive chain scission. A method of overcoming this difficulty was discovered by Thomas and Sparks of Standard Oil Development Company in 1937. It was found that copolymers of isobutene containing small amounts of conjugated diene can be vulcanized with sulphur. The copolymer with isoprene, which is the preferred diene, is known as butyl rubber (or IIR) and has become an important commercial material. The first commercial butyl plant came on stream in 1943 in the USA.

2.11.2 Preparation

Commercial butyl rubbers generally contain 1–3 mole % isoprene. In a typical process, a solution of the monomers in methyl chloride is pre-cooled to $-100°C$ and then fed continuously into the base of a reactor together with a solution of aluminium chloride (initiator) in the same diluent. The reaction is highly exothermic and in order to maintain the temperature at $-100°C$ the reactor is fitted with a powerful stirrer and cooling coils containing liquid ethylene. The polymer forms as a granular slurry which is displaced continuously from the top of the reactor. The slurry passes into a flash tank where it is vigorously agitated with hot water. This treatment deactivates the initiator and extracts water-soluble products. At the same time the diluent and unreacted monomers are flashed off and are purified and re-used. Also at this point a lubricant and an antioxidant are introduced into the polymer. The lubricant, usually zinc stearate, prevents agglomeration of wet crumb and the antioxidant, usually phenyl-β-naphthylamine, reduces degradation during drying operations. The slurry then passes to a screen where wet polymer crumb is isolated. The crumb is dried at about 95–170°C, fed through an extruder and then hot milled. The object of these last operations is to compact the polymer and remove all traces of residual water.

In butyl rubber the isoprene is present as random 1,4-units and the copolymer thus has the following form:

$$\sim CH_2-\underset{\underset{CH_3}{|}}{\overset{\overset{CH_3}{|}}{C}}-CH_2-\underset{\underset{CH_3}{|}}{\overset{\overset{CH_3}{|}}{C}}-CH_2-\overset{\overset{CH_3}{|}}{C}=CH-CH_2-CH_2-\underset{\underset{CH_3}{|}}{\overset{\overset{CH_3}{|}}{C}}-CH_2-\underset{\underset{CH_3}{|}}{\overset{\overset{CH_3}{|}}{C}}\sim$$

2.11.3 Properties

Unvulcanized butyl rubber resembles polyisobutene of comparable molecular weight in many properties. It is soluble in hydrocarbons and chlorinated hydrocarbons and shows considerable cold flow.

The unvulcanized material may also be halogenated and both chlorinated and brominated butyl rubber are commercially available. These modified products are of interest in that they are more easily vulcanized than the straight copolymer and are more compatible with conventional diene rubbers. Halogenation is carried out by treating a solution of the rubber in an inert solvent, such as carbon tetrachloride, with the free halogen at room temperature. In commercial products, halogenation is normally continued until the polymer contains about one halogen atom per isoprene unit; this amounts to about 1% chlorine or 2% bromine. During halogenation, substitution and some chain scission occur. Substitution appears to occur predominantly at the isoprene units giving various allylic halide structures.

Since the substitution is little affected by light or by the presence of antioxidants or oxygen, a free radical reaction is unlikely; the following ionic mechanism has been suggested [9]:

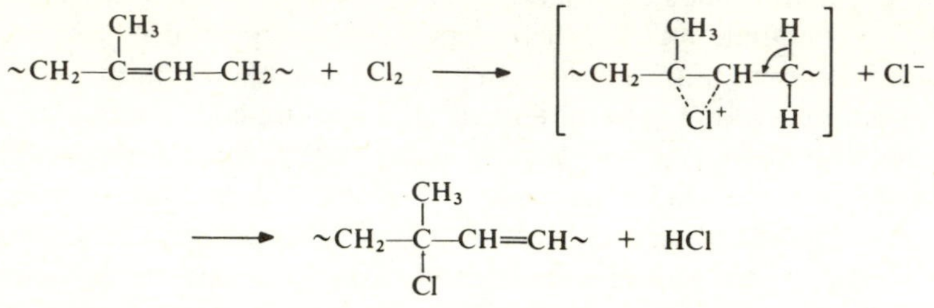

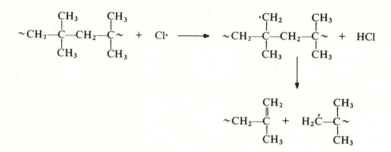

Chain scission, on the other hand, probably proceeds by a radical mechanism involving attack of halogen atoms on the methyl groups of isobutene units:

Vulcanization rates are higher than for normal butyl rubber because the presence of allylic halide increases the reactivity of the unsaturated sites. Brominated butyl rubber shows higher vulcanization rates than chlorinated butyl rubber. Halogenated butyl rubbers can be vulcanized with various reagents, e.g. diamines, dihydroxy aromatic compounds and zinc oxide.

Vulcanized butyl rubber broadly resembles vulcanized natural rubber in physical characteristics, but shows better resistance to abrasion, repeated flexing and tear; it also has lower resilience. As is to be expected from its almost completely saturated hydrocarbon structure, vulcanized butyl rubber has relatively good resistance to degradation by heat, oxygen and ozone. The outstanding property of butyl rubber, however, is its very low permeability to gases. Because of this last characteristic, butyl rubber is very widely used for tyre inner tubes. Other applications include automobile damping fixtures, radiator hose and window seals, cable insulation and engine mounting pads.

2.12 POLYBUTENE

Polybutene (polybutylene) (PB) is manufactured by stereospecific Ziegler-Natta polymerization of 1-butene:

Commercial products are predominantly isotactic (98–99.5%) and of high molecular weight ($M_w = 2.3 \times 10^5$–7.5×10^5). An unusual feature of the polymer is that it can exist in several distinct crystalline forms. Crystallization from the melt yields the metastable Form II, which changes to the stable Form I over a period of 5 to 7 days at room temperature. Other crystalline forms maybe obtained from solution. Form I has a crystalline melting point of 135°C and a density of 0.95 g/cm³. Form II has a crystalline melting point of 124°C and a density of 0.89 g/cm³.

Polybutene has properties to be expected from a crystalline polyolefin. It has a melting point and stiffness between those of high density and low density polyethylene. Its thermal stability lies between that of polyethylene and polypropylene. The polymer exhibits resistance to oxidizing and chemical environments broadly similar to that shown by polypropylene; like polypropylene, polybutene is immune from environmental stress cracking. An outstanding property of polybutene is its high creep resistance, due probably to the prevention of slippage of the polymer chains by the large number of ethyl side chains and the high molecular weight of commercial polymers. The main commercial application of polybutene is in extruded pipe, which has good resistance to rupture under pressure. Such pipe finds use in hot and cold water plumbing and for the transport of abrasive or corrosive materials.

2.13   POLY(4-METHYL-1-PENTENE)

Poly(4-methyl-1-pentene) was first introduced in 1965 by Imperial Chemical Industries Ltd (UK) but since 1975 the polymer has been manufactured solely by Mitsui Petrochemical Industries Ltd. Polymerization is carried out using a Ziegler-Natta catalyst such as titanium trichloride/diethylaluminium chloride in a hydrocarbon diluent at atmospheric pressure and 30–60°C:

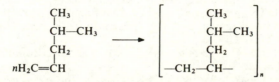

The commercial material contains a comonomer, possibly 1-hexene, which enhances clarity. The polymer configuration is predominantly isotactic.

Generally, this material has the basic physical properties to be expected from a crystalline polyolefin but in some respects it offers significant improvements over other polyolefins. One outstanding property of poly(4-methyl-1-pentene) is the very low specific gravity, which at 0.83 is the lowest of current polymers. The crystalline melting point is 240°C and the Vicat softening temperature is 179°C; these high values mean that a useful form stability is maintained up to about 200°C. The transparency of the polymer is of a high order, being comparable to poly(methyl methacrylate) and polystyrene. Poly(4-methyl-1-pentene) exhibits resistance to oxidizing and other chemical environments broadly similar to that shown by polypropylene; however, poly(4-methyl-1-pentene) does undergo environmental stress cracking comparable to low density polyethylene. The permeability of poly(4-methyl-1-pentene) to gases and water vapour is considerably higher than that for other polyolefins. Poly(4-methyl-1-pentene) may be extruded and injection moulded using standard equipment. The material has been used in several applications where transparency and heat resistance are required, e.g. medical and laboratory ware.

REFERENCES

1. Hambling, J. K. (1969) *Chem. Brit.*, **5**, 345.
2. Aubrey, D. W. (1968) in *Addition Polymers: Formation and Characterization*, (ed. D. A. Smith), Butterworths, London, ch. 7.
3. Friedlander, H. N. (1964) in *Crystalline Olefin Polymers, Part I*, (eds R. A. V. Raff and K. W. Doak), Interscience Publishers, New York.
4. Boor, J. (1979) *Ziegler-Natta Catalysts and Polymerizations*, Academic Press, New York, ch. 12.
5. Batleman, H. L. (1975) *Plastics Engineering*, **31**, No. 4, 73.
6. Al-Malaika, S. and Scott, G. (1983) in *Degradation and Stabilization of Polyolefins*, (ed. N. S. Allen), Applied Science Publishers, London, ch. 6.
7. Baker, G. (1980) *Plastics Engineering*, **36**, No. 2, 45.
8. Reed, D. (Jan., 1986) *Eur. Rubb. J.*, 16.
9. Baldwin, F. P. *et al.* (May 1961) *Rubber and Plastics Age*, 500.

BIBLIOGRAPHY

Renfrew, A. and Morgan, P. (eds) (1960) *Polythene – The Technology and Uses of Ethylene Polymers*, 2nd edn, Iliffe Books Ltd, London.
Sittig, M. (1961) *Polyolefin Resin Processes*, Gulf Publishing Co., Houston.
Topchiev, A. V. and Krentsel, B. A. (1962) *Polyolefins* (Translated by A. D. Norris), Pergamon Press, Oxford.
Gibson, R. O. (1964) *The Discovery of Polythene*, The Royal Institute of Chemistry, London, Lecture Series.
Raff, R. A. V. and Doak, K. W. (1965, 1964) *Crystalline Olefin Polymers, Part 1, Part 2*, Interscience Publishers, New York.
Frank, H. P. (1968) *Polypropylene*, Gordon and Breach Science Publishers, New York.
Miller, S. A. (ed.) (1968) *Ethylene and its Industrial Derivatives*, Benn Bros. Ltd, London.

Ritchie, P. D. (ed.) (1968) *Vinyl and Allied Polymers*, Vol. I, Iliffe Books Ltd, London.
Rubin, I. D. (1968) *Poly(1-Butene) – Its Preparation and Properties*, Gordon and Breach Science Publishers, New York.
Kresser, T. O. J. (1969) *Polyolefin Plastics*, Van Nostrand Reinhold Co., New York.
Sittig, M. (1976) *Polyolefin Production Processes – Latest Developments*, Noyes Data Corporation, Park Ridge.
Ahmed, M. (1982) *Polypropylene Fibres – Science and Technology*, Elsevier Scientific Publishing Co., Amsterdam.
Seymour, R. B. and Cheng, T. (eds.) (1986) *History of Polyolefins*, D. Reidel Publishing Company, Dordrecht.

Petroleum Chemicals
Staff of Royal Dutch/Shell Group of Companies (1983) *The Petroleum Handbook*, 6th edn, Elsevier, Amsterdam.
List, H. L. (1986) *Petroleum Technology*, Prentice-Hall, Englewood Cliffs.
Wiseman, P. (1986) *Petrochemicals*, Ellis Horwood Limited, Chichester.

3

POLYSTYRENE AND STYRENE COPOLYMERS

3.1 SCOPE

In this chapter the homopolymer, polystyrene, is considered together with styrene-acrylonitrile copolymers, acrylonitrile-butadiene-styrene copolymers and styrene-α-methylstyrene copolymers. The important styrene-butadiene copolymers are described with other diene polymers in Chapter 20. The use of styrene in the cross-linking of unsaturated polyesters is described in Chapter 11.

3.2 POLYSTYRENE

3.2.1 Development

Commercial interest in polystyrene began in the 1930s when the material was found to have good electrical insulation characteristics, and limited production was started by I. G. Farbenindustrie (Germany) and Dow Chemical Company (USA) shortly before the Second World War. During the war enormous quantities of styrene were produced in the USA for use in the synthetic rubber programme which was undertaken when supplies of natural rubber were no longer available. Thus when the war ended and natural rubber supplies were resumed, a large styrene capacity was available for civilian use and new outlets were sought. Applications for polystyrene were vigorously explored and the material was quickly adopted in many fields. Polystyrene is now one of the major commercial plastics, being very extensively used in such diverse applications as domestic appliances, food containers, packaging, toys and, in expanded form, thermal insulation.

3.2.2 Raw materials

The bulk of commercial styrene is prepared from benzene by the following route:

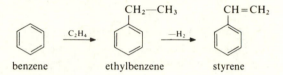

In the first stage, a Friedel-Crafts reaction is commonly carried out by treating benzene with ethylene in the liquid phase at 90–100°C at slightly above atmospheric pressure. The catalyst is aluminium chloride (with ethyl chloride as catalyst promoter). A molar excess of benzene is used to reduce the formation of polyethylbenzenes; the molar ratio of reactants is generally about 1:0.6. The reactants are fed continuously into the bottom of a reactor whilst crude product is removed from near the top. The product is cooled and allowed to separate into two layers; the lower layer, which consists of an aluminium chloride-hydrocarbon complex, is removed and returned to the reactor. The remaining ethylbenzene is then separated by distillation from polyethylbenzenes and benzene, which are recycled (since dealkylation also occurs under the reaction conditions). In newer plants, ethylbenzene is produced in a gas phase process. An excess of benzene is treated with ethylene at about 420°C and 1.2–2 MPa (12–20 atmospheres) in the presence of a zeolite catalyst.

The second stage of the styrene process involves the dehydrogenation of ethylbenzene. The reaction is carried out in the vapour phase at temperatures of 600–650°C over catalysts based on either ferric or zinc oxides with lesser amounts of other metallic oxides such as chromic, cupric and potassium oxides. The reaction is favoured by low pressure and in order to reduce the partial pressure of the ethylbenzene the feed is mixed with superheated steam before passage over the catalyst. Normally, a conversion of 35–40% per pass is achieved. The product is cooled and allowed to separate into two layers; the aqueous layer is discarded. The organic layer consists of styrene (about 37%), ethylbenzene (about 61%) and benzene, toluene and tar (about 2%). The separation of styrene by distillation is difficult because of the susceptibility of the monomer to polymerization at quite moderate temperatures and because the boiling point of styrene (145°C) is rather close to that of ethylbenzene (136°C). It is necessary therefore to use specially designed columns and to add a polymerization inhibitor (commonly sulphur) before distillation and to distil under reduced pressure. In a typical process, a four-column distillation train is used. In the first column benzene and toluene are removed at atmospheric pressure; in the second and third columns ethylbenzene is removed at about 5 kPa (35 mmHg); in the fourth column styrene is separated from sulphur and tar, also at about 5 kPa. Finally, an inhibitor is added to the styrene; *tert*-butyl catechol is preferred for this purpose rather than sulphur which leads to discoloration of the final polymer. Styrene is a colourless liquid with a characteristic odour.

3.2.3 Preparation

Styrene may be polymerized by means of all four techniques outlined in section 1.5, i.e. by bulk, solution, suspension and emulsion polymerization. Each of these methods is practised commercially, but solution polymerization is now the most extensively used. The four processes are described below.

(a) Bulk polymerization

In a common type of process, styrene is partially polymerized batch-wise by heating the monomer (without added initiator) in large vessels at about 80°C for 2 days until about 35% conversion is attained. The viscous solution of polymer in monomer is then fed continuously into the top of a tower which is some 25 feet high. The top of the tower is maintained at a temperature of about 100°C, the centre at about 150°C and the bottom at about 180°C. As the feed material traverses the temperature gradient, polymerization occurs and fully polymerized material emerges from the base of the tower. The reaction is controlled by a complex array of heating and cooling jackets and coils with which the tower is fitted. The molten material is fed into an extruder, extruded as filament and then cooled and chopped into granules. Since the product contains few impurities, it has high clarity and good electrical insulation properties. The polymer has a broader molecular weight distribution than polymer prepared at one temperature.

(b) Solution polymerization

Continuous solution processes have found wide commercial utilization, the main advantage over bulk methods being a lessening of the problems associated with the movement and heat transfer of viscous masses. However, the technique does require the added steps of solvent removal and recovery. Typically, a mixture of monomer, solvent (3–12% ethylbenzene) and initiator is fed into a train of three polymerization reactors, each with several heating zones. The reaction temperature is progressively increased, rising from 110–130°C in the first reactor to 150–170°C in the last. The polymer solution is then extruded as fine strands into a devolatilizing vessel. In this vessel, which is at a temperature of 225°C, removal of solvent and unreacted monomer takes place, being aided by the large surface area of the strands. The molten material is fed into an extruder, extruded as filament, cooled and chopped. It may be noted that this type of process is commonly regarded as a continuous bulk process since the amount of solvent used is so small.

(c) Suspension polymerization

Suspension processes simplify the heat transfer problems associated with bulk methods and, unlike solution methods, they do not involve solvent removal and recovery. The disadvantages of the suspension technique are that it requires the added step of drying and it does not readily lend itself to continuous operation. Typically, polymerization is carried out batch-wise in a stirred reactor, jacketed for heating and cooling. A typical formulation might be as follows [1]:

Styrene (inhibitor free)	100 parts by weight	
Water (demineralized)	70	
Tricalcium phosphate	0.8	(suspending agent)
Dodecylbenzene sulphonate	0.003	(suspending agent)
Benzoyl peroxide	0.2	(initiator)

Reaction temperature is about 90°C. When polymerization is complete, the product, in the form of a slurry, is washed with hydrochloric acid and water to remove suspending agent, centrifuged, dried in warm air (at about 60°C), extruded and chopped.

(d) Emulsion polymerization

Emulsion processes are not used for making solid grades of polystyrene. This is because these processes lead to polymer containing large quantities of soap residues which impair the electrical insulation properties and optical clarity. Emulsion polymerization does, however, find limited application in the production of polystyrene latex used in water-based surface coatings. The techniques employed are very similar to those used for other polymer latices, e.g. poly(vinyl acetate) latex (Chapter 5).

3.2.4 Structure

It is shown in section 1.6.3 that, on theoretical grounds, vinyl polymers may be expected to have a head-to-tail structure. For polystyrene there is experimental evidence that this is indeed the case. When polystyrene is pyrolysed *in vacuo* at 290–320°C the following fractions are obtained [2]:

Monomer	Dimer	Trimer	Tetramer	Residue
40	20	24	4	12%

Separation of the dimer and trimer fraction yields the following compounds:

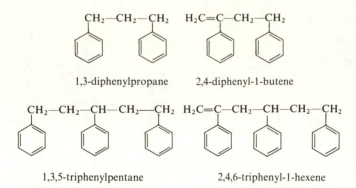

1,3-diphenylpropane   2,4-diphenyl-1-butene

1,3,5-triphenylpentane   2,4,6-triphenyl-1-hexene

Since the arrangement of styrene skeletons in these compounds is head-to-tail, it is reasonable to conclude that the same arrangement predominates in polystyrene itself:

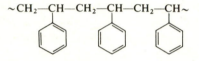

Commercial polymers normally have an average molecular weight (M_v) in the range 50 000–200 000.

Polystyrene produced by free radical polymerization techniques is largely syndiotactic but the polymer is non-crystalline. Isotactic polystyrene has been prepared by the use of Ziegler-Natta catalysts and n-butyllithium. Isotactic polystyrene has a high crystalline melting point of 230°C, which makes it a difficult material to process; also it is less transparent and more brittle than the non-crystalline polymer. For these reasons isotactic polystyrene has not achieved commercial importance.

3.2.5   Properties

Straight polystyrene (commonly referred to as crystal polystyrene) is a hard, rigid, rather brittle material. It has a relatively low softening point and does not withstand the temperature of boiling water. (See Table 3.1 for some comparative properties.) Polystyrene is highly transparent, transmitting about 90% of visible light; it also has a high refractive index of 1.59 which gives it particular brilliance. Since it is a hydrocarbon, polystyrene has good electrical insulation characteristics; also, it shows low moisture absorption and the electrical properties are maintained in humid conditions. Polystyrene is a low cost material and has good mouldability; these factors, together with its transparency and colourability, are the principal reasons for its widespread application.

Table 3.1 Comparative properties of typical commercial grades of styrene-containing polymers

	Polystyrene (general purpose)	Polystyrene (medium impact)	Polystyrene (high impact)	SAN	ABS (medium impact)	ABS (high impact)
Specific gravity	1.04	1.04	1.05	1.08	1.04	1.04
Yield tensile strength (MPa)	42	31	18	69	40	37
(lbf/in^2)	6100	4500	2600	10 000	5800	5400
Ultimate elongation (%)	2.0	25	60	2.5	25	25
Impact strength, Izod (J/m)	19	64	110	27	270	400
(ft lbf/in)	0.35	1.2	2.0	0.5	5.0	7.5
Softening point, Vicat (°C)	100	100	100	110	105	106

Polystyrene is readily soluble in a wide range of solvents. It is dissolved by aromatic hydrocarbons such as benzene and toluene; by chlorinated hydrocarbons such as carbon tetrachloride, chloroform, trichloroethylene and o-dichlorobenzene; by ketones such as methyl ethyl ketone (but not acetone); by esters such as ethyl acetate and amyl acetate; and by a few oils such as pine oil. Some non-solvents such as alcohols, heptane and kerosene cause crazing and cracking, particularly if internal stresses are present in the sample.

Polystyrene is unaffected by most acids, alkalis and aqueous solutions but strong oxidizing acids attack the polymer. Oxidation of polystyrene occurs in air on exposure to ultraviolet light and/or elevated temperature. Prolonged exposure results in deterioration of mechanical properties but before these effects become appreciable the polymer generally becomes discoloured. The yellowing of polystyrene in sunlight is a limitation of the material although considerable improvement may be effected by the addition of stabilizers such as o-hydroxybenzophenones and o-hydroxyarylbenzotriazoles, e.g.

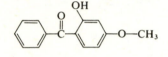

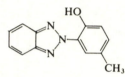

2-hydroxy-4-methoxybenzophenone   2-(2'-hydroxy-5'-methylphenyl)benzotriazole

These compounds absorb strongly in the region 3000–4000 Å and are light stable. Their mode of action is mainly through competitive absorption of the ultraviolet energy responsible for the polymer degradation. The yellowing of polystyrene is accompanied by the formation of carbonyl groups and it is generally supposed that a free radical reaction occurs which is analogous to that described previously for polyethylene (section 2.3.3). Photodecomposition of an intermediate hydroperoxide results in cleavage of the polymer chain and formation of a ketone:

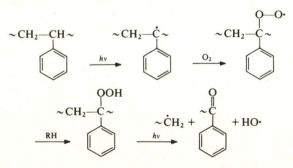

It has also been suggested [3] that, in addition to the above, a hydroperoxide group may form on the benzene nucleus to give a quinone-type compound:

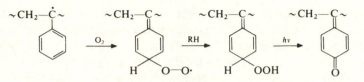

A further possibility is that α-diketones are also formed [4]:

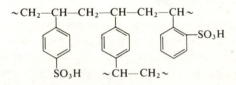

(See Reference 5 for a review of photodecomposition of polystyrene.)

The benzene nucleus in polystyrene undergoes normal aromatic reactions and thus the polymer may be, for example, alkylated, halogenated, nitrated and sulphonated. A frequent practical difficulty in carrying out such reactions is one of finding a reaction medium in which the polymer is soluble; also, chain scission, cross-linking and discoloration often accompany these reactions. Chemically-modified polystyrenes have found little commercial application but sulphonated styrene-divinylbenzene copolymers (which are cross-linked) find use as cationic exchange resins; the nature of such resins is illustrated by the following structure:

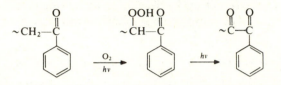

3.2.6 Rubber-modified polystyrenes

A serious limitation of polystyrene in many applications is its brittleness and a number of attempts have been made to improve the polymer in this respect. The most successful approach to this problem has been the addition of rubbery materials (usually 5–15%) to the polymer and rubber-modified polystyrene (commonly referred to as impact polystyrene) is an important commercial material. In fact, production of impact polystyrene currently exceeds that of crystal polystyrene. Several rubbers have been tried for the preparation of impact polystyrene but cis-1,4-polybutadiene is the most commonly used. The method of mixing the polystyrene and rubber has a profound effect on the properties of the product. Simple blending of the two polymers in, for example, an internal mixer or two-roll mill gives a product which has only slightly better impact strength than straight polystyrene. Likewise, the mixing of the polymers in latex form and isolation of the product by coagulation results in only marginal improvement. However, much better results are obtained if the polystyrene is prepared in the presence of the rubber. In one commercial process, for example, rubber is dissolved in a styrene-solvent mixture and then the styrene is polymerized in solution, using techniques similar to those outlined in section 3.2.3(b). The products of such methods contain not only polystyrene and straight rubber but also a graft polymer in which polystyrene side chains are attached to the rubber:

$$\sim CH_2-CH=CH-CH_2\sim \ + \ I\cdot \ \longrightarrow \ \sim \overset{\cdot}{C}H-CH=CH-CH_2\sim \ + \ IH$$

$$\sim \overset{\cdot}{C}H-CH=CH-CH_2\sim \ + \ nH_2C=CH \longrightarrow \ \sim \underset{\underset{n}{\left[CH_2-CH-\right]}}{CH}-CH=CH-CH_2\sim$$

where $I\cdot$ is an initiator fragment.

Compared to straight polystyrene, rubber-modified polystyrenes have much improved impact strength, but they have reduced clarity, softening point and tensile strength. (See Table 3.1.)

3.2.7   Expanded polystyrene

Expanded polystyrene has become very important as a thermal insulating material and to a lesser extent as a packaging material.

It may be prepared in various ways but most commercial methods make use of 'expandable beads'. In one process, styrene is polymerized in suspension in much the same way as described previously (section 3.2.3(c)) except that a low-boiling hydrocarbon, such as n-pentane, is added to the system. This modification results in the formation of polystyrene beads containing 5–8% of the volatile hydrocarbon. Alternatively, a slurry of preformed polystyrene beads may be treated under heat and pressure with the volatile hydrocarbon. The impregnated beads are then expanded, commonly by treatment with steam. When the beads are heated in steam, they soften and volatilization of the low-boiling hydrocarbon and diffusion of steam into the beads cause the beads to expand to about forty times their original size. (At this stage the beads are not fused together.) The beads are then allowed to stand so that air permeates into them and they attain atmospheric pressure at room temperature. The beads are subsequently loaded into a mould through which steam is passed; they again expand (by a small amount) and, being enclosed in the mould, are consolidated into a block.

By such methods, it is possible to prepare polystyrene foam with density as low as 16 kg/m³ (1 lb/ft³). This material has extremely low thermal conductivity, being comparable to expanded cork and glass wool in this respect.

3.3   STYRENE-ACRYLONITRILE COPOLYMERS

Many copolymers of styrene have been investigated, very often in the hope of finding a material which has greater heat resistance and toughness than straight polystyrene whilst retaining the low cost, rigidity and transparency of the homopolymer. Up to the present time results have been rather disappointing and, with the exception of acrylonitrile-containing materials, styrene copolymers have not achieved much commercial importance as rigid materials.

Commercial styrene-acrylonitrile copolymers (SAN) generally contain 20–30% acrylonitrile. They are random amorphous copolymers and are produced by bulk or suspension polymerization using techniques similar to those described previously for the homopolymer (section 3.2.3).

Compared to straight polystyrene, styrene-acrylonitrile copolymers have a higher softening point and improved impact strength. (See Table 3.1.) They are also transparent but tend to have a slight yellow tint. Because of the polar nature of acrylonitrile, the copolymers are more resistant to hydrocarbons and oils than polystyrene. The higher the acrylonitrile content the greater the heat resistance, impact strength and chemical resistance but the ease of moulding declines. Styrene-acrylonitrile copolymers have found some use in applications with somewhat more stringent requirements than can be met by straight polystyrene, e.g. appliance knobs, refrigerator compartments, mixer bowls, cassette cases and syringes.

3.4   ACRYLONITRILE-BUTADIENE-STYRENE COPOLYMERS

Although the impact strength of acrylonitrile-styrene copolymers is higher than that of polystyrene, it is still sufficiently low to make it a limiting factor in many applications. Understandably, therefore, the addition of rubbery materials to acrylonitrile-styrene copolymers has been extensively investigated and as a result materials based on acrylonitrile, butadiene and styrene have become commercially important. Such materials are commonly referred to as ABS polymers. It should be noted that commercial ABS polymers are not random copolymers of acrylonitrile, butadiene and styrene but are what have become known as multipolymers or polyblends. A few other related multipolymers are also available and are described in this section.

3.4.1   Preparation

Two basic processes are used commercially to prepare ABS polymers, namely blending and grafting. These processes give rise to materials which are rather different to each other. Of the two processes, grafting is now the more important.

(a)   Blending

The products obtained by this method are mechanical blends of styrene-acrylonitrile copolymers and acrylonitrile-butadiene rubbers. The preferred method of preparation is by blending latices of the two copolymers and coagulating the mixture. A wide range of products is possible, depending on the composition of each copolymer and the relative amounts of each employed. A typical blend would consist of the following (solids):

70 parts of styrene-acrylonitrile copolymer (70:30)
40 parts of acrylonitrile-butadiene rubber (35:65)

It has been found that non-cross-linked acrylonitrile-butadiene rubbers are compatible with styrene-acrylonitrile copolymers and the mixtures show little improvement in impact strength and have low softening points. However, if the rubber is sufficiently cross-linked so as to be not completely soluble in the copolymer then the mixtures have high impact strengths and high softening points. A convenient method of preparing a suitably cross-linked acrylonitrile-butadiene rubber is to take an emulsion polymerization to high conversion; alternatively, a small amount of divinylbenzene can be added to the emulsion recipe. After the two latices have been mixed, coagulation is brought about by the addition of either an acid or a salt. The resulting crumb is washed, filtered, dried, extruded and chopped into granules.

An alternative method of preparing a blend of the two copolymers is by mixing the solids on a two-roll mill. In this case, a non-cross-linked acrylonitrile-butadiene rubber may be used as starting material. The rubber is firstly cross-linked by milling with a peroxide and then the styrene-acrylonitrile copolymer is added.

The physical nature of these blends does not appear to be the same as that of rubber-modified polystyrenes. When this type of ABS polymer is treated with a solvent such as methyl ethyl ketone the sample swells and only partially breaks up; this indicates that rubber networks permeate the styrene-acrylonitrile copolymer matrix. When rubber-modified polystyrenes are treated with a solvent such as toluene, complete disintegration into fine particles occurs.

(b)   Grafting

In this method of preparing ABS polymers, acrylonitrile and styrene are polymerized in the presence of a polybutadiene latex. A wide range of products is possible, depending on the relative quantities of reactants. A typical recipe might be as follows [6]:

Polybutadiene latex (solids)	34 parts by weight
Acrylonitrile	24
Styrene	42
Water	200
Sodium disproportionated rosin	2.0 (surfactant)
Mixed tertiary mercaptans	1.0 (transfer agent)
Potassium persulphate	0.2 (initiator)

The reaction is carried out at about 50°C. The solid product is then isolated from the latex by the techniques described in the previous section.

ABS polymers prepared in this way consist of a continuous matrix of styrene-acrylonitrile copolymer, dispersed particles of polybutadiene and a boundary layer of polybutadiene grafted with acrylonitrile and styrene.

3.4.2   Properties

As has already been mentioned, the range of possible ABS polymers is very large since both the ratio of the three monomers and the manner in which they are assembled in the final product can be varied considerably. Thus commercial ABS polymers are available with appreciable differences in properties, but they are generally characterized by high impact strengths and softening points as high as, or higher than, straight polystyrene. (See Table 3.1 for typical values.)

The ABS polymers prepared by grafting contain varying amounts of ungrafted polybutadiene (which has a low glass transition temperature) and therefore have good low-temperature impact strength. Also, the ABS polymers prepared by grafting are superior to those obtained by blending in that moulded specimens commonly have a better surface appearance.

In recent years, special grades of ABS polymers have been produced in which a fourth monomer is present. For example, heat-resistant grades contain α-methylstyrene (see section 3.5) and transparent grades incorporate methyl methacrylate.

ABS polymers can be injection moulded and extruded and have been used in a great variety of applications which require toughness, rigidity and good appearance, e.g. automobile fascia panels and radiator grilles, household appliances, business machines, telephones and pipe and pipe fittings.

3.4.3   Related multipolymers

ABS polymers have poor resistance to outdoor ultraviolet light and exposure results in significant changes in appearance and mechanical properties. Improvements can be effected by replacing the polybutadiene by other rubbers and a few styrene-acrylonitrile multipolymers of this kind are commercially available. In acrylic-styrene-acrylonitrile multipolymers (ASA) an acrylate rubber (see section 6.4) is used whilst in acrylonitrile-ethylene-propylene-styrene multipolymers (AES) an ethylene-propylene-diene rubber (see section 2.8) is utilized. Acrylonitrile-chlorinated polyethylene-styrene multipolymers (ACS) contain elastomeric chlorinated polyethylene (see section 2.3.3) and in addition to showing good weatherability have excellent flame-retardant properties.

3.5   STYRENE-α-METHYLSTYRENE COPOLYMERS

α-Methylstyrene is obtained from benzene and propylene by the following route:

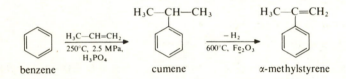

α-Methylstyrene is not readily polymerized by free radical initiators but it is susceptible to cationic polymerization. Polymerization can be effected, for example, in ethyl chloride at $-130°C$ with aluminium chloride but poly(α-methylstyrene) prepared in this manner is a clear brittle solid which has little commercial application. Polymers with much lower molecular weight (300–600) obtained by using alkaline earth catalysts, however, are viscous liquids and have found some use as plasticizers in surface coatings and adhesives.

Styrene-α-methylstyrene copolymers are marketed for extrusion and injection moulding. Compared to polystyrene, these copolymers have the advantage of higher softening points (104–106°C); they too are transparent and colourless.

REFERENCES

1. Grim, J. M. (1955) U.S. Patent 2715118.
2. Staudinger, H. and Steinhofer, A. (1935) *Ann.*, **517**, 35.
3. Achhammer, B. G. *et al.* (1951) *J. Res. Nat. Bur. Std.*, **47**, 116.
4. Reiney, M. J. *et al.* (1953) *J. Res. Nat. Bur. Std.*, **51**, 155.

5. Savides, C. *et al.* (1968) in *Stabilization of Polymers and Stabilizer Processes,* (ed. R. F. Gould), American Chemical Society, Washington, ch. 20.
6. Nelb, R. G. (1964) in *Manufacture of Plastics,* (ed. W. M. Smith), Reinhold Publishing Corporation, New York, ch. 11.

BIBLIOGRAPHY

Boundy, R. H. and Boyer, R. F. (eds.) (1952) *Styrene: Its Polymers, Copolymers and Derivatives*, Reinhold Publishing Corporation, New York.
Teach, W. C. and Kiessling, G. C. (1960) *Polystyrene*, Reinhold Publishing Corporation, New York.
Basdekis, C. H. (1964) *ABS Plastics*, Reinhold Publishing Corporation, New York.
Bishop, R. B. (1971) *Practical Polymerization for Polystyrene*, Cahners Books, Boston.
Brighton, C. A., Pritchard, G. and Skinner, G. A. (1979) *Styrene Polymers: Technology and Environmental Aspects*, Applied Science Publishers Ltd, London.

POLY(VINYL CHLORIDE) AND RELATED POLYMERS

4.1 SCOPE

In this chapter the homopolymer, poly(vinyl chloride) (PVC) is considered together with vinyl chloride-vinyl acetate copolymers and other vinyl chloride copolymers of lesser importance. Also discussed are the commercially important copolymers of vinylidene chloride.

4.2 POLY(VINYL CHLORIDE)

4.2.1 Development

Although the polymerization of vinyl chloride was reported as early as 1872, the polymer remained merely a laboratory curiosity for over 50 years. The principal reason for this was that the material could be processed in the melt only at temperatures where decomposition was appreciable. Commercial interest in poly(vinyl chloride) was revealed in patents filed independently in 1928 by Carbide and Carbon Chemical Corporation (USA), E. I. du Pont de Nemours and Company (USA) and I. G. Farbenindustrie (Germany). In each case the patents described vinyl chloride-vinyl acetate copolymers which could be processed at temperatures not sufficiently high to cause serious decomposition. In 1930 B. F. Goodrich Chemical Company (USA) discovered an alternative approach when it was found that poly(vinyl chloride) could be plasticized with high-boiling liquids like tritolyl phosphate to give material not unlike the copolymers. Thus the way was opened for the development of vinyl chloride polymers, and commercial production began in Germany and the USA in 1933. In the main, the plasticized homopolymer has found greater application than the copolymers. During the Second World War there was great demand for poly(vinyl chloride) for cable insulation. After the war the civilian use of poly(vinyl chloride) expanded rapidly, aided by the development of new types of materials such as paste polymers, easy-processing polymers and rigid poly(vinyl chloride). An important event was the discovery in 1974 that vinyl chloride is a human carcinogen, causing angiosarcoma of the liver. This has resulted in significant changes in process operation to ensure acceptable plant hygiene, environmental protection and low levels of residual monomer in the polymer. Polymer with a residual monomer content of less than 5 ppm is now generally available while polymer intended for use in contact with foodstuffs has a monomer content of less than 1 ppm.

In terms of tonnage, poly(vinyl chloride) is (with polyethylene, polypropylene and polystyrene) one of the four most important plastics currently in use. It is extremely widely used in such applications as bottles, building accessories, cable insulation, flooring, packaging film and pipe.

4.2.2 Raw materials

There are two principal industrial routes for the preparation of vinyl chloride (often designated as VCM for vinyl chloride monomer), namely from acetylene and from ethylene. Ethylene is cheaper than acetylene and although some vinyl chloride is still manufactured from acetylene, new capacities are unlikely to be based on this feedstock (except in countries such as South Africa which have coal-based economies).

Acetylene itself is manufactured in two ways. The first method is based on calcium carbide, which is obtained by heating a mixture of coke and lime at about 3000°C in an electric furnace; treatment of the carbide with water yields acetylene:

$$CaO + 3C \longrightarrow CaC_2 + CO$$
$$CaC_2 + 2H_2O \longrightarrow C_2H_2 + Ca(OH)_2$$

Depending on the purity of the carbide used, the acetylene contains impurities such as ammonia, arsine, hydrogen sulphide and phosphine. Scrubbing with water removes all these impurities except phosphine which is removed with acid dichromate.

The second method of production of acetylene is based on the cracking either of methane in natural gas or of higher hydrocarbons in, for example, a naphtha feedstock. Various processes have been devised for these cracking operations, which in the case of methane, may be represented as:

$$2CH_4 \rightleftharpoons C_2H_2 + 3H_2$$

The formation of acetylene is favoured by high temperatures. Probably the most widely used commerical processes are those which employ partial oxidation as the means of providing the energy for the main reaction. Thus the methane is mixed with oxygen (insufficient to oxidize all the methane) and ignited. Part of the methane is oxidized according to the reaction:

$$2CH_4 \;+\; 3O_2 \longrightarrow 2CO \;+\; 4H_2O$$

This reaction provides a flame temperature of about 1500°C in a specially designed burner and the remainder of the methane is subjected to cracking. The cracked gases, which contain about 8% by volume of acetylene, are cooled and scrubbed with water to remove carbon. The scrubbed gases are then compressed to about 1 MPa (10 atmospheres) and the acetylene is extracted with a selective solvent such as methylpyrrolidone. The solution passes to a tower in which the pressure is released and pure acetylene is recovered.

Ethylene is obtained from cracking operations as described in section 2.2.

(a) Preparation of vinyl chloride from acetylene

This method involves the addition of hydrogen chloride to acetylene:

$$HC{\equiv}CH \;+\; HCl \longrightarrow H_2C{=}CHCl$$

The reaction is carried out in the vapour phase in a multi-tubular reactor packed with a catalyst of mercuric chloride on an activated charcoal support. The reaction is highly exothermic and cooling is applied to keep the temperature at 90–140°C. The pressure used is about 0.15 MPa (1.5 atmospheres). The gases from the reactor are cooled and washed with aqueous sodium hydroxide to remove unreacted hydrogen chloride. The product is then liquefied by cooling to $-40°C$ and pure vinyl chloride is obtained by fractional distillation. Provided pure reactants are used, this preparation of vinyl chloride is clean and easily accomplished. Vinyl chloride is a colourless gas (b.p. $-14°C$) with a pleasant, sweet odour. It is conveniently handled, under slight pressure, as a liquid, which may be stored without the addition of a polymerization inhibitor.

(b) Preparation of vinyl chloride from ethylene

The traditional ethylene route to vinyl chloride consists of the following two-part process:

$$H_2C{=}CH_2 \;+\; Cl_2 \longrightarrow ClH_2C{-}CH_2Cl \longrightarrow H_2C{=}CHCl \;+\; HCl$$

The first part of the process consists of the addition of chlorine to ethylene to give ethylene dichloride. Chlorination is commonly carried out by allowing the gases to react in a solvent, conveniently ethylene dichloride. A metal halide catalyst such as ferric chloride is used and the reaction temperature is kept down to 50–70°C to minimize the formation of more highly chlorinated products. A pressure of about 0.4 MPa (4 atmospheres) is used. In the second part of the process ethylene dichloride is dehydrochlorinated. Typically, this reaction is carried out by pyrolysis of ethylene dichloride at about 500°C and 2.5 MPa (25 atmospheres). The exit gases are quenched in a stream of unreacted ethylene dichloride; the gases which do not condense are scrubbed with water to recover hydrogen chloride. Vinyl chloride is obtained from the liquid mixture by distillation under pressure and then purified by redistillation.

In the late 1950s processes for producing ethylene dichloride from ethylene by oxychlorination rather than by direct chlorination were developed. In these processes ethylene is treated with a mixture of oxygen and hydrogen chloride in the presence of a catalyst consisting of a mixture of cuprous and cupric chlorides:

$$H_2C{=}CH_2 \ + \ 2HCl \ + \ \tfrac{1}{2}O_2 \ \longrightarrow \ ClH_2C{-}CH_2Cl \ + \ H_2O$$

The catalyst mixture is supported on alumina or silica and is either packed into a tubular reactor or used in a fluidized bed. Reaction conditions are 200–350°C and 0.2–1 MPa (2–10 atmospheres) and high yields of ethylene dichloride are obtained. The ethylene dichloride is then pyrolysed as described previously (and the hydrogen chloride produced is recycled).

It will be noted that an important feature of the oxychlorination route is that no by-products are formed (other than water) whereas in direct chlorination half the chlorine consumed appears as hydrogen chloride, for which there may be no outlet. Often, producers operate a balanced facility in which the hydrogen chloride obtained from a direct chlorination process is used in an oxychlorination process.

4.2.3 Preparation

In commercial practice, poly(vinyl chloride) is mainly prepared by suspension polymerization whilst bulk and emulsion polymerization are used to a lesser extent. The homopolymer is seldom made by solution methods.

(a) Bulk polymerization

The only commercially successful bulk polymerization process for poly(vinyl chloride) is that developed by Pechiney St. Gobain (now Rhone-Poulenc Industries) (France). This process is conducted in two stages, permitting better control of particle morphology than is possible with a one-stage process. The first stage is carried out in a stainless steel reactor, jacketed for heating and cooling and fitted with a reflux condenser and high speed agitator. About half of the monomer required for the final amount of polymer is fed into the reactor together with an acyl peroxide or peroxydicarbonate initiator. In the first stage, polymerization is carried out at about 60–75°C and 0.5–1.2 MPa (5–12 atmospheres) for a short time (about 20 minutes) to give a conversion of about 8%. At this point the product consists of small particles of polymer dispersed in liquid monomer (since the polymer is insoluble in the monomer). The size of the polymer particles is determined principally by the rate of agitation and must be carefully controlled since it affects the final processing properties of the polymer. A mean particle diameter of about 10^{-5} cm is usual for the first stage product (pre-polymer). For the second stage of the process, the seed is transferred into a larger reactor, jacketed for heating and cooling and fitted with a reflux condenser and low speed agitator. Additional monomer is added to the reactor together with a further quantity of initiator such as diisopropyl peroxydicarbonate. In the second stage, polymerization is carried out at a constant pressure of about 1 MPa (10 atmospheres) while the temperature rises from about 55°C to 75°C. Reaction proceeds for 3–5 hours until a conversion of about 80% is reached. At this point the product is in the form of a powder containing absorbed monomer. Unreacted monomer is distilled off and recycled and the remaining product is degassed *in vacuo* using steam or nitrogen as a carrier. The final product consists of particles (about 10^{-2} cm in diameter) which are agglomerates of smaller particles (about 10^{-4} cm in diameter).

(b)   Suspension polymerization

The principal characteristics of suspension polymerization have been described in the previous discussion of polystyrene (section 3.2.3(c)). Typically, the suspension polymerization of vinyl chloride is carried out batch-wise in a stirred reactor, jacketed for heating and cooling. The reactor is also connected to a vacuum line. A typical basic formulation might be as follows:

Vinyl chloride	100 parts by weight
Water	150
Poly(vinyl alcohol)	0.1 (suspending agent)
Sodium carbonate	0.02 (pH regulator)
Di(2-ethylhexyl) peroxydicarbonate	0.04 (initiator)

All the ingredients, except the monomer, are added to the reactor which is then evacuated. Liquid vinyl chloride is then drawn in from weighing tanks and the reactor is sealed. The reactants are heated to about 50°C and the pressure in the reactor rises to about 0.7 MPa (7 atmospheres). This temperature is maintained for about 6 hours after which time the pressure begins to drop as the last of the monomer is consumed. When the pressure reaches about 0.07 MPa (0.7 atmospheres) (which corresponds to about 90% conversion) the residual monomer is vented off and recycled. The slurry is then treated at 80–120°C with large quantities of steam in a column to strip residual monomer. The slurry is discharged to a centrifuge where the polymer is separated and washed. Finally, the polymer is dried in hot air at 70–100°C and screened to remove any oversize particles. The final product consists of particles (about 10^{-2} cm in diameter) which are agglomerates of smaller particles (about 10^{-4} cm in diameter).

The nature of the suspending system used in suspension polymerization (sometimes called granular polymerization) of vinyl chloride is of technological significance in that it determines the nature of the polymer particles obtained. It is generally desirable to produce porous particles. Such particles facilitate removal of residual monomer, readily absorb additives such as stabilizers, lubricants and plasticizers, and make for easier fabrication into the finished article. Widely used suspending agents include partially hydrolysed grades of poly(vinyl alcohol) (see section 5.4.1) and cellulose derivatives such as methyl cellulose and sodium carboxymethyl cellulose (see section 13.7).

(c)   Emulsion polymerization

Poly(vinyl chloride) prepared by emulsion techniques contains soap residues and, as a result, the heat and colour stabilities and the electrical insulation properties are rather poor compared to those of suspension polymer. Nevertheless, emulsion polymer is manufactured for pastes which find use in non-critical applications. There is also some direct use of poly(vinyl chloride) latices for coating and impregnating paper and textiles. Emulsion polymerization is carried out in a pressure reactor of the type used for suspension polymerization. A typical basic formulation might be as follows:

Vinyl chloride	100 parts by weight	
Water	200	
Ammonium stearate	3	(emulsifier)
Ammonium persulphate	0.3	(initiator)

For the above recipe a reaction temperature of about 50°C is appropriate but if a redox initiating system (see section 20.5.3(a)) is used rapid polymerization may be achieved at about 20°C. Reaction is allowed to continue, usually for 1–2 hours, until the pressure in the reactor drops and then the resulting latex is stripped of residual monomer and spray dried. The polymer obtained by this type of process is in the form of particles (about 10^{-3} cm in diameter) which are agglomerates of smaller primary particles (about 10^{-4} cm in diameter).

4.2.4   Structure

It is shown in section 1.6.3 that, on theoretical grounds, vinyl polymers may be expected to have a head-to-tail structure. For poly(vinyl chloride) there is experimental evidence that this is indeed the case. When poly(vinyl chloride) in solution in dioxan is treated with zinc dust the resulting polymer is saturated and has a small chlorine content [1]. These findings are consistent with a head-to-tail structure of the initial poly(vinyl chloride), dechlorination of which may be represented as:

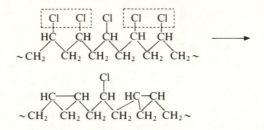

Since dechlorination involves random pairs of chlorine atoms it is to be expected that some unreacted chlorine atoms would become isolated, as shown above, and complete dechlorination does not occur. The amount of chlorine which is not removed (about 15% of that originally present) agrees closely with the quantity predicted statistically for a completely head-to-tail polymer. If poly(vinyl chloride) contained head-to-head units then dechlorination would lead to unsaturation and if the polymer were completely head-to-head/tail-to-tail, complete dechlorination would be possible:

$$\sim CH_2-\underset{\fbox{Cl}}{CH}-\underset{\fbox{Cl}}{CH}-CH_2-CH_2-\underset{\fbox{Cl}}{CH}-\underset{\fbox{Cl}}{CH}-CH_2-CH_2-\underset{\fbox{Cl}}{CH}-\underset{\fbox{Cl}}{CH}-CH_2 \sim \longrightarrow$$

$$\sim CH_2-CH=CH-CH_2-CH_2-CH=CH-CH_2-CH_2-CH=CH-CH_2 \sim$$

It may be noted that recent studies using NMR techniques indicate that the number of intact head-to-head units in poly(vinyl chloride) is very small (less than 0.2 per 1000 monomer units).

There is also evidence that commercial poly(vinyl chloride) is branched. At one time it was believed, on the basis of infrared spectroscopic studies, that there were about 20 branches per polymer chain. Recent NMR spectroscopic studies, however, have shown the presence of some 6 branches per 1000 monomer units. The following types of branches have been identified:

chloromethyl	4 per 1000 monomer units
2-chloroethyl	<0.5
2,4-dichlorobutyl	1
long branches	<0.5

Chloromethyl branches arise through an occasional head-to-head addition followed by 1,2-chlorine migration and subsequent chain growth:

$$\sim CH_2\text{—}\overset{\cdot}{C}HCl \; + \; ClHC{=}CH_2 \longrightarrow \sim CH_2\text{—}CHCl\text{—}CHCl\text{—}\overset{\cdot}{C}H_2CH_2Cl$$

$$\longrightarrow \sim CH_2\text{—}CHCl\text{—}\overset{\cdot}{C}H\text{—}CH_2Cl \xrightarrow{\;H_2C{=}CHCl\;} \sim CH_2\text{—}CHCl\text{—}\underset{\overset{|}{CH_2Cl}}{CH}\text{—}CH_2\text{—}\overset{\cdot}{C}HCl \text{ etc.}$$

Production of dichlorobutyl branches has been attributed to a back-biting mechanism analogous to that proposed for low density polyethylene (section 2.3.1) [2]:

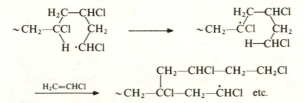

Long branches arise during polymerization by transfer of hydrogen from within the polymer to growing radicals. The most labile hydrogen atoms in poly(vinyl chloride) are those in the –CHCl– groups since the resultant free radical has resonance stability:

$$\sim CH_2\text{—}CHCl\sim \; + \; \sim CH_2\text{—}\overset{\cdot}{C}HCl \longrightarrow \sim CH_2\text{—}\overset{\cdot}{C}Cl\sim \; + \; \sim CH_2\text{—}CH_2Cl$$

$$\sim CH_2\text{—}\overset{\cdot}{C}Cl\sim \; + \; nH_2C{=}CHCl \longrightarrow \sim CH_2\text{—}\underset{\overset{|}{[CH_2\text{—}CHCl\text{—}]_n}}{\overset{\cdot}{C}Cl}\sim$$

Thus tertiary chlorine atoms are formed at points of branching and it is thought that they feature in polymer degradation reactions (See section 4.2.6.).

For most applications, commerical poly(vinyl chloride) has an average molecular weight (M_n) in the range of 40 000–80 000 but for material to be used in rigid applications the range is generally 30 000–50 000. X-ray studies show that commercial poly(vinyl chloride) is substantially amorphous although a small amount (about 5%) of crystallinity is present. NMR studies indicate that ordinary poly(vinyl chloride) is about 55% syndiotactic in structure, the rest being atactic. Syndiotacticity may be increased by lowering polymerization temperature: for example, polymer prepared at $-50°C$ is 66% syndiotactic. A completely syndiotactic polymer has been obtained by use of a urea canal complex [3]. Compared to the atactic material the syndiotactic polymers are more difficult to process and give brittle products; they have not, therefore, become of commercial importance.

4.2.5 Properties

Straight poly(vinyl chloride) is a colourless rigid material. It has a relatively high density and low softening point. (See Table 4.1 for some comparative properties.) Poly(vinyl chloride) has a higher dielectric constant and power

Table 4.1 Comparative properties of typical commercial grades of vinyl chloride polymers

	Unplasticized p.v.c.	Vinyl chloride-vinyl acetate copolymer	Plasticized p.v.c.*	Chlorinated p.v.c.
Specific gravity	1.40	1.35	1.31	1.52
Softening point, Vicat (°C)	80	70	Flexible	105
Tensile strength (MPa)	59	48	19	59
(lbf/in²)	8500	7000	2700	8500
Elongation at break (%)	5	5	300	

* Plasticized with 50 parts of diisooctyl phthalate per 100 parts of polymer.

factor than polyethylene owing to the polar carbon-chlorine bond. However, at temperatures below the glass transition temperature (80°C) the power factor is still comparatively low due to the immobility of the dipole. The high chlorine content of poly(vinyl chloride) renders it flame retarding.

Poly(vinyl chloride) has very limited solubility. The most effective solvents are those which appear to be capable of some form of interaction with the polymer. It has been suggested [4] that poly(vinyl chloride) is a weak proton donor and effective solvents are proton acceptors. Thus the polymer is soluble at room temperature in oxygen-containing solvents such as ethers, e.g. dioxan, tetrahydrofuran; ketones, e.g. cyclohexanone, methyl isobutyl ketone; and nitrocompounds, e.g. nitrobenzene. Poly(vinyl chloride) is also soluble in chlorinated solvents of similar solubility parameter, e.g. ethylene dichloride. Plasticizers for poly(vinyl chloride) (which may be regarded as non-volatile solvents) have solubility parameters similar to that of the polymer and are also proton acceptors. They have too large a molecular size to dissolve the polymer at room temperature but they may be incorporated at elevated temperatures to give mixtures stable at room temperature. Many materials can be used as plasticizers for poly(vinyl chloride) but the more important include alkyl phthalates, e.g. diisooctyl phthalate; aryl phosphates, e.g. tritotyl phosphate; esters of aliphatic acids, e.g. dibutyl sebacate; and epoxidized oils, e.g. epoxidized soya bean oil. Generally, 40–60 parts of plasticizer per 100 parts of polymer are used for most common applications. Poly(vinyl chloride) containing this level of plasticizer is a flexible rubber-like material. It may be noted here that whilst in the past most poly(vinyl chloride) has been used in a plasticized form, unplasticized poly(vinyl chloride) (UPVC) now accounts for about 60% of the current consumption of the polymer. This change has come about largely through the development of better stabilizing systems (see later) and improved processing equipment. When properly compounded and processed, unplasticized poly(vinyl chloride) has excellent weatherability and corrosion resistance and it is very widely used for a variety of building accessories.

Poly(vinyl chloride) is unaffected by acids, alkalis and aqueous solutions; even strong oxidizing agents such as chromic and nitric acids have little action.

Uncompounded poly(vinyl chloride) has relatively poor heat and light stability compared to other polymers. Exposure to temperatures above 70°C and/or ultraviolet light has a number of adverse affects on the properties of the polymer. The first physical manifestation of degradation is usually a change in colour. Thus when the initially colourless polymer is heated in air it turns in sequence, yellow, orange, brown, black. Further heating causes a general deterioration in mechanical and electrical properties. The commercial success of poly(vinyl chloride) has been very largely due to the discovery of stabilizers, the incorporation of which alleviates the effects of degradation and enables technologically useful materials to be made. Many compounds have been found to exert a stabilizing effect on poly(vinyl chloride) and several are used commercially. The more important stabilizers include lead compounds such as basic lead carbonate and dibasic lead phthalate; metal soaps (carboxylates) such as barium, cadmium and zinc laurates, octoates and stearates; cadmium-barium phenates; organotin compounds such as dibutyltin dilaurate and maleate; and epoxidized oils such as epoxidized linseed and soya bean oils. Stabilizing systems are very commonly made up of mixtures of stabilizers which have a synergistic effect on each other. The mechanisms of degradation and stabilization are considered in the next sections of this chapter.

Chemically, poly(vinyl chloride) is a rather inert material. However, it may be halogenated and chlorinated poly(vinyl chloride) is produced commercially; this material is described in section 4.2.8.

4.2.6 Degradation

As mentioned previously, the exposure of poly(vinyl chloride) to either ultraviolet light or heat leads to degradation, as a result of which discoloration and changes in mechanical properties can occur. The degradation of poly(vinyl chloride) has been the subject of extensive investigation and much information has accumulated. However, the reactions which take place are complex and are not completely understood at the present time. Whilst thermal degradation and photodegradation of poly(vinyl chloride) have much in common, they are distinct processes and are considered separately.

(a) Thermal degradation

In the absence of oxygen, the thermal degradation of poly(vinyl chloride) involves dehydrochlorination to give polyene sequences followed by cross-linking. When the polymer is subjected to heat in the presence of oxygen, dehydrochlorination again occurs but at an accelerated rate. During thermo-oxidative degradation, chain scission also takes place, competing with cross-linking. These various processes are now considered.

(i) *Dehydrochlorination.* It has been firmly established that dehydrochlorination begins in the early stages of degradation and can continue until only traces of chlorine remain in the polymer. In fact, determination of hydrogen chloride evolved is a common method of following the degradation process.

The finding that dehydrochlorination can continue until the polymer contains virtually no chlorine indicates that elimination does not occur in a random fashion. If random elimination occurred the resulting polymer would have an appreciable chlorine content since isolated chlorine atoms would remain, as illustrated below:

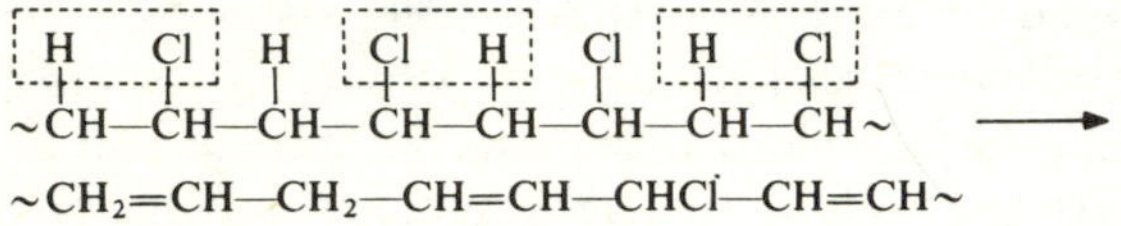

$$\sim CH-CH-CH-CH-CH-CH-CH-CH\sim \longrightarrow$$

$$\sim CH_2=CH-CH_2-CH=CH-CHCl-CH=CH\sim$$

Thus the degradation of poly(vinyl chloride) must involve the liberation of hydrogen chloride from successive units of the polymer; such a process is often referred to as 'unzipping'. It has been widely postulated that sequential elimination occurs because of allylic activation, i.e. the carbon-carbon double bond formed in a dehydrochlorination reaction renders the adjacent unit particularly susceptible to dehydrochlorination, as illustrated below:

$$\sim CH=CH-CH_2-CHCl-CH_2-CHCl-CH_2-CHCl \sim \longrightarrow$$

$$\sim CH=CH-CH=CH-CH_2-CHCl-CH_2-CHCl \sim \longrightarrow$$

$$\sim CH=CH-CH=CH-CH=CH-CH_2-CHCl \sim \longrightarrow$$

$$\sim CH=CH-CH=CH-CH=CH-CH=CH \sim \longrightarrow$$

Over the years, several different schemes for dehydrochlorination have been proposed. These schemes may be classified as involving ionic, uni-molecular and free radical mechanisms. At the present time, the ionic mechanism is perhaps the most plausible.

The main support for the occurrence of ionic processes comes from observations that degradation is catalysed by hydrogen chloride, other acids and strong bases and that the rate of dehydrochlorination in solution is influenced by the dielectric constant of the solvent. An ion-pair mechanism is consistent with these facts [5]:

$$\sim CH=CH-CHCl-CH_2-CHCl-CH_2\overset{Cl^-}{\underset{+}{N}} \longrightarrow \ \sim CH=CH-\overset{+}{C}H-CH_2-CHCl-CH_2 \sim$$

$$\longrightarrow \ \sim CH=CH-CH=CH-\overset{Cl^-}{\underset{+}{C}H}-CH_2 \sim \ \text{etc.}$$

The unimolecular theory largely rests on the basis of negative evidence which opposes other theories. An elimination mechanism involving a cyclic transition state has been proposed [6]:

A possible free radical mechanism, based on pyrolysis/mass spectrometry studies [7, 8], is as follows:

$$\sim CH=CH-CH_2-CHCl-CH_2-CHCl \sim \ + \ Cl\cdot \longrightarrow$$
$$\sim CH=CH-\dot{C}H-CHCl-CH_2-CHCl \sim \ + \ HCl$$

$$\downarrow$$

$$\sim CH=CH-CH=CH-CH_2-CHCl \sim \ + \ Cl\cdot$$

$$\sim CH=CH-CH=CH-CH_2-CHCl \sim + Cl\cdot \longrightarrow$$
$$\sim CH=CH-CH=CH-\dot{C}H-CHCl \sim \ + \ HCl$$

$$\downarrow$$

$$\sim CH=CH-CH=CH-CH=CH \sim \ + \ Cl\cdot$$

Regardless of the uncertainty of the mechanism of dehydrochlorination, a great deal of work has been directed toward identifying the dehydrochlorination initiation site. The initial step in dehydrochlorination is generally considered to occur at structural abnormalities in the polymer on the grounds that the group $-CH_2-CHCl-$ is quite stable in low molecular weight compounds such as 1,3,5-trichlorohexane up to at least 300°C. Several possible initiation sites have been suggested at various times but recent progress in establishing the nature and abundance of irregular structures in poly(vinyl chloride) has made many of these suggestions improbable. It is now considered likely that tertiary chloride is the most important labile structure in the polymer. As seen previously, tertiary chlorine is associated with dichlorobutyl branches and with long branches. Thus commercial poly(vinyl chloride) contains about one tertiary chlorine per 1000 monomer units, which is sufficient to account for the instability of the polymer. By the use of specially prepared polymers containing increased amounts of branching, a linear relationship between the rate of dehydrochlorination and total content of dichlorobutyl and long branches has been established [9].

It will be noticed that dehydrochlorination of poly(vinyl chloride) by an unzipping process results in sequences of conjugated double bonds. In thermally degraded poly(vinyl chloride) the average number of double bonds per sequence is generally in the range 6–14. It is these conjugated double bonds which are responsible for colour formation in degraded polymer. When the sequence length is longer than 8 units, the polyene absorbs radiation in the visible region and discoloration appears.

In addition to polyene formation, dehydrochlorination may also lead to cross-linking. Thus it has been found that when poly(vinyl chloride) is heated in nitrogen there is a continuous increase in molecular weight and the polymer becomes insoluble [10]. Cross-linking occurs by Diels-Alder cycloaddition between two polyenes:

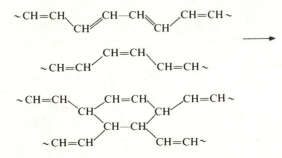

(ii) *Oxidation.* Poly(vinyl chloride) degrades more rapidly when heated in air than when heated in an inert atmosphere. The rate of dehydrochlorination increases substantially. It is generally supposed that this acceleration is due to peroxy radicals, which are formed by the oxidation of a hydrocarbon polymer (see section 2.3.2). The peroxy radicals abstract hydrogen from intact monomer units and thus initiate further dehydrochlorination:

$$\sim CH_2-CHCl-CH_2-CHCl\sim \xrightarrow[-ROOH]{ROO\cdot} \sim \overset{\cdot}{C}H-CHCl-CH_2-CHCl\sim$$

$$\xrightarrow[-Cl\cdot]{} \sim CH=CH-CH_2-CHCl\sim \xrightarrow[-HCl,-Cl\cdot]{Cl\cdot} \sim CH=CH-CH=CH\sim \quad etc.$$

The polymeric hydroperoxides formed in this sequence apparently undergo decomposition leading to chain scission since it has been found that there is a preliminary decline in molecular weight when poly(vinyl chloride) is heated in air. This decline in molecular weight is followed by an increase, presumably through the Diels-Alder cycloaddition cross-linking described above.

It has also been found that compared to purely thermal degradation, polymer discoloration is not as severe during thermo-oxidative degradation.

In the latter case, the polyene sequences are shorter as a result of reaction between the polyenes and oxygen, possibly as follows:

$$\sim CH{=}CH{-}CH{=}CH \sim \quad + \quad O_2 \longrightarrow \quad \sim CH{-}CH{=}CH{-}CH \sim$$

$$\underset{\displaystyle O{-}\!\!-\!\!-\!\!-\!\!-\!\!-\!\!O}{\quad\quad\quad\quad\quad\quad\quad\quad\quad\quad\quad\quad\quad\qquad \big| \quad\quad\quad\quad\quad\quad \big| }$$

(b) Photodegradation

When unstabilized poly(vinyl chloride) is exposed to ultraviolet radiation of wavelengths above 250 nm, dehydrochlorination occurs to give polyene sequences. Simultaneously, the polymer chains undergo scission and cross-linking. Photolysis in the presence of oxygen also results in dehydrochlorination together with chain scission and cross-linking.

(i) *Dehydrochlorination.* As in thermal degradation, the main feature of photolysis of poly(vinyl chloride) is dehydrochlorination to give conjugated polyene sequences. Depending on irradiation time, sequence lengths of 2–20 conjugated double bonds are formed, resulting in coloured material. Based mainly on ESR studies, it is now generally agreed that photolytic dehydrochlorination of poly(vinyl chloride) proceeds by the radical chain reaction described above under thermal degradation.

Although there is general agreement on the mechanism of photolytic dehydrochlorination, there is less certainty as to the initiation site. Whilst oxygenated structures such as carbonyl and hydroperoxide groups (which are usually present in commercial poly(vinyl chloride)) may represent initiation sites, it seems more likely that unsaturated structures (which are also present in commercial polymer) serve as the major initiation sites. Isolated alkene linkages absorb only ultraviolet radiation of wavelength below 220 nm and so cannot be responsible for the initiation of degradation by sunlight or by artificial light of wavelengths above 250 nm. If the double bonds are conjugated, however, absorption shifts steadily to longer wavelengths as the sequence length increases. Thus any dienes, trienes and longer polyenes present in the polymer would absorb the light and photo-cleavage would produce free radicals capable of initiating dehydrochlorination. The most likely bond to be cleaved in poly(vinyl chloride) is an allylic C–Cl bond, as shown in the following example:

$$\sim CH{=}CH{-}CH{=}CH{-}\underset{\displaystyle \underset{Cl}{|}}{CH}{-}CH_2 \sim \quad \overset{h\nu}{\longrightarrow} \quad \sim CH{=}CH{-}CH{=}CH{-}\dot{C}H{-}CH_2 \sim \quad + \quad Cl\cdot$$

Support for the conjugated alkene photosensitized degradation of poly(vinyl chloride) comes from the observation that photolysis of thermally degraded polymer (which contains conjugated polyene sequences) gives a substantially increased rate of dehydrochlorination [11].

In addition to polyene formation, photodehydrochlorination of poly(vinyl chloride) also leads to cross-linking and chain scission. When a solution of the polymer is irradiated there is a decrease in viscosity (indicating scission) and a broadening of the molecular weight distribution at both ends of the curve (indicating cross-linking) [12]. The mechanisms of these processes have not been fully elucidated but it is generally supposed that they are free radical in nature. Cross-linking is usually regarded as arising by combination of polymeric radicals. The radicals which are involved in polyene formation (which have labile β-chlorine) are probably too short-lived to undergo combination. Radicals with no β-chlorine, however, would be more likely to have a lifetime sufficiently long to allow combination. Such radicals are formed in the initiation step described above and coupling of the following kind may be envisaged:

$$\sim CH{=}CH{-}\overset{\cdot}{C}H{-}CH_2 \sim \qquad\qquad \sim CH{=}CH{-}\underset{\displaystyle |}{CH}{-}CH_2 \sim$$
$$\longrightarrow$$
$$\sim CH{=}CH{-}\overset{\cdot}{C}H{-}CH_2 \sim \qquad\qquad \sim CH{=}CH{-}CH{-}CH_2 \sim$$

The same type of radical may also undergo scission in a competitive process:

$$\sim CH{=}CH{-}\overset{\cdot}{C}H{-}CH_2{-}CHCl \sim \longrightarrow \sim CH{=}CH{-}CH{=}CH_2 + Cl H\overset{\cdot}{C} \sim$$

(ii) *Oxidation.* When poly(vinyl chloride) undergoes photolysis in air, zip-dehydrochlorination is again the major process. In fact, this process does not appear to be much affected by the presence of oxygen. At the same time oxygenated structures appear in the chain, being mostly carbonyl and hydroperoxide groups. Concomitant with these reactions, cross-linking and chain scission occur. The mechanisms of these various oxidative processes have not been fully elucidated but it is generally agreed that alkyl peroxy radicals are the major precursors. The scheme shown in Fig. 4.1 accounts for the principal features observed when poly(vinyl chloride) undergoes photo-oxidation but it must be emphasized that the scheme is somewhat speculative and far from complete.

4.2.7 Stabilization

As mentioned previously, many materials have been found which lessen the effects of thermal and photolytic degradation of poly(vinyl chloride). The manner in which these stabilizers interfere with the degradation processes is complex and is not fully understood, despite a good deal of attention to the matter. However, although details obviously vary according to the type of stabilizer used, a number of common features have become clear. Thus most stabilizers appear to function through the operation of at least one (and generally more than one) of the following processes.

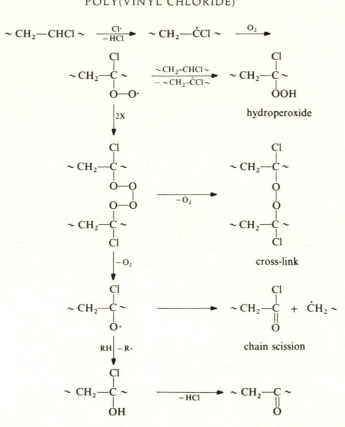

Fig. 4.1 Scheme for photo-oxidation of poly(vinyl chloride) (after [13]).

(a) Reaction with hydrogen chloride

As seen in the preceding section, the primary reaction occurring in the degradation of poly(vinyl chloride) is dehydrochlorination. The hydrogen chloride formed exerts a catalytic effect on the rate of reaction and all poly(vinyl chloride) stabilizers have the capacity to function as hydrogen chloride acceptors. For example, cadmium carboxylates and dibutyltin dicarboxylates react as follows:

$$Cd(OCOR)_2 \ + \ 2HCl \longrightarrow CdCl_2 \ + \ 2RCOOH$$

$$Bu_2Sn(OCOR)_2 \ + \ 2HCl \longrightarrow Bu_2SnCl_2 \ + \ 2RCOOH$$

(b) Reaction with polymer

Using radiotracer techniques it has been shown that stabilizer ligands are attached to the polymer during the degradation of poly(vinyl chloride). In most cases, allylic chlorine appears to be the reactive site in the polymer. For example, dibutyl tin mercaptides react as follows:

$$\underset{\sim}{CH}=CH-\underset{\underset{Cl}{|}}{CH}-CH_2\sim \quad + \quad Bu_2Sn(SCH_2-OCO-R)_2 \longrightarrow$$

$$\sim CH=CH-\underset{\underset{S-CH_2-OCO-R}{|}}{CH}-CH_2\sim \qquad + \qquad Bu_2Sn\underset{S-CH_2-OCO-R}{\overset{Cl}{<}}$$

Thus the zip-dehydrochlorination which normally follows allylic activation does not occur.

(c) Decomposition of hydroperoxide

It has been shown in the preceding section that hydroperoxides are involved in the oxidative degradation of poly(vinyl chloride). Many sulphur compounds are known to decompose hydroperoxides into non-radical products and it is likely that organotin mercaptide stabilizers can also react in this way. For example, dimethyltin bis(isooctylthioglycolate) decomposes the model compound, *tert*-butyl hydroperoxide as follows [14]:

$$(CH_3)_2Sn(SCH_2-OCO-C_8H_{17})_2 \quad + \quad (CH_3)_3C-OOH \longrightarrow$$

$$(CH_3)_2Sn=O \quad + \quad (C_8H_{17}-COO-CH_2S-)_2 \quad + \quad (CH_3)_3C-OH$$

(d) Reaction with polyenes

As seen in the preceding section, the discoloration of degraded poly(vinyl chloride) is due to the formation of polyene sequences. Some stabilizers are capable of restoring the original colour of the polymer, indicating reaction with the polyenes present. For example, dibutyltin maleate undergoes a Diels-Alder reaction with polyenes [15]. Thus the sequence of conjugated double bonds is shortened.

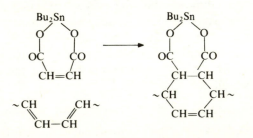

4.2.8   Chlorinated poly(vinyl chloride)

Depending on the conditions used, various products may be obtained by the post-chlorination of poly(vinyl chloride). Two types of product are made commercially, one by chlorination at low temperature and one by chlorination at high temperature.

(i) *Low temperature chlorination.* In one process, an aqueous dispersion of poly(vinyl chloride) (containing a swelling agent such as chloroform) is heated to about 50°C, illuminated and treated with chlorine. Under these conditions, substitution occurs mainly at the unsubstituted methylene groups so that *s*-dichloroethylene units are formed. Commercial chlorinated poly(vinyl chloride) has a chlorine content of about 63–68% and is thus, in effect, a copolymer of vinyl chloride and *s*-dichloroethylene (poly-*s*-dichloroethylene would have 73% Cl):

$$\sim CHCl{-}CHCl{-}CHCl{-}CHCl{-}CH_2{-}CHCl{-}CHCl{-}CHCl\sim$$

The properties of chlorinated poly(vinyl chloride) of this type are in the main similar to those of unplasticized poly(vinyl chloride). (See Table 4.1 for some comparative properties.) The most significant result of chlorination is the elevation of softening point, as a result of which the maximum service temperature is about 100°C compared to about 65°C for unplasticized poly(vinyl chloride). Chlorination also leads to a deterioration in heat stability and an increase in melt viscosity and thus chlorinated poly(vinyl chloride) is rather more difficult to process; nevertheless, the material can be satisfactorily extruded and injection moulded. The major application for chlorinated poly(vinyl chloride) is in plumbing applications, particularly hot and cold water distribution and hot, corrosive effluent handling.

(ii) *High temperature chlorination.* In this case, chlorination is conducted in solution at about 100°C. A suitable solvent for the poly(vinyl chloride) is tetrachloroethane. Under these conditions, substitution occurs extensively at the $-CH_2-$ group and there is also chain scission. The most significant effect of this type of chlorination is that the product is soluble in low cost solvents such as acetone, butyl acetate and methylene chloride. Such solutions find application in adhesives and protective coatings; they are also used for spinning fibre which, because of its resistance to corrosive liquids, is utilized mainly for chemical filter cloth. Poly(vinyl chloride) chlorinated in this manner has a low softening point, low impact strength and poor colour stability and is not used for the production of articles by conventional moulding operations.

4.2.9 Vinyl chloride-vinyl acetate copolymers

The principal effects of introducing vinyl acetate into a vinyl chloride polymer are to increase solubility and to improve moulding characteristics by lowering the temperature at which the material can be manipulated and by increasing flow. Thus vinyl chloride-vinyl acetate copolymers are used for non-flexible mouldings for which the processing characteristics of the homopolymer are not satisfactory. The most important of these applications are gramophone records, in which exceptionally good flow is required, and flooring compositions, in which high loadings of filler result in high melt viscosity. Vinyl chloride-vinyl acetate copolymers are also used for fibres and in surface coatings, where the increased solubility of the copolymers is utilized. Copolymers intended for moulding applications are generally prepared by suspension polymerization whilst those intended for surface coatings are usually made by solution polymerization.

(i) *Suspension polymerization.* Typical copolymers designed for ease of moulding contain about 10 mole % vinyl acetate. They are prepared by processes essentially similar to that described previously for the homopolymer. The copolymers have an increased tendency toward agglomeration during the polymerization cycle and it is usual to have much higher levels of suspending agents in the system. Compared to the homopolymer, these vinyl chloride-vinyl acetate copolymers have reduced softening points, tensile strength, abrasion resistance, chemical resistance and heat stability. (See Table 4.1 for some comparative properties.)

(ii) *Solution polymerization.* The vinyl acetate content of most copolymers designed for coatings is in the range 2–10 mole %. Solution polymerization is carried out under pressure in a solvent for the copolymer which is then isolated by precipitation, washing and drying. Alternatively, polymerization is conducted in a solvent such as butane in which the copolymer is insoluble; the polymer is then obtained as a slurry. In one process, polymerization is carried out at 30°C using the active initiator, acetyl benzoyl peroxide. Solution polymerization has the advantage over suspension polymerization of allowing more effective skew feeding of monomers to give a more uniform copolymer composition. Sometimes, in order to promote adhesion, polar groups are introduced into the polymer by incorporating about 1% maleic acid or methacrylic acid or by partial hydrolysis of the acetate groups. Usually copolymers intended for coatings have rather low molecular weights ($M_v \simeq 10\,000$). The copolymers are soluble in chlorinated hydrocarbons, esters and ketones and diluents such as toluene or xylene may be added for cost reduction and evaporation rate control. Coatings based on vinyl chloride-vinyl acetate copolymers have excellent resistance to water, acids, alkalis, aliphatic hydrocarbons, oils and fats; they are non-flammable, tough and flexible (which allows postforming of coated metal sheet). These coatings are used in such applications as industrial plant exposed to corrosive environments, marine equipment, food cans and collapsible tubes.

4.2.10   Other vinyl chloride copolymers

Many copolymers in which vinyl chloride is the major component have been investigated but none has equalled the commercial importance of the vinyl acetate copolymers described above. Of these other copolymers which have been commercialized, the following may be mentioned.

(a) Vinylidene chloride copolymers

These copolymers generally contain 5–12% vinylidene chloride and have properties similar to the vinyl acetate copolymers described in section 4.2.9. They have been suggested for use in specialized calendering applications and as a filler polymer in rigisols. Vinylidene chloride-vinyl chloride copolymers in which vinylidene chloride is the major component are described in section 4.3.

(b) Acrylonitrile copolymers

The copolymer containing 60% vinyl chloride and 40% acrylonitrile is used for the production of fibres, which are spun from an acetone solution. Fabrics made from this fibre are non-flammable and have good chemical resistance; they are used for industrial garments and filter cloths.

(c) Ethylene copolymers

Copolymers containing 15–30% ethylene show increased film flexibility and are used in latex form for can coating and as a binder for non-woven fabrics.

4.3 VINYLIDENE CHLORIDE COPOLYMERS

Vinylidene chloride is readily polymerized but the homopolymer does not have sufficient thermal stability to withstand melt processing. The homopolymer is not therefore of commercial importance. However, by the copolymerization of vinylidene chloride with lesser amounts (generally 10–30%) of a vinyl monomer, processable copolymers may be obtained. The comonomers most widely used for this purpose are vinyl chloride and acrylonitrile.

4.3.1 Raw materials

Vinylidene chloride is prepared from either vinyl chloride or ethylene dichloride by the following route:

$$H_2C=CHCl \xrightarrow{Cl_2}$$
vinyl chloride

$$ClH_2C-CHCl_2 \xrightarrow{-HCl} H_2C=CCl_2$$
1,1,2-trichloroethane   vinylidene chloride

$$H_2CCl-CH_2Cl \xrightarrow[-HCl]{Cl_2}$$
ethylene dichloride

In the first alternative, trichloroethane is prepared by the liquid phase chlorination of vinyl chloride at 30–50°C under pressure. In the second alternative, trichloroethane is obtained by liquid phase chlorination of ethylene dichloride at about 60°C in the presence of aluminium chloride as catalyst. The trichloroethane is then dehydrochlorinated by agitating with an aqueous suspension of calcium hydroxide at about 50°C. Crude vinylidene chloride distills off as it is formed and is then purified by distillation under pressure. The dehydrochlorination of trichloroethane may also be accomplished by pyrolysis at 400°C. Vinylidene chloride is a colourless liquid (b.p. 32°C). It is rather difficult material to handle since it readily polymerizes on standing. Polymerization occurs rapidly on exposure to air, water or light but even storage under an inert atmosphere does not completely prevent polymer formation. The monomer is therefore commonly inhibited with a phenol, such as *p*-methoxyphenol, which is removed by distillation or alkali-washing before polymerization.

4.3.2   Preparation

Vinylidene chloride copolymers are prepared by suspension and emulsion processes. Suspension processes are generally preferred for the production of melt-processable polymer whilst latices are used directly for barrier coatings. The procedures used are essentially as described for the preparation of poly(vinyl chloride) (section 4.2.3). The polymerization rate for vinylidene chloride/vinyl chloride mixtures is markedly less than for either monomer alone so somewhat higher temperatures and initiator levels are used together with longer reaction times. Molecular weights are generally relatively low ($M_w \simeq 20\,000$–$50\,000$).

4.3.3   Properties

Poly(vinylidene chloride) has a very regular structure:

$$\sim CH_2-\underset{\underset{\displaystyle Cl}{|}}{\overset{\overset{\displaystyle Cl}{|}}{C}}-CH_2-\underset{\underset{\displaystyle Cl}{|}}{\overset{\overset{\displaystyle Cl}{|}}{C}}-CH_2-\underset{\underset{\displaystyle Cl}{|}}{\overset{\overset{\displaystyle Cl}{|}}{C}}\sim$$

It is therefore a highly crystalline material and has a crystalline melting point of about 200°C. At this temperature the polymer has a high rate of decomposition. Copolymerization reduces the molecular regularity and thus lowers the softening point, which for 85% vinylidene chloride – 15% vinyl chloride copolymer is about 140°C. Since crystallization is thermodynamically favoured even in the presence of liquids of similar solubility parameter and since there is little interaction between the polymer and any liquid, there are no effective solvents at room temperature for the homopolymer. The copolymers, however, are soluble in ethers and ketones and the solutions may be used for coating applications. The copolymers are resistant to most other organic solvents and to acids; there is some attack by alkalis.

Even in the copolymers there is extensive crystallization and so the materials have high specific gravity (1.7 for 85% vinylidene chloride–15% vinyl chloride copolymer) and low permeability to moisture vapour and gases. The copolymers also have high tensile strengths, namely about 70 MPa (10 000 lb/in^2) for unoriented material and up to 290 MPa (40 000 lb/in^2) for oriented filament. It may be noted that the copolymers are not readily plasticized as all but small amounts of plasticizer exude from the finished product.

The main application of the vinylidene chloride–vinyl chloride copolymers is for packaging film, which is made by extrusion and biaxial stretching. The film has very good clarity, toughness and moisture and gas impermeability. The copolymers are also used for filaments, which are made by melt extrusion and drawing. The filaments are used for such applications as car upholstery, garden chair fabrics and filter cloths where toughness, durability and chemical resistance are required.

The main use of the vinylidene chloride–acrylonitrile copolymers is as coatings for materials such as cellophane, paper and polyethylene. The coatings confer moisture and gas impermeability and they are heat-sealable.

REFERENCES

1. Marvel, C. S. *et al.* (1939) *J. Am. Chem. Soc.*, **61**, 3241.
2. Bovey, F. A. and Tiers, G. V. D. (1962) *Chem. Ind.*, 1826.
3. White, D. M. (1960) *J. Am. Chem. Soc.*, **82**, 5678.
4. Small, P. A. (1953) *J. Appl. Chem.*, **3**, 71.
5. Starnes, W. H. (1981) in *Developments in Polymer Degradation – 3*, (ed. N. Grassie), Applied Science Publishers, London, p. 135.
6. Imoto, M. and Nakaya, T. (1965) *Kogyo Kagaku Zasshi*, **68**, 2285.
7. Stromberg, R. R. *et al.* (1959) *J. Polymer Sci.*, **35**, 355.
8. Winkler, D. E. (1959) *J. Polymer Sci.*, **35**, 3.
9. Hjertberg, T. and Sorvik, E. M. (1983) *Polymer*, **24**, 673.
10. Druesedow, D. and Gibbs, C. F. (1953) *Nat. Bur. Std. Circ.*, **525**, 95.
11. Owen, E. D. and Williams, J. I. (1974) *J. Polym. Sci., Polym. Chem. Ed.*, **12**, 1933.
12. Balandier, M. and Decker, C. (1978) *Eur. Polym. J.*, **14**, 995.
13. Decker, C. (1984) in *Degradation and Stabilisation of PVC*, (ed. E. D. Owen), Elsevier Applied Science Publishers, London, ch. 3.
14. Wirth, H. O. and Andreas, H. (1977) *Pure Appl. Chem.*, **49**, 627.
15. Troitskaya, L. S. and Troitski, B. B. (1968) *Plast. Massy*, 12.

BIBLIOGRAPHY

Smith, W. M. (1958) *Vinyl Resins*, Reinhold Publishing Corporation, New York.

Kaufman, M. (ed.), (1962) *Advances in PVC Compounding and Processing*, Maclaren & Sons Ltd, London.

Chevassus, F. and de Broutelles, R. (1963) *The Stabilization of Polyvinyl Chloride*, Edward Arnold Ltd, London.

Geddes, W. C. (1966) *The Mechanism of PVC Degradation*, Technical Review No. 31, RAPRA, Shawbury.

Gould, R. F. (ed.) (1968) *Stabilization of Polymers and Stabilizer Processes*, American Chemical Society, Washington.

Koleske, J. V. and Wartman, L. H. (1969) *Poly(vinyl chloride)*, Gordon and Breach Science Publishers, New York.

Sarvetnick, H. A. (1969) *Polyvinyl Chloride*, Van Nostrand Reinhold Company, New York.

Mathews, G. (1972) *Vinyl and Allied Polymers*, Vol. 2, Iliffe Books, London.

Henson, J. H. L. and Whelan, A. (eds) (1973) *Developments in PVC Technology*, John Wiley & Sons, New York.

Wessling, R. A. (1977) *Polyvinylidene Chloride*, Gordon and Breach Science Publishers, New York.

Whelan, A. and Craft, J. L. (eds) (1977) *Developments in PVC Production and Processing - 1*, Applied Science Publishers Ltd, London.

Sittig, M. (1978) *Vinyl Chloride and PVC Manufacture*, Noyes Data Corporation, Park Ridge.

Burgess, R. H. (ed.) (1982) *Manufacture and Processing of PVC*, Applied Science Publishers Ltd, London.

Butters, G. (1982) *Particulate Nature of PVC*, Applied Science Publishers Ltd, London.

Owen, E. D. (ed.) (1984) *Degradation and Stabilization of PVC*, Elsevier Applied Science Publishers, London.

Titow, W. V. (1984) *PVC Technology*, 4th edn, Elsevier Applied Science Publishers, London.

Nass, L. I. and Heiberger, C. A. (eds) (1986) *Encyclopedia of PVC*, Vol. 1, 2nd edn, Marcel Dekker, Inc., New York.

Wypych, J. (1986) *Polyvinyl Chloride Stabilization*, Elsevier, Amsterdam.

POLY(VINYL ACETATE) AND RELATED POLYMERS

5.1   SCOPE

In this chapter poly(vinyl acetate) and some vinyl acetate copolymers are described. The important copolymers, ethylene-vinyl acetate and vinyl chloride-vinyl acetate are dealt with in Chapters 2 and 4 respectively. Also considered are poly(vinyl alcohol) and poly(vinyl acetal)s, which are derivatives of poly(vinyl acetate).

5.2   POLY(VINYL ACETATE)

5.2.1   Raw materials

Until the mid-1960s, vinyl acetate manufacture was based almost entirely on acetylene. Since that time, however, more economical ethylene-based processes have been developed and now very few manufacturers use the acetylene route.

(i) *Ethylene route.* This route involves the oxidation of a mixture of ethylene and acetic acid:

$$H_2C{=}CH_2 \; + \; CH_3COOH \; + \; \tfrac{1}{2}O_2 \; \longrightarrow \; H_2C{=}CH{-}O{-}\overset{\displaystyle O}{\overset{\displaystyle \|}{C}}{-}CH_3 \; + \; H_2O$$

Initially, liquid phase processes were developed but because of severe corrosion problems, gas-phase processes are now more common. Typically, a mixture of ethylene and acetic acid is oxidized over a palladium catalyst at 0.5–1 MPa (5–10 atmospheres) and 150–200°C. The exit gases are quenched and vinyl acetate is separated by distillation. A yield of about 95% is obtained.

(ii) *Acetylene route*. This route is based on the reaction of acetylene and acetic acid:

$$HC{\equiv}CH \ + \ CH_3COOH \ \longrightarrow \ H_2C{=}CH{-}O{-}\overset{\displaystyle O}{\overset{\|}{C}}{-}CH_3$$

Further reaction can occur between vinyl acetate and acetic acid to give ethylidene diacetate:

$$H_2C{=}CH{-}OCOCH_3 \ + \ CH_3COOH \ \longrightarrow \ H_3C{-}CH(OCOCH_3)_2$$

Formation of this by-product is minimized by using a molar excess of acetylene, short reaction times and low temperatures.

The reaction may be carried out in the vapour or liquid phase, the former being generally preferred for industrial processes. In a typical vapour phase preparation, a mixture of acetylene (in excess) and acetic acid vapour is passed through a reaction tube at 190–220°C. The tube contains a catalyst such as zinc acetate or zinc silicate and about 50% of the acetic acid is converted per pass. The exit vapours are cooled; acetylene is recycled and the liquid stream is distilled to give vinyl acetate and acetic acid, which is also recycled.

In the liquid phase process, acetylene is passed into acetic acid containing mercuric sulphate as catalyst at a temperature of 75–80°C. The reactor is fitted with a condenser held at 72–74°C; this permits the passage of vinyl acetate vapour (which is swept from the reactor by a current of acetylene) but returns acetic acid. The exit vapours are cooled; acetylene is recycled and the vinyl acetate is purified by distillation.

Vinyl acetate is a colourless liquid (b.p. 72.5°C) with a pleasant, sweet odour. Before shipment, the monomer is usually inhibited by such materials as cupric acetate or hydroquinone.

5.2.2 Preparation

The first commercial production of poly(vinyl acetate) began in Germany in 1920.

Vinyl acetate may be readily polymerized by bulk, solution, suspension and emulsion techniques. In commercial practice, emulsion polymerization is the predominant method, the resulting latices being mainly used directly in water-based products such as paints and adhesives. The other polymerization techniques [1] find limited use in the preparation of poly(vinyl acetate) which is required in solid form.

(a) Emulsion polymerization

Polymerization is normally carried out batch-wise in a stirred reactor, jacketed for heating and cooling. A typical formulation might be as follows:

Vinyl acetate (inhibitor free)	100 parts by weight
Water	100
Hydroxyethylcellulose	2.5 (protective colloid)
Poly(ethylene glycol) ether of lauryl alcohol	2.5 (surfactant)
Sodium dodecylbenzenesulphonate	0.1 (surfactant)
Sodium bicarbonate	0.5 (buffer)
Potassium persulphate	0.5 (initiator)

Reaction temperature is about 75–80°C and reaction time about 2 hours. The reaction is highly exothermic and in order to achieve better control and a product with smaller particle size it is common practice to polymerize firstly only a portion of the monomer to give a 'seed latex' and then add the remaining monomer slowly over 2–4 hours. The buffer is added to the system to minimize hydrolysis of the vinyl acetate. As mentioned previously, the resulting latex is normally used as such and the solid polymer is not isolated.

5.2.3 Structure

It is shown in section 1.6.3 that, on theoretical grounds, vinyl polymers may be expected to have a head-to-tail structure. For poly(vinyl acetate) there is experimental evidence that this is in fact the case. It has been found that poly(vinyl alcohol) has a mainly head-to-tail structure (see section 5.4.2); since poly(vinyl alcohol) is derived from poly(vinyl acetate), it follows that the latter also has a mainly head-to-tail structure:

$$
\begin{array}{ccccc}
CH_3 & & CH_3 & & CH_3 \\
| & & | & & | \\
CO & & CO & & CO \\
| & & | & & | \\
O & & O & & O \\
| & & | & & | \\
\sim CH_2{-}CH{-}CH_2{-}CH{-}CH_2{-}CH\sim
\end{array}
$$

When vinyl acetate is polymerized to degrees of conversion above about 30%, appreciable branching occurs because of chain transfer. Since a growing radical may abstract a hydrogen atom from either the chain or the methyl group, two kinds of branches are possible:

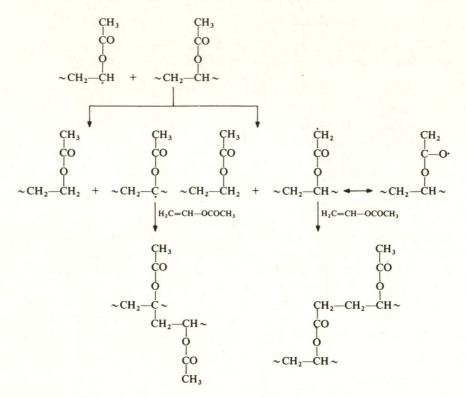

The radical resulting from transfer to the methyl group is stabilized by delocalization and, in practice, most branching occurs at the side-groups; branching from chain methenyl groups occurs to a small extent. (See also section 5.4.2.)

5.2.4   Properties

Poly(vinyl acetate) is too brittle and shows too much cold flow to be useful in bulk form for rigid applications. The glass transition temperature is 28°C, above which temperature the polymer is soft but not very rubber-like. Poly(vinyl acetate) mostly finds application in film form in surface coatings and adhesives. In order to promote film formation in such applications and to give the film flexibility over a wider temperature range, it is usual to lower the glass transition temperature of the polymer to below room temperature by either plasticization or copolymerization. Plasticization is readily accomplished by mixing a plasticizer, typically dibutyl phthalate (about 10–15% of the polymer weight), into the latex. Copolymerization is discussed in section 5.3.

Poly(vinyl acetate) is soluble in a wide range of solvents. It is soluble at room temperature in aromatic hydrocarbons such as benzene and toluene;

chlorinated hydrocarbons such as carbon tetrachloride, chloroform and dichloroethylene; lower alcohols such as methanol and ethanol; esters such as ethyl acetate and butyl acetate; and ketones such as acetone and methyl isobutyl ketone. The polymer is resistant to greases and oils.

Poly(vinyl acetate) swells and softens on prolonged immersion in water and the polymer is readily hydrolyzed to poly(vinyl alcohol) by acids and alkalis. (See section 5.4.)

Extended exposure to temperatures above about 70°C results in discoloration and deterioration of mechanical properties, accompanied by elimination of acetic acid. Ester decomposition proceeds as follows:

The double bond formed promotes decomposition of the next unit and so the molecular process passes along the chain to give long conjugated sequences. The overall reaction may be simply represented as follows:

$$\sim CH_2-\underset{\underset{\underset{CH_3}{CO}}{\overset{O}{\overset{|}{O}}}}{CH}-CH_2-\underset{\underset{\underset{CH_3}{CO}}{\overset{O}{\overset{|}{O}}}}{CH}-CH_2-\underset{\underset{\underset{CH_3}{CO}}{\overset{O}{\overset{|}{O}}}}{CH}\sim \xrightarrow{-CH_3COOH} \sim CH=CH-CH=CH-CH=CH\sim$$

5.3   VINYL ACETATE COPOLYMERS

As mentioned in the previous section, improved film formation and film flexibility in poly(vinyl acetate) may be achieved by the addition of such materials as dibutyl phthalate.

A limitation of external plasticizers of this kind is that they may eventually be lost by evaporation or by migration into the substrate, leaving an imperfect and brittle film. This limitation may be overcome by the use of copolymers and these are now widely used in surface coatings and other applications. Comonomers which may be employed for this purpose include butyl acrylate, 2-ethylhexyl acrylate, diethyl fumarate, diethyl maleate and vinyl esters of fatty acids (e.g. a branched C_{10} fatty acid). Typically, the copolymers contain 15–20% by weight of such comonomers. These copolymers are readily prepared by the emulsion polymerization techniques described previously for the homopolymer.

5.4   POLY(VINYL ALCOHOL)

5.4.1   Preparation [2, 3]

Vinyl alcohol has not been isolated in the free state; the keto tautomer, acetaldehyde, is much the more stable form and is always obtained:

$$\underset{\substack{|\\ \text{OH}}}{\text{H}_2\text{C}=\text{CH}} \rightleftharpoons \underset{\substack{\| \\ \text{O}}}{\text{H}_3\text{C}-\text{CH}}$$

Thus poly(vinyl alcohol) cannot be prepared from its monomer by the usual techniques, although the polymerization of acetaldehyde with sodium amalgam at -80 to $-20°\text{C}$ has been found to give poly(vinyl alcohol) of low molecular weight [4]. For commercial purposes, poly(vinyl alcohol) is obtained exclusively from poly(vinyl acetate).

Poly(vinyl acetate) is readily hydrolysed by treating an alcoholic solution with aqueous acid or alkali. Acid hydrolysis results in traces of acid in the poly(vinyl alcohol) which are difficult to remove and which lead to instability of the polymer; alkaline hydrolysis results in contamination of the product by a large amount of sodium acetate which is also difficult to remove and which has little intrinsic value. These difficulties are avoided if poly(vinyl alcohol) is prepared from poly(vinyl acetate) by alcoholysis using a small amount of base as catalyst. The reaction is commonly carried out by treating poly(vinyl acetate) with methanol in the presence of sodium methoxide:

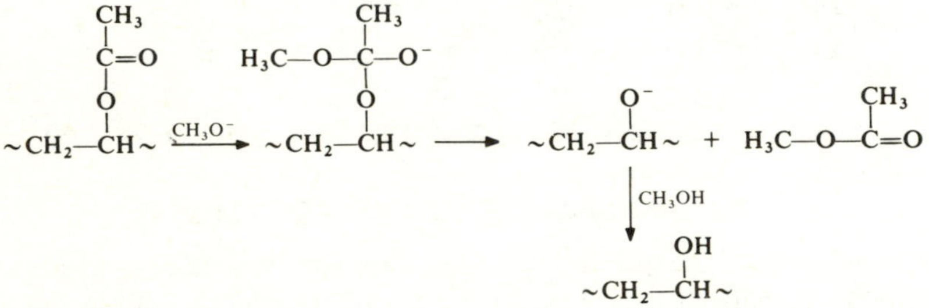

The preferred methods of preparing poly(vinyl acetate) for conversion to poly(vinyl alcohol) are solution and suspension polymerization. The former technique has the advantage that if polymerization is conducted in methanol the resulting solution can be used directly without the need for isolating the polymer; this method is the most suitable for continuous processes. Bulk polymerized poly(vinyl acetate) tends to give low molecular weight poly(vinyl alcohol) of poor colour.

In one continuous process, a solution of poly(vinyl acetate) in methanol (about 20%) is mixed with the catalyst solution in a high speed in-line mixer. The mixture then passes through a 'gelling zone' on a conveyor belt. Typically, the material is kept at $40°\text{C}$ for 10 minutes in this zone during which time the alcoholysis reaction occurs; poly(vinyl alcohol) is insoluble in methanol and a gel is produced. The gel is chopped up and neutralized with acetic acid to stop reaction; the liquid content (which is mainly methanol and methyl acetate) is then expressed and recovered. The residual solid is washed with methanol, dried and pulverized.

It is possible to control the extent to which acetate groups are replaced by hydroxyl groups by changing the reaction conditions. In particular, the catalyst concentration and the time of reaction have a major effect on the degree of alcoholysis. The most common commercial types of poly(vinyl alcohol) are the so-called partially hydrolysed grades in which 87–89% of the acetate groups have been replaced and the completely hydrolysed grades in which 99–100% of the acetate groups have been replaced. The degree of alcoholysis has an effect on the properties of the polymer (see section 5.4.3).

5.4.2   Structure

It follows that poly(vinyl alcohol) has a substantially head-to-tail structure since neither periodic acid nor lead tetraacetate, reagents which cleave 1,2-diols, are consumed to any measurable extent by the polymer. However, after treatment with these reagents the polymer does have a reduced solution viscosity, indicating the presence of a few head-to-head linkages (1–2% of the total linkages) [5]:

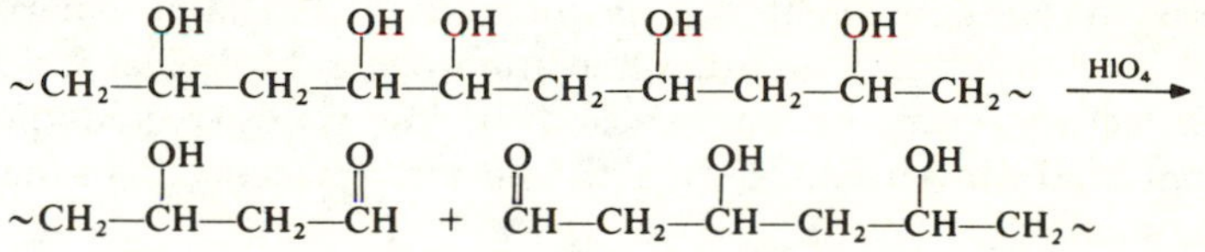

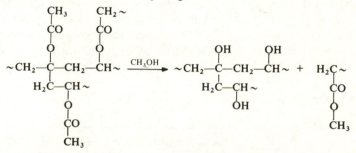

As mentioned previously (section 5.2.3.), two types of branching may occur in poly(vinyl acetate). The alcoholysis of poly(vinyl acetate) causes removal of branches resulting from chain transfer to the acetate group but not those arising from transfer to chain hydrogen:

Thus it is found that poly(vinyl alcohol) has a degree of polymerization considerably lower than that of the poly(vinyl acetate) from which it was derived. There appears to be little branching in poly(vinyl alcohol), from which it may be deduced that most of the branching in poly(vinyl acetate) arises from transfer to the acetate group. Generally speaking, poly(vinyl alcohol) is commercially available in four molecular weight ranges which are commonly referred to as super-high-, high-, medium- and low-viscosity poly(vinyl alcohol)s. The corresponding average molecular weights (M_v) are 250 000–300 000; 170 000–220 000; 120 000–150 000; and 25 000–35 000 respectively.

Since poly(vinyl acetate), as normally prepared, is atactic poly(vinyl alcohol) is also atactic. However, although poly(vinyl acetate) is amorphous poly(vinyl alcohol) exhibits crystallinity. The hydroxyl group is small enough to fit into a crystal lattice which is essentially the same as that of polyethylene.

5.4.3   Properties

The physical properties of poly(vinyl alcohol) are somewhat dependent on the degree of alcoholysis. (See Table 5.1 for some comparative properties.) Thus completely hydrolysed poly(vinyl alcohol) has a higher tensile strength and tear resistance than the partially hydrolysed material, in which crystallinity and hydrogen bonding are less extensive. Physical properties are also affected by environmental humidity. Water acts as a plasticizer and poly(vinyl alcohol) conditioned at, for example, 50% relative humidity shows a decrease in tensile strength but an increase in elongation compared to material conditioned in a moisture-free atmosphere. Molecular weight also has an effect on physical properties and the low viscosity grade polymers have appreciably lower tensile strength and tear resistance than the higher viscosity grades.

An important characteristic of poly(vinyl alcohol) is its solubility in water. As the acetate groups of poly(vinyl acetate) are replaced by hydroxyl groups the water sensitivity of the polymer increases. Maximum sensitivity occurs at a degree of alcoholysis of about 88% and polymers with a degree of alcoholysis in the range 87–89% are readily soluble in cold water. At higher degrees of alcoholysis, hydrogen bonding becomes more appreciable and results in a reduction in the ease of solubility. Thus, completely hydrolysed grades of poly(vinyl alcohol) are dissolved in water only by heating to above 85°C.

Table 5.1   Comparative properties of typical commercial grades of poly(vinyl alcohol)

	Partially hydrolysed, high viscosity grade	Completely hydrolysed, high viscosity grade
Specific gravity	1.3	1.3
Tensile strength		
(dry) (MPa)	120	150
(1bf/in^2)	18 000	22 000
(at 50% R.H.) (MPa)	72	83
(1bf/in^2)	10 500	12 000

Poly(vinyl alcohol) is resistant to a very wide range of organic solvents. In general, resistance to organic solvents increases with degree of alcoholysis. All grades are substantially unaffected by most aliphatic and aromatic hydrocarbons, chlorinated hydrocarbons, higher monohydric alcohols, esters, ethers and ketones. Lower monohydric alcohols have some effect on the partially hydrolysed grades of polymer but negligible effect on completely hydrolysed grades. The only effective solvents are those capable of hydrogen bonding with the hydroxyl groups. Examples of effective solvents are poly-hydroxy compounds such as ethylene glycol and glycerol; amides such as formamide and acetamide; and amines such as diethylenetriamine and tri-ethylenetetramine. The two last-named solvents are among the few which dissolve poly(vinyl alcohol) at room temperature; the other solvents are effective only at elevated temperature. Glycerol is commonly used as a plasticizer for poly(vinyl alcohol).

The intermolecular hydrogen bonding present in poly(vinyl alcohol) results in a high crystalline melting point (about 230°C for completely hydrolysed grades) and appreciable decomposition occurs before this temperature is reached. Thus the polymer can be processed only in solution (as in film casting or solution spinning of fibre) or if plasticized (when compression moulding and extrusion are possible). If poly(vinyl alcohol) is heated above about 120°C, water is eliminated to give conjugated double bonds which impart colour; there is also a reduction in solubility, possibly due to the formation of ether cross-links.

The secondary hydroxyl groups in poly(vinyl alcohol) are reactive and many derivatives have been prepared. The most important commercial derivatives are the acetals, which are described in the following section. Among the other derivatives which have found application are the acid sulphates (as ion exchange resins), the hydroxyethyl ethers (which have better low temperature flexibility than the parent alcohols) and the thiols (which find use in the isolation of metals such as silver, mercury and platinum by the formation of insoluble mercaptides). These derivatives are prepared as follows:

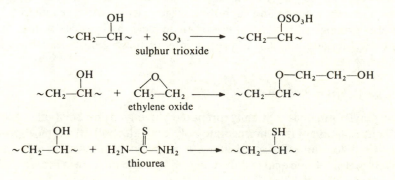

It will be appreciated that poly(vinyl alcohol) has an unusual combination of properties. In particular, it has much greater tensile strength than is normally associated with water-soluble materials; at the same time it has outstanding chemical resistance. This combination results in a wide variety of applications. Thus poly(vinyl alcohol) film is used for water-soluble packages for materials such as bath salts, bleaches, disinfectants and insecticides; film is also employed as a release agent in the production of reinforced plastics. Poly(vinyl alcohol) may be spun into fibres (which are insolubilized by treatment with formaldehyde), which have been developed particularly in Japan. Aqueous poly(vinyl alcohol) solutions are used in the formulation of adhesives, paper treatments and textile sizes.

5.5   POLY(VINYL ACETAL)S

Treatment of poly(vinyl alcohol) with aldehydes or ketones results in the formation of poly(vinyl acetal)s and poly(vinyl ketal)s, of which only the former products are of commercial significance. At the present time, poly(vinyl formal) and poly(vinyl butyral) are the most important members of this group; poly(vinyl acetal) itself now finds little application.

The preparation of a poly(vinyl acetal) from poly(vinyl alcohol) may be represented as follows:

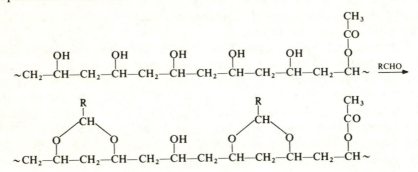

Thus, in practice, a poly(vinyl acetal) contains acetal groups, hydroxyl groups (which become isolated and cannot undergo reaction with the aldehyde) and acetate groups (which are due to incomplete hydrolysis of the parent poly(vinyl acetate)). Differing proportions of these groups lead to various grades of materials.

5.5.1   Poly(vinyl formal)

Poly(vinyl formal) is generally prepared directly from poly(vinyl acetate) without isolation of the intermediate poly(vinyl alcohol). In a typical process, poly(vinyl acetate) is dissolved in acetic acid and then formalin and a small quantity of sulphuric acid (catalyst) are added. The mixture is heated at 70°C for 24 hours. Water is then added to the product with rapid agitation and the poly(vinyl formal) is precipitated as granules, which are washed with alkali and dried.

Product specifications for some typical commercial grades of poly(vinyl formal) are given in Table 5.2. As the formal content increases there is a progressive increase in softening point, impact strength and hardness. The materials, being amorphous, are soluble in solvents of similar solubility parameter, such as dichloroethylene and dioxan; as the acetate content increases, the polymers become more soluble in ketones, esters and glycol ethers.

The principal use of poly(vinyl formal) is in the electrical insulation of magnet wire. For this purpose, polymers with low hydroxyl content (5–6%) and low acetate content (10–13%) are used in admixture with a resol phenolic resin. On heating at about 175°C, cross-linking occurs; this reaction probably involves the formation of ether links by condensation of methylol groups in the resol and hydroxyl groups in the poly(vinyl formal). These coatings are extremely tough; of particular significance is the fact that they are able to withstand high-speed winding operations.

5.5.2 Poly(vinyl butyral)

The main application of poly(vinyl butyral) is as interlayers in safety glass laminates. In this application, a pure (colourless, light-stable) product with high hydroxyl content (for good adhesion) is required. Such material is preferably prepared from poly(vinyl alcohol) rather than directly from poly(vinyl acetate). In a typical process, completely hydrolysed poly(vinyl alcohol) is suspended in ethanol and then butyraldehyde and a small quantity

Table 5.2 Product specifications for some typical commercial grades of poly(vinyl acetal)s

	Poly(vinyl formal)		Poly(vinyl butyral)	
	Grade 1	Grade 2	Grade 1	Grade 2
Average molecular weight (M_w)	21 000	34 000	64 000	50 000
Hydroxyl content (as % poly(vinyl alcohol))	5–7	5–6	18–21	9–13
Acetate content (as % poly(vinyl acetate))	40–50	10–13	3	3
Formal content (as % poly(vinyl formal)), approx.	50	83	–	–
Butyral content (as % poly(vinyl butyral)), approx.	–	–	80	88

of sulphuric acid (catalyst) are added. The mixture is heated at 80°C for 6 hours. Water is then added to the product with rapid agitation and the poly(vinyl butyral) is precipitated as granules, which are washed with alkali and dried.

Product specifications for some typical commercial grades of poly(vinyl butyral) are given in Table 5.2. Poly(vinyl butyral) is generally softer than poly(vinyl formal) with lower softening point, hardness and tensile strength. In commercial polymers the hydroxyl content exerts a major influence on the solubility characteristics. Most grades are soluble in alcohols, glycol ethers and dioxan and as the hydroxyl content decreases the polymers become more soluble in ketones and esters.

Poly(vinyl butyral) used in safety glass generally has 78–80% butyral content, 18–19% hydroxyl content and less than 2% acetate content. For this application the polymer is plasticized with an ester such as dibutyl sebacate to the extent of about 30 parts of plasticizer per 100 parts of polymer; the compound is then sheeted by calendering. To produce safety glass, the sheet is placed between two pieces of glass and the laminate is then bonded under heat and pressure.

REFERENCES

1. Vona, J. A. *et al.* (1964) in *Manufacture of Plastics*, (ed. W. M. Smith), Reinhold Publishing Corporation, New York. ch. 4.
2. Dickstein, J. and Bouchard, R. (1964) in *Manufacture of Plastics*, (ed. W. M. Smith), Reinhold Publishing Corporation, New York, ch. 5.
3. Hackel, E. (1968) in S. C. I. Monograph No. 30, *Properties and Applications of Polyvinyl Alcohol*, Society of Chemical Industry, London, p. 1.
4. Imoto, T. and Matsubara, T. (1962) *J. Polymer Sci.*, **56**, S4.
5. Marvel, C. S. and Denoon, C. E. (1938) *J. Am. Chem. Soc.*, **60**, 1045; Flory, P. J. and Leutner, F. S. (1948) *J. Polymer Sci.*, **3**, 880; (1950); **5**, 267.

BIBLIOGRAPHY

S. C. I. Monograph No. 30, (1968) *Properties and Applications of Polyvinyl Alcohol*, Society of Chemical Industry, London.
Pritchard, J. G. (1970) *Poly(vinyl alcohol): Basic Properties and Uses*, Macdonald & Co. (Publishers) Ltd, London.
Llewellyn, I. and Williams, H. (1972) in *Vinyl and Allied Polymers*, Vol. 2, (ed. G. Matthews), Iliffe Books, London, ch. 18, (Poly(vinyl acetate)).
Finch, C. A. (ed.) (1973) *Polyvinyl Alcohol*, John Wiley and Sons, London.
Modi, T. W. (1980) in *Handbook of Water-Soluble Gums and Resins*, (ed. R. L. Davidson), McGraw Book Company, New York, ch. 20, (Poly(vinyl alcohol)).
El-Asser, M. S. and Vanderhoff, J. W. (eds.) (1981) *Emulsion Polymerization of Vinyl Acetate*, Applied Science Publishers, London.

6

ACRYLIC POLYMERS

6.1 SCOPE

For the purposes of this chapter, acrylic polymers are defined as polymers based on acrylic acid and its homologues and their derivatives. The principal commercial polymers in this class are based on acrylic acid itself (I) and methacrylic acid (II); esters of acrylic acid (III) and of methacrylic acid (IV); acrylonitrile (V); acrylamide (VI); cyanoacrylates (VII); and copolymers of these compounds. Acrylic-ethylene copolymers are described in Chapter 2. The important styrene-acrylonitrile and acrylonitrile-butadiene-styrene co-polymers are discussed in Chapter 3 whilst acrylonitrile-butadiene copoly-mers are dealt with in Chapter 20.

$$H_2C{=}CH{-}COOH \qquad H_2C{=}\overset{\overset{\displaystyle CH_3}{|}}{CH}{-}COOH \qquad H_2C{=}CH{-}COOR$$

$$\text{(I)} \qquad\qquad\qquad \text{(II)} \qquad\qquad\qquad \text{(III)}$$

$$H_2C{=}\overset{\overset{\displaystyle CH_3}{|}}{C}{-}COOR \qquad H_2C{=}CH{-}CN \qquad H_2C{=}CH{-}CO{-}NH_2$$

$$\text{(IV)} \qquad\qquad\qquad \text{(V)} \qquad\qquad\qquad \text{(VI)}$$

$$H_2C{=}\overset{\overset{\displaystyle CN}{|}}{C}{-}COOR$$

$$\text{(VII)}$$

6.2   RAW MATERIALS

6.2.1   Acrylic acid

There are three routes to acrylic acid which have commercial significance; they are based on propylene, acetylene and ethylene respectively. At the present time, most acrylic acid is produced via the propylene route.

(i) *Propylene route.* This route involves the two-stage oxidation of propylene:

$$H_2C=CH-CH_3 \xrightarrow{O_2} H_2C=CH-CHO \xrightarrow{O_2} H_2C=CH-COOH$$

propylene   acrolein   acrylic acid

A mixture of propylene, air and steam is fed into a reactor containing a catalyst at about 320°C to give acrolein. This intermediate is not isolated but is passed directly to a second reactor, also containing a catalyst, at about 280°C. The effluent is cooled by contact with cold aqueous acrylic acid. Acrylic acid is extracted from the solution with a solvent and then separated by distillation. Because of the ready availability of low cost propylene, this route has become the preferred route for the production of acrylic acid.

(ii) *Acetylene route.* This route involves the reaction of acetylene, carbon monoxide and water:

$$HC{\equiv}CH \ + \ CO \ + \ H_2O \longrightarrow H_2C=CH-COOH$$

In one process, the reaction is conducted in solution in tetrahydrofuran at about 200°C and 6–20 MPa (60–200 atmospheres). Nickel bromide is used as catalyst. The solution of acrylic acid in tetrahydrofuran, after separation of the unconverted acetylene and carbon monoxide in a degassing column, passes to a distillation tower where tetrahydrofuran is taken overhead and acrylic acid is the bottom product. The reaction between acetylene, carbon monoxide and water may also be carried out by using nickel carbonyl as the source of carbon monoxide. In this case, milder reaction conditions are possible. Owing to the high cost of acetylene, this route is now of little commercial importance.

(iii) *Ethylene route.* This route consists of the following sequence:

$$\underset{\substack{\text{ethylene}\\\text{oxide}}}{\overset{O}{\overset{\triangle}{CH_2-CH_2}}} \xrightarrow{HCN} \underset{\substack{\text{ethylene}\\\text{cyanohydrin}}}{\overset{OH \quad CN}{\underset{}{CH_2-CH_2}}} \xrightarrow{-H_2O} \underset{\text{acrylonitrile}}{[H_2C=CH-CN]} \xrightarrow[-NH_3]{H_2O} \underset{\text{acrylic acid}}{H_2C=CH-COOH}$$

The production of ethylene oxide is described in section 9.4.1. The addition of hydrogen cyanide to ethylene oxide takes place at 55–60°C in the presence of a basic catalyst such as diethylamine. The reaction is exothermic and is carried out in solution to facilitate control; the solvent is conveniently ethylene cyanohydrin. The reaction mixture is neutralized and ethylene cyanohydrin is separated by distillation. The second stage of the synthesis involves the dehydration and hydrolysis of ethylene cyanohydrin; these reactions are carried out in one step by heating the cyanohydrin with aqueous sulphuric acid at about 175°C. (It is possible, of course, that the intermediate in this conversion may be acrylonitrile, as shown, or β-hydroxy-propionic acid or both.) At one time this was the standard route for the preparation of acrylic acid but it has been largely displaced by the more economical propylene route.

Acrylic acid is a colourless liquid with a pungent odour, b.p. 141°C.

6.2.2 Methacrylic acid

The most common route for the preparation of methacrylic acid is from acetone as follows:

$$H_3C-CO-CH_3 \xrightarrow{\text{HCN}} H_3C-\underset{\underset{\text{acetone}}{\overset{\text{OH}}{|}}}{\overset{\overset{CH_3}{|}}{C}}-CN \xrightarrow{H_2SO_4} \left[H_2C=\underset{}{\overset{CH_3\overset{+}{N}H_2.H_2SO_4}{|}}C-C=O \right]$$

acetone   acetone cyanohydrin   methacrylamide sulphate

$$\xrightarrow[-NH_3]{H_2O} H_2C=\underset{}{\overset{CH_3}{|}}C-COOH$$

methacrylic acid

In a typical process, acetone is treated with hydrogen cyanide at 140°C in the presence of ammonia as catalyst. The acetone cyanohydrin produced is treated with concentrated sulphuric acid at 100°C to form methacrylamide sulphate. This intermediate is not isolated but is directly converted to methacrylic acid by treatment with water at about 90°C.

A competitive route now in commercial operation involves the two stage oxidation of isobutene with air. The reaction proceeds via methacrolein:

$$H_2C=\underset{}{\overset{CH_3}{|}}C-CH_3 \xrightarrow{O_2} H_2C=\underset{}{\overset{CH_3}{|}}C-CHO \xrightarrow{O_2} H_2C=\underset{}{\overset{CH_3}{|}}C-COOH$$

Methacrylic acid is a colourless liquid with a pungent odour, b.p. 160°C.

6.2.3 Esters of acrylic acid

Various esters of acrylic acid are commercially available but methyl, ethyl and butyl acrylates are the most widely used. These esters are obtained by esterification of acrylic acid. The appropriate alcohol, acrylic acid and a catalyst (generally sulphuric acid) are fed into a reactor at about 80°C and the ester separated by distillation. Higher acrylates are conveniently prepared by ester interchange.

6.2.4 Esters of methacrylic acid

Various esters of methacrylic acid are commercially available but methyl methacrylate is by far the most important. The ester can be made from free methacrylic acid but the standard method is from acetone cyanohydrin without isolation of intermediates (cf. section 6.2.2). In a typical process, acetone cyanohydrin is treated with concentrated sulphuric acid at 100°C to form methacrylamide sulphate which is fed directly into aqueous methanol. The methyl methacrylate is separated by steam distillation and purified by distillation.

Methyl methacrylate is a colourless liquid with a characteristic sweet odour, b.p. 100.5°C. For shipping and storage, hydroquinone or p-methoxyphenol are commonly added as inhibitors.

Higher methacrylates are generally prepared by ester interchange.

6.2.5 Acrylonitrile

There are three routes to acrylonitrile which have commercial significance; they are based on propylene, acetylene and ethylene respectively. At the present time, virtually all acrylonitrile is produced via the propylene route.

(i) *Propylene route.* This route involves the ammoxidation of propylene:

$$H_2C{=}CH{-}CH_3 \; + \; NH_3 \; + \; \tfrac{3}{2}O_2 \; \longrightarrow \; H_2C{=}CH{-}CN \; + \; 3H_2O$$

In a typical process, approximately stoichiometric quantities of propylene, ammonia and oxygen (as air) are fed into a reactor containing a fluidized antimony-iron oxide catalyst. The reaction is conducted at 400–500°C and 0.1–0.3 MPa (1–3 atmospheres). The exit gases are scrubbed with water and acrylonitrile is obtained from the aqueous solution by a series of distillations. In a variation of this process, propylene is treated with nitric oxide (which may be regarded as a product of ammonia oxidation) at about 700°C in the presence of a silver catalyst:

$$2H_2C{=}CH{-}CH_3 \; + \; 3NO \; \longrightarrow \; 2H_2C{=}CH{-}CN \; + \; \tfrac{1}{2}N_2 \; + \; 3H_2O$$

The propylene route does not give particularly high yields of acrylonitrile but it is attractive in that raw material costs are low and hydrogen cyanide is obtained as a valuable by-product. This route therefore now dominates acrylonitrile production.

(ii) *Acetylene route.* The reaction involved is the addition of hydrogen cyanide to acetylene:

$$HC{\equiv}CH \; + \; HCN \; \longrightarrow \; H_2C{=}CH{-}CN$$

A catalyst consisting of cuprous chloride in hydrochloric acid is used and reaction is carried out in the liquid phase. In a typical process, hydrogen cyanide and a large excess of acetylene, slightly above atmospheric pressure, are passed through the catalyst solution at 80–90°C. The reactor effluent is cooled and passed into water, whereby the acrylonitrile is extracted. The acrylonitrile solution is stripped with steam and the acrylonitrile is purified by passage through a series of distillation columns, the last of which operates under reduced pressure. In the past this route has accounted for the major part of acrylonitrile production but it has been largely superseded by the more economical propylene route.

(iii) *Ethylene route.* This route to acrylonitrile is as follows:

$$H_2C{=}CH_2 \xrightarrow{\;O_2\;} \underset{\substack{\text{ethylene}\\\text{oxide}}}{\overset{O}{CH_2{-}CH_2}} \xrightarrow{\;HCN\;} \underset{\substack{\text{ethylene}\\\text{cyanohydrin}}}{\overset{OH\quad CN}{CH_2{-}CH_2}}$$
$$\xrightarrow{\;-H_2O\;} \underset{\text{acrylonitrile}}{H_2C{=}CH{-}CN}$$

(ethylene)

The production of ethylene cyanohydrin is described in section 6.2.1. The dehydration reaction may be carried out in the vapour phase by passing the cyanohydrin over activated alumina at 300°C or in the liquid phase by heating the cyanohydrin (b.p. 221°C) with a catalyst such as magnesium carbonate or sodium formate at 200°C. This route is not economical to run and is now of little commercial importance.

Acrylonitrile is a colourless liquid with a slightly pungent odour, b.p. 77°C.

6.2.6 Acrylamide

Acrylamide is prepared by the graded hydrolysis of acrylonitrile:

$$H_2C{=}CH{-}CN \;+\; H_2O \longrightarrow H_2C{=}CH{-}CO{-}NH_2$$

At one time the hydrolysis was generally carried out with sulphuric acid but now catalytic methods are preferred. In one process, 7% aqueous acrylonitrile is fed into a reactor containing a copper-chromium oxide catalyst at 85°C. Acrylamide is separated from the solution by crystallization. Acrylamide is a white crystalline solid, m.p. 84–85°C.

6.2.7 Cyanoacrylates

2-Cyanoacrylates are synthesized by condensing the appropriate alkyl cyanoacetate with formaldehyde in the presence of a basic catalyst to give a low molecular weight poly(alkyl 2-cyanoacrylate) which is depolymerized by heating:

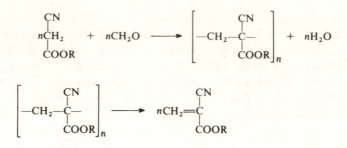

6.3   POLY(ACRYLIC ACID) AND POLY(METHACRYLIC ACID)

These two polymers are considered together since they are generally quite similar. In particular, they both have the significant property of being soluble in water. Mainly because of their water solubility, the polymers have found use in a host of miscellaneous applications such as textile sizes, suspending agents for inorganic pigments and thickeners for polymer latices used for textile finishes and paints. The polymers cannot be melt processed and find little use in the massive state.

Poly(acrylic acid) and poly(methacrylic acid) may be prepared by direct polymerization of the appropriate monomer:

$$n H_2C{=}CH{-}COOH \longrightarrow \left[-CH_2{-}\overset{\displaystyle COOH}{\underset{}{CH}}- \right]_n$$

$$n H_2C{=}\overset{\displaystyle CH_3}{\underset{}{C}}{-}COOH \longrightarrow \left[-CH_2{-}\overset{\displaystyle COOH}{\underset{\displaystyle CH_3}{C}}- \right]_n$$

Conventional free radical techniques are applicable and commonly solution polymerization is carried out, using water as the solvent. In a typical process a water-soluble initiator such as potassium persulphate is added to an aqueous solution of the acid together with an activator, e.g. sodium thiosulphate, and a transfer agent, e.g. mercaptosuccinic acid; a reaction temperature of 50–100°C is used. Alternatively, if solid polymer is required, it is convenient to conduct the polymerization in a solvent such as benzene in which the polymer is insoluble; in this case benzoyl peroxide is a suitable initiator. Salts of acrylic and methacrylic acids may also be polymerized in aqueous solution by treatment with free radical initiators. In addition to the polymerization of the appropriate monomer, the hydrolysis of a suitable polymer represents a further method of preparation of polymeric acids and salts (see sections 6.4, 6.6.3 and 6.8).

Polymers of acrylic and methacrylic acids and their salts are hard, brittle, transparent materials. Poly(acrylic acid) and poly(methacrylic acid) are soluble in such polar solvents as methanol, ethanol, ethylene glycol, dioxan, dimethylformamide and water but are insoluble in non-polar solvents such as aliphatic and aromatic hydrocarbons. Monovalent metal and ammonium salts of these polymers are generally soluble in water.

Poly(acrylic acid) and poly(methacrylic acid) undergo reactions characteristic of carboxylic acids. Neutralization of aqueous solutions of the polyacids by bases such as sodium hydroxide is of interest since it causes viscosity changes which demonstrate the presence of two different molecular conformations. Ordinarily, aqueous solutions of the polyacids have low viscosities since the polymer is tightly coiled, being only slightly ionized. As sodium hydroxide is added more and more carboxyl groups become ionized; mutual repulsion of charges forces the polymer chains to uncoil and the extended chains lead to higher solution viscosity. If hydrochloric acid is added to aqueous solutions of the polyacids there is a decrease in viscosity as the ionization of the polyacid is reduced from its initial low value and the chains become more tightly coiled. Interestingly, addition of base and acid to a filament of polymer causes reversible elongation and contraction and the filament can be made to perform mechanical work [1]. Poly(methacrylic acid) shows a much greater effect than poly(acrylic acid).

A further reaction shown by poly(acrylic acid) and poly(methacrylic acid) is with ammonia to form cyclic imides. Polymethacrylimide is prepared commercially by heating a copolymer of methacrylic acid and methacrylonitrile with an ammonia-producing compound (e.g. urea or ammonium hydrogen carbonate) at about 200°C:

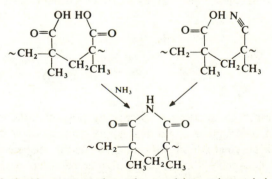

Although imidation is not complete, the resulting polymer is in the form of a rigid cellular product. Such foams have good solvent resistance and may be used under load up to 160°C. Suggested applications include engine covers and aircraft landing-gear doors.

Diesters of acrylic and methacrylic acids are of interest in that they may be used as thermosetting resins. For example, one commercial material is produced by esterification of methacrylic acid with the addition product of ethylene oxide (section 9.4.1(a)) and bisphenol A (section 18.3.1):

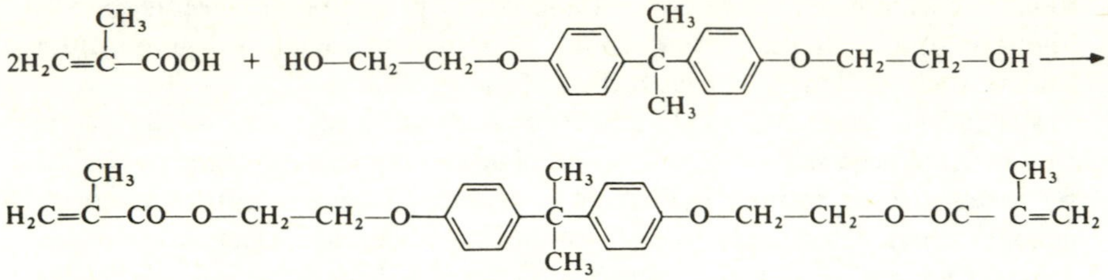

Such materials may be cross-linked through the terminal unsaturated sites and find use as laminating resins. Compared to polyester resins, the acrylic resins exhibit less shrinkage on cure.

6.4 POLYACRYLATES

Polymers of the lower n-alkyl acrylates have found some commercial use. The lowest member of the series, poly(methyl acrylate) has poor low temperature properties and is water-sensitive and its use is therefore restricted to such applications as textile sizes and leather finishes. Ethyl acrylate and butyl acrylate are the most commonly used major components of commercial acrylate rubbers. At the present time, most commercial acrylate rubbers are copolymers of either ethyl or butyl acrylate with an alkoxy acrylate such as methoxyethyl acrylate (VIII) and ethoxyethyl acrylate (IX). The presence of a proportion of longer side-chains gives a rubber with improved low temperature flexibility.

$$H_2C=CH—COO—CH_2—CH_2—O—CH_3 \quad H_2C=CH—COO—CH_2—CH_2—O—CH_2—CH_3$$
$$(VIII) \qquad\qquad\qquad\qquad (IX)$$

$$H_2C=CH—O—CH_2—CH_2Cl \qquad H_2C=CH—O—CO—CH_2Cl$$
$$(X) \qquad\qquad\qquad\qquad (XI)$$

The most important method of preparing polyacrylates is by emulsion polymerization. The high rate and heat of polymerization of acrylates make control of bulk polymerization impractical whilst in suspension polymerization there is a tendency for the soft polymer beads to coalesce. Solution polymerization results in solutions of high viscosity, which are difficult to handle unless the molecular weight of the polymer or the degree of conversion are restricted. This high viscosity arises because during polymerization of acrylates, extensive branching and cross-linking occur due to the ease with which tertiary hydrogen is abstracted from the polymer chain. Such abstraction is favoured by the stability of the resulting radical, e.g.

$$
\begin{array}{ccc}
\text{C}_2\text{H}_5 & & \left[\text{C}_2\text{H}_5 \right. \\
| & & | \\
\text{O} & & \text{O} \\
| & \longrightarrow & | \\
\text{CO} & & \text{CO} \\
| & & | \\
n\text{H}_2\text{C}=\text{CH} & & \left. -\text{CH}_2-\text{CH}- \right]_n
\end{array}
$$

$$
\begin{array}{ccccc}
\text{C}_2\text{H}_5\text{O}-\text{C}=\text{O} & & \text{C}_2\text{H}_5\text{O}-\text{C}=\text{O} & & \text{C}_2\text{H}_5-\text{C}-\text{O} \cdot \\
| & \xrightarrow{\text{R}\cdot} & | & \longleftrightarrow & \parallel \\
\sim\text{C}_2\text{H}-\text{CH}\sim & & \sim\text{CH}_2-\dot{\text{C}}\sim & & \sim\text{CH}_2-\text{C}\sim
\end{array}
$$

where R· represents a polymeric free radical.

With emulsion polymerization, high molecular weight polymer is produced rapidly and easily and the latex has a much lower viscosity than the corresponding solution. In a typical process potassium persulphate is used as initiator with a reaction temperature of 80–90°C. Since acrylate monomers are easily hydrolysed under basic conditions it is important that polymerization is conducted under neutral or acid conditions. For this reason, fatty acid salts such as potassium oleate are not satisfactory emulsifying agents; however, salts of strong acids such as long chain sulphonic acids may be used. When the polyacrylate is intended for such uses as textile sizing the latex is used directly; for the production of rubbers the latex is coagulated and the crumb washed and dried.

As with other rubbers, acrylate rubbers must be vulcanized before they exhibit useful technological properties. It is possible to cross-link straight polyacrylates by various means. For example, treatment with a peroxide leads to abstraction of tertiary hydrogen and subsequent cross-linking, e.g.

$$
\begin{array}{ccc}
\text{C}_2\text{H}_5\text{O}-\text{CO} & & \text{C}_2\text{H}_5\text{O}-\text{CO} \\
| & & | \\
\sim\text{CH}_2-\dot{\text{C}}\sim & & \sim\text{CH}_2-\text{C}\sim \\
& \longrightarrow & | \\
\sim\text{CH}_2-\dot{\text{C}}\sim & & \sim\text{CH}_2-\text{C}\sim \\
| & & | \\
\text{C}_2\text{H}_5\text{O}-\text{CO} & & \text{C}_2\text{H}_5\text{O}-\text{CO}
\end{array}
$$

Various basic compounds such as sodium hydroxide and sodium metasilicate also bring about vulcanization; in these cases, cross-linking is accompanied by the liberation of alcohol which indicates the following type of reaction:

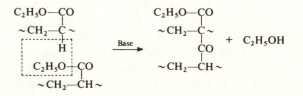

In general, however, vulcanizates of straight polyacrylates are weak and do not have good ageing properties. Consequently, commercial acrylate rubbers are invariably vulcanized through reactive sites introduced into the polymer by copolymerization.

At one time reactive sites were obtained by the use of 2-chloroethyl vinyl ether (X) but because of processing problems this comonomer has been replaced by compounds such as vinyl chloroacetate (XI) which give more reactive cross-linking sites. Whereas the earlier acrylate rubbers were vulcanized with amines or ammonium salts, now the most commonly used system is the so-called sulphur-soap system which consists of sulphur and potassium or sodium stearate.

The important characteristics of acrylate rubbers are good oil resistance and good heat resistance; carbon black reinforced materials age very well in air up to about 200°C. It is these properties which account for the major use of acrylate rubbers, i.e. in oil seals for automobiles. Polyacrylates are fairly readily hydrolysed to poly(sodium acrylate) by heating with aqueous sodium hydroxide for a few hours.

6.5 POLYMETHACRYLATES

Polymers of a few n-alkyl methacrylates have found commercial application. By far the most important of these polymers is poly(methyl methacrylate). This is an established major plastics material and is considered in detail in the next section.

As mentioned in section 1.2, the physical properties of polymeric materials are largely determined by molecular weight, strength of intermolecular forces, regularity of polymer structure and flexibility of the molecule. The influence of these factors is clearly demonstrated by the difference in properties of various polymethacrylates. In a series of poly(n-alkyl methacrylate)s, as the size of the alkyl group increases the polymer molecules become further spaced apart and intermolecular attraction is reduced. Thus as the side-chain length increases, the softening point decreases and the polymers become rubbery at progressively lower temperatures. However, when the number of carbon atoms in the side-chain exceeds 12, side-chain crystallization becomes significant and the polymers become less rubbery. The effect of side-chain length is illustrated in Fig. 6.1. As is shown in Table 6.1, poly(alkyl methacrylate)s in which the alkyl group is branched have higher softening points than the unbranched isomers. This effect arises because of the better packing which is possible with the branched isomers and because the lumpy branched structures impede rotation about the carbon-carbon bond in the main chain. Similarly, the α-methyl group present in polymethacrylates reduces chain flexibility and the lower polymethacrylates have higher softening points than the corresponding polyacrylates. (In the higher polyacrylates, side-chain crystallization is the more pronounced effect.) (Fig. 6.1.)

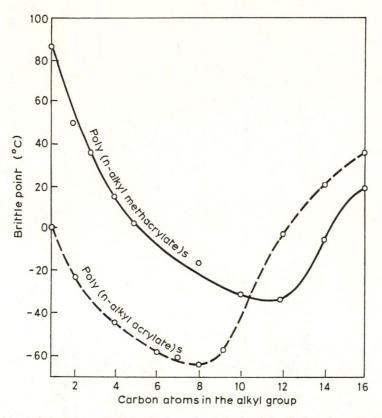

Fig. 6.1   Brittle points of poly(n-alkyl acrylate)s and poly(n-alkyl methacrylate)s [2].

Some other comparative properties of various poly(alkyl methacrylate)s are given in Table 6.2 and the influence of the alkyl group is again illustrated.

The nature of the alkyl group in poly(alkyl methacrylate)s also determines the mode of thermal decomposition of the polymer. When poly(methyl methacrylate) is heated above about 200°C, an unzipping reaction results in the almost quantitative production of monomer (see section 6.6.3). Polymers of higher esters, however, tend to degrade only partially to monomer, poorer yields of monomer generally being obtained in passing up the homologous series and from primary to secondary and tertiary esters. Thus poly(n-butyl methacrylate) yields about 50% monomer whilst poly(*tert*-butyl methacry-late) gives about 1% monomer and almost quantitative amounts of isobutene and poly(methacrylic anhydride) [4]. The decomposition of poly(*tert*-butyl methacrylate) involves ester decomposition to give poly(methacrylic acid) which then loses water to form the anhydride:

Table 6.1 Softening points of poly(alkyl methacrylate)s [3]

$$\left[-CH_2-\underset{\underset{COOR}{|}}{\overset{\overset{CH_3}{|}}{C}}- \right]_n$$

R	Softening point (°C)	R	Softening point (°C)	R	Softening point (°C)
H_3C-	125				
H_3C-CH_2-	65				
$H_3C-CH_2-CH_2-$	38	$H_3C-\overset{\overset{CH_3}{|}}{CH}-CH-$	95		
$H_3C-CH_2-CH_2-CH_2-$	33	$H_3C-CH_2-CH_2-\overset{\overset{CH_3}{|}}{CH}-$	62	$H_3C-\overset{\overset{CH_3}{|}}{CH}-CH_2-$	70

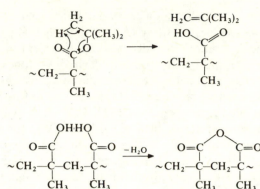

It may be noted here that the thermal degradation of polyacrylates does not follow the same pattern as that of the polymethacrylates. None of the polyacrylates gives appreciable yields of monomer or olefin; the principal products are polymeric fragments which arise by chain scission accompanied by transfer of tertiary hydrogen atoms.

Poly(methyl methacrylate) is by far the predominant polymethacrylate in use in rigid applications, mainly because of its relatively high softening point and strength. However, a few other polymethacrylates have found limited commercial utilization. The n-butyl, n-octyl and n-nonyl esters are used as leather finishes and poly(lauryl methacrylate) is used to depress the pour-point and improve viscosity-temperature characteristics of lubricating oils.

Table 6.2 Typical values for various properties of some poly(alkyl methacrylate)s [3]

	Methyl	Ethyl	n-Propyl	n-Butyl	Isobutyl
Specific gravity	1.19	1.11	1.06	1.05	1.02
Tensile strength					
(MPa)	62	34	28	10	23
(1bf/in^2)	9000	5000	4000	1500	3400
Elongation at					
break (%)	4	7	5	230	2

6.6 POLY(METHYL METHACRYLATE)

6.6.1 Development

Initial researches into acrylic polymers were concerned rather more with polyacrylates than polymethacrylates. The first acrylic polymer to be produced commercially was poly(methyl acrylate); production was begun in 1927 by Rohm and Haas AG in Germany. In about 1930, Hill of Imperial Chemical Industries Ltd (UK) prepared cast poly(methyl methacrylate) and found it to be a potentially useful material but high raw material costs prohibited commercial development. (At this time, methyl methacrylate was obtained by dehydration of the hydroxyisobutyric ester.) In 1932 Crawford (also of Imperial Chemical Industries Ltd) devised a synthesis based on the cheap raw materials, acetone and hydrogen cyanide, which is described in section 6.2.4. Poly(methyl methacrylate) thus became a feasible proposition and commercial production began in 1934. Cast sheet poly(methyl methacrylate) was much used during the Second World War for aircraft glazing. Since the war, sundry other uses of the sheet material have been developed; these include display signs, lighting fittings and bathroom fittings. Poly(methyl methacrylate) may also be melt processed and, in fact, a recent trend has been the increasing displacement of cast sheet by material fabricated by more cost-effective means. For example, street lamp housings are now injection moulded and extruded sheet now approaches cast sheet in quality. Poly(methyl methacrylate) is also extensively used for the production of dentures.

6.6.2 Preparation

The polymerization of methyl methacrylate:

$$nH_2C=\underset{\underset{COOCH_3}{|}}{\overset{\overset{CH_3}{|}}{C}} \longrightarrow \left[-CH_2-\underset{\underset{COOCH_3}{|}}{\overset{\overset{CH_3}{|}}{C}}- \right]_n$$

is readily accomplished by bulk, solution, suspension and emulsion techniques. Of these methods, bulk and suspension polymerization are mainly used for the production of the homopolymer.

(a) Bulk polymerization

Techniques which involve a combination of bulk polymerization and casting are extensively used in the manufacture of poly(methyl methacrylate) sheet. In most processes, the first step is the preparation of a partially polymerized material. Typically, monomer is stirred with benzoyl peroxide (0.5%) at 90°C for about 10 minutes to give a syrup which is then cooled to room temperature. Colourant, plasticizer and ultraviolet absorber, if required, are added at this point. At this stage the degree of conversion of monomer to polymer is about 20%; the use of such a syrup reduces shrinkage in the casting cell and also lessens leakage from the cell. The syrup is then poured into a casting cell, consisting of two glass plates separated by a rubber gasket. The plates are held together by spring-loaded clamps so that the plates continuously move together in response to the shrinkage of approximately 20% which occurs on conversion of monomer to polymer. The filled cell is then passed through a heating tunnel wherein the temperature is maintained at 40°C for 15 hours and then 95°C for 1 hour. The sheet is then cooled and removed from the cell. With castings of thickness greater than about 2 cm, the exothermic reaction may result in local temperatures above the boiling point of the monomer (100.5°C) and bubbles may form. In such cases, polymerization may be carried out under pressure so that the boiling point of the monomer is raised.

Rod is also manufactured by casting. In one process, syrup is contained in vertical aluminium tubes which are very slowly lowered into a bath at 40°C. As the lowest portion of syrup polymerizes it contracts and the syrup above moves downwards. In this way a homogeneous rod, free from voids, is obtained. Dentures are normally made from a polymer-monomer dough in a plaster mould. Bulk polymerization carried out in the preparation of sheet, as described above, results in polymer of very high average molecular weight ($\simeq 10^6$).

(b) Suspension polymerization

Suspension polymerization of methyl methacrylate is used mainly for the production of injection moulding and extrusion grades of polymer. Suspension polymer is also used in the preparation of polymer-monomer doughs for dentures. Polymerization is carried out batch-wise in a stirred reactor, jacketed for heating and cooling; the reactor is capable of withstanding a pressure of 0.3–0.4 MPa (3–4 atmospheres). A typical basic formulation might be as follows:

Methyl methacrylate	100 parts by weight	
Water	200	
Magnesium carbonate	3	(suspending agent)
Benzoyl peroxide	0.2	(initiator)

Reaction is carried out under nitrogen. Typically, the mixture is initially heated to about 80°C but the exothermic reaction causes the temperature to rise to about 120°C, with accompanying increase in pressure. Polymerization is rapid and is complete in about 1 hour. The suspension is cooled and acidified with sulphuric acid to remove the suspending agent. The beads of polymer are then filtered off, washed and dried in air at about 80°C. The dried beads may be used for moulding without further treatment or they may be compounded with additives (e.g. colourants), extruded and granulated.

Suspension polymerized poly(methyl methacrylate) normally has an average molecular weight of about 60 000.

6.6.3 Properties

Poly(methyl methacrylate) is a hard, rigid transparent substance. Values for some physical properties of the material are given in Tables 6.1 and 6.2. Straight poly(methyl methacrylate) is somewhat tougher than polystyrene but is less tough than the ABS polymers. An outstanding property of poly(methyl methacrylate) is its clarity. The material absorbs very little visible light but there is about 4% reflection at each polymer-air interface for normal incident light. Thus the transmission of normal incident light through a sheet of the polymer is about 92%. Poly(methyl methacrylate) is a polar material and has a rather high dielectric constant and power factor; it is a good electrical insulator at low frequencies but is less satisfactory at high frequencies.

Poly(methyl methacrylate) prepared by free radical polymerization is amorphous and is therefore soluble in solvents of similar solubility parameter. Effective solvents include aromatic hydrocarbons such as benzene and toluene; chlorinated hydrocarbons such as chloroform and ethylene dichloride; and esters such as ethyl acetate and amyl acetate. Some organic materials, although not solvents for the polymer, cause crazing and cracking, e.g. aliphatic alcohols and amines. Poly(methyl methacrylate) has very good resistance to attack by water, alkalis, aqueous inorganic salts and most dilute acids. Some dilute acids such as hydrocyanic and hydrofluoric acids, however, do attack the polymer, as do concentrated oxidizing acids. Poly(methyl methacrylate) has much better resistance to hydrolysis than poly(methyl acrylate), probably by virtue of the shielding presented by the α-methyl group. Poly(methyl methacrylate) may be converted to poly(sodium methacrylate) only by rather drastic treatment with, for example, molten sodium hydroxide.

A further outstanding property of poly(methyl methacrylate) is its good outdoor weathering, in which respect the material is markedly superior to most other thermoplastics. After several years under tropical conditions the colour change is extremely small. When poly(methyl methacrylate) is heated above the glass transition temperature ($105°C$) it becomes rubbery and sheet material is easily manipulated at $150–160°C$. Above about $200°C$ decomposition becomes appreciable and at $350–450°C$ a nearly quantitative yield of monomer is readily obtained. Thus the recovery of monomer from scrap polymer is a feasible proposition.

6.7   ACRYLIC COPOLYMERS

Several acrylic copolymers have been mentioned in the preceding sections, but generally the products described contain a preponderance of one monomer with minor amounts of other monomer(s). There also exists a large number of acrylic copolymers in which one monomer does not predominate to any great extent. Such copolymers find widespread use in surface coatings. Two distinct types of copolymers are used in surface coatings, namely thermoplastic acrylics and thermosetting acrylics and it is these materials which are the subject of this section.

6.7.1 Thermoplastic acrylics

These are linear copolymers which are generally based on approximately equal amounts of methyl methacrylate (which confers hardness) and an alkyl acrylate (which confers flexibility). Two forms of such copolymers are commonly produced, namely acrylic solutions and acrylic latices.

(i) *Acrylic solutions* are prepared directly by solution polymerization. Commonly, polymerization is conducted in an aromatic hydrocarbon such as benzene or toluene or a ketone such as methyl ethyl ketone. The initiator is generally benzoyl peroxide (0.2–1% by weight of the monomer). The reaction is carried out at 90–110°C under slight pressure in a jacketed kettle fitted with a heavy-duty stirrer. Since reaction is vigorous, it is common practice to add the monomer gradually to the hot solvent over a period of 1–4 hours. This technique gives polymer with a wide molecular weight distribution; normally, conditions are chosen to give an average molecular weight of about 90 000. The final product is a 40–60% solution of polymer in solvent which is used directly in the preparation of surface coatings. The main area of application for such coatings is in the automotive industry, where they are used extensively for refinishing.

(ii) *Acrylic latices* are prepared by emulsion polymerization. Most commonly a semi-continuous process is carried out using a reactor fitted with a stirrer and jacketed for heating and cooling. A typical formulation might be as follows:

Initial charge		
Water (deionized)	68.0	parts by weight
Sodium lauryl sulphate	0.2	(surfactant)
Ammonium persulphate	0.5	(initiator)

Monomer charge	
Methyl methacrylate	50.0
Butyl acrylate	49.0
Methacrylic acid	1.0
Water (deionized)	30.5
Sodium lauryl sulphate	0.2

The initial charge is heated to 85°C and the monomer charge is fed in continuously over 2.5 hours. The temperature is then raised to 95°C to complete the reaction. The addition of monomer over this time span permits better control of the exothermic reaction and also leads to a more uniform distribution of comonomer units in the final copolymer. In this formulation, methacrylic acid is included to give improved latex stability and adhesion. The final product is used directly in such applications as latex paints, adhesives for high speed packaging, paper coatings and inks. About half of the current output of acrylic latices is used to prepare domestic paints. Compared to latex paints based on vinyl acetate copolymers, straight acrylic latex paints show improved gloss and exterior durability: they are, however, more costly.

Non-aqueous dispersions (NAD) in which the dispersing medium is not water but an organic solvent may also be made by emulsion polymerization. Although such systems combine the advantages of high solids contents and adjustable drying rates, they find limited commercial use.

6.7.2   Thermosetting acrylics

Thermosetting acrylics are designed to give cross-linked films on heating and they are generally terpolymers. The first monomer confers hardness and rigidity to the copolymer, the second monomer contributes to the flexibility of the copolymer and the third monomer provides pendant reactive groups which are the sites for subsequent cross-linking. Table 6.3 shows examples of each of the three classes of monomer which are used commercially.

Table 6.3   Comonomers used in the preparation of thermosetting acrylic copolymers

Monomers conferring hardness	Monomers conferring flexibility	Monomers conferring reactive sites
Acrylonitrile	Butyl acrylate	Acrylic acid
Methyl methacrylate	Ethyl acrylate	Butoxymethylacrylamide
Styrene	2-Ethylhexyl acrylate	Glycidyl acrylate
Vinyltoluene		Hydroxyethyl acrylate

Most commonly, the copolymers are applied as solutions and it is therefore convenient to prepare the solutions directly from the monomer by solution polymerization. Solvents used include butanol and xylene. In order that the solution has an acceptable viscosity for application and a satisfactory solids content (usually 40–60%), the molecular weight of the polymer is kept down to about 20 000–30 000 by the use of relatively high initiator concentrations (about 2% on the monomer) and high temperatures (about 100–140°C). A typical formulation might be as follows:

Methyl methacrylate	41.3 parts by weight	
2-Ethylhexyl methacrylate	27.5	
Hydroxyethyl methacrylate	31.2	
Xylene	50.0	
Ethoxyethanol	25.0	
tert-Butyl perbenzoate	2.0	(initiator)

The monomers and initiator are added to the solvent at 140°C under nitrogen over 3 hours. Heating is then continued for a further 2 hours to complete the reaction.

It will be noted from Table 6.3 that the sites used for subsequent cross-linking reactions include carboxyl, *N*-methylol ether, epoxy and hydroxy-alkyl groups. Although some copolymers are self-reactive and cross-link merely on heating, it is general practice to add cross-linking agents to effect cure. A large number of cross-linking agents, which may themselves be polymeric, is used for this purpose. The choice of cross-linking agent is governed by the service conditions envisaged for the coating and by the nature of the reactive groups in the copolymer. For example, copolymers with carboxyl or *N*-methylol ether groups are commonly cured by the addition of epoxy resins (see Chapter 18), e.g.

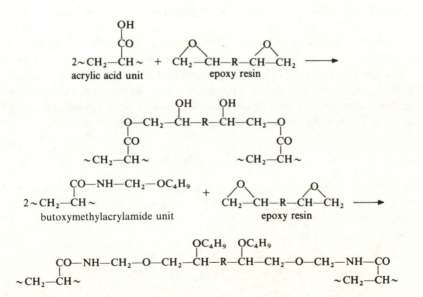

Thermosetting acrylic copolymers containing epoxy groups are usually cross-linked by polyfunctional amines, e.g.

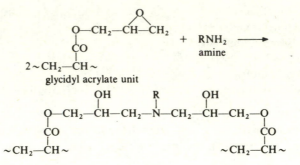

Copolymers with hydroxyalkyl groups are commonly cured by butylated melamine- or urea-formaldehyde resins (see Chapter 15), e.g.

O—CH₂—CH₂—OH
|
CO
|
2 ~CH₂—CH~   +   C₄H₉O—CH₂—NH—R—NH—CH₂—OC₄H₉
hydroxyethyl acrylate unit   butylated aminoresin

O—CH₂—CH₂—O—CH₂—NH—R—NH—CH₂—O—CH₂—CH₂—O
| |
CO   CO
| |
~CH₂—CH~   + 2C₄H₉OH   ~CH₂—CH~

It will be appreciated from the various examples given above that the range of thermosetting acrylic copolymers is very large. By careful selection of the monomers from which the base polymer is prepared and of the cross-linking agent, acrylic finishes can be tailored to meet many requirements. Such finishes range from the very flexible coatings needed for strip metal coating to the hard, chemically resistant coatings for domestic appliances.

6.8   POLYACRYLONITRILE

Approximately 70% of the commercial output of acrylonitrile is polymerized (with minor amounts of comonomers) to give polymers which are used for textile fibres:

$$n\text{H}_2\text{C}=\overset{\overset{\displaystyle \text{CN}}{|}}{\text{CH}} \longrightarrow \left[-\text{CH}_2-\overset{\overset{\displaystyle \text{CN}}{|}}{\text{CH}}- \right]_n$$

The most important methods for the preparation of polyacrylonitrile are solution polymerization and suspension polymerization. The former method is particularly convenient, since when a solvent for the polymer is used, the resulting solution may be utilized directly for fibre spinning. Concentrated aqueous solutions of inorganic salts such as calcium thiocyanate, sodium perchlorate and zinc chloride make suitable solvents; suitable organic solvents include dimethylacetamide, dimethylformamide and dimethylsulphoxide. Emulsion polymerization suffers from the disadvantage that the monomer has appreciable water-solubility and the formation of polymer in the aqueous phase can lead to coagulation of the latex. This tendency is reduced by the addition of ethylene dichloride to the system.

Fibres prepared from straight polyacrylonitrile are difficult to dye and, in order to improve dyeability, commercial fibres invariably contain a minor proportion (about 10%) of one or two comonomers such as methyl methacrylate, vinyl acetate and 2-vinylpyridine.

The average molecular weight (M_W) of commercial polyacrylonitrile is generally in the range 80 000–170 000.

In polyacrylonitrile appreciable electrostatic forces occur between the dipoles of adjacent nitrile groups on the same polymer molecule. This intramolecular interaction restricts bond rotation and leads to a stiff chain. As a result, polyacrylonitrile has a very high crystalline melting point (317°C) and is soluble in only a few solvents such as dimethylacetamide and dimethylformamide and in aqueous solutions of inorganic salts. Polyacrylonitrile cannot be melt processed since extensive decomposition occurs before any appreciable flow occurs and fibres are therefore spun from solution. In one process, for example, a solution of the polymer in dimethylformamide is extruded into a coagulating bath of glycerol and the fibre formed is drawn and wound.

As mentioned above, polyacrylonitrile is unstable at elevated temperatures. On heating above about 200°C, polyacrylonitrile yields a red solid with very little formation of volatile products. When the red residue is heated at about 350°C there is produced a brittle black material of high thermal stability. The first step in these changes consists of a nitrile polymerization reaction whilst the second step involves aromatization to form a condensed polypyridine *ladder polymer*:

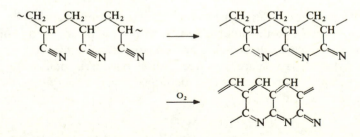

Continued heating at high temperatures (1500–3000°C) results in the elimination of all elements other than carbon to leave a carbon fibre with graphitic

crystalline structure of great strength. Polyacrylonitrile fibres have become
the most important source for carbon fibres.

Polyacrylonitrile is hydrolysed by heating with concentrated aqueous
sodium hydroxide to poly(sodium acrylate).

6.9   NITRILE RESINS

As mentioned in the preceding section, polyacrylonitrile cannot be melt
processed because extensive decomposition occurs before any appreciable
flow takes place. Various copolymers in which acrylonitrile is the major
component are, however, melt-processable and several have been produced
commercially. Such copolymers are commonly referred to as nitrile resins.
The main reason for the development of these copolymers was their low gas
permeability and carbonated-beverage containers represented a large potential market. However, concern about the carcinogenicity of residual acrylonitrile has rendered uncertain the future use of nitrile resins as containers for
foodstuffs.

Owing to the uncertain situation, several companies have withdrawn from
this field and at the present time only one type of nitrile resin is commercially
available. This product is a copolymer of acrylonitrile and methyl acrylate
(70:30 parts by weight) which is formed in the presence of a butadiene-
acrylonitrile rubber (10–15%). The product thus contains graft copolymer
and has a structure resembling that of ABS polymers previously described
(section 3.4.1(b)).

In addition to their outstanding gas-barrier properties, nitrile resins have
good resistance to most organic solvents (but not acetone and methyl ethyl
ketone) and acids and bases. Nitrile resins may be melt-processed by all the
usual methods, e.g. injection moulding, blow moulding and sheet extrusion.
Applications include containers for agricultural, automotive and household
products.

6.10   POLYACRYLAMIDE

Polyacrylamide resembles poly(acrylic acid) and poly(methacrylic acid)
(section 6.3) in being water-soluble and, as with those polymers, it is mainly
this property which results in some limited commercial utilization.

Polyacrylamide is prepared by free radical polymerization, using techniques essentially similar to those described for poly(acrylic acid) and
poly(methacrylic acid):

$$n\mathrm{H_2C}{=}\mathrm{CH}{-}\mathrm{CONH_2} \longrightarrow \left[-\mathrm{CH_2}{-}\underset{\mathrm{CH}}{\overset{\mathrm{CONH_2}}{|}}- \right]_n$$

It may be noted that whereas this reaction gives a vinyl polymer, a different type of polymer is obtained when polymerization is initiated by a strong base. In this case, a polyamide (nylon 3) is formed. Active initiators for this type of polymerization include alkoxides (RO^-) and reaction probably takes place according to the following scheme, which involves the rearrangement of a carbanion to a more stable amide anion:

Initiation

$$RO^- \ + \ H_2C=CH-CO-NH_2 \ \longrightarrow \ RO-CH_2-\overset{-}{C}H-CO-NH_2$$

$$\longrightarrow \ RO-CH_2-CH_2-CO-\overset{-}{N}H$$

Propagation

$$RO-CH_2-CH_2-CO-\overset{-}{N}H \ + \ H_2C=CH-CO-NH_2 \ \longrightarrow$$

$$RO-CH_2-CH_2-CO-NH-CH_2-\overset{-}{C}H-CO-NH_2 \ \longrightarrow$$

$$RO-CH_2-CH_2-CO-NH-CH_2-CH_2-CO-\overset{-}{N}H \ \text{etc.}$$

Polyacrylamide is a hard, brittle material. It is readily soluble in cold water but solubility in organic compounds is generally very limited. The polymer undergoes reactions characteristic of the amide group; for example, alkaline hydrolysis introduces carboxylic groups and reaction with formaldehyde gives methylol groups.

Polyacrylamide has found use as a flocculant in the processing of minerals and in water treatment. Copolymers of acrylamide and acrylic acid are used to increase the dry strength of paper.

6.11 POLY(2-CYANOACRYLATE)S

Methyl and ethyl 2-cyanoacrylates form the basis of so-called 'super glues'. The presence of two electron-withdrawing groups polarizes the carbon-carbon double bond and makes the monomer very susceptible to anionic polymerization. Weak bases such as water bring about rapid polymerization and, in practice, a trace of moisture on a substrate leads to very fast bonding:

Initiation

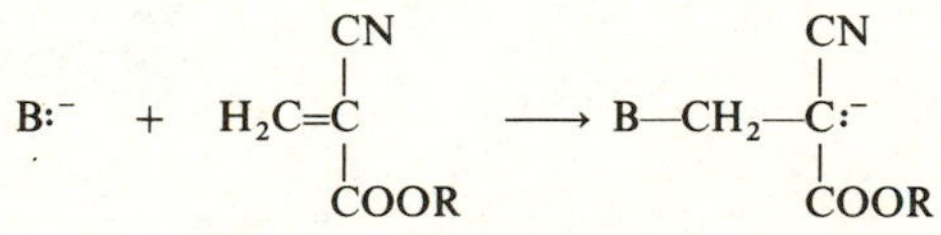

$$B:^- \ + \ H_2C=\underset{\underset{COOR}{|}}{\overset{\overset{CN}{|}}{C}} \ \longrightarrow \ B-CH_2-\underset{\underset{COOR}{|}}{\overset{\overset{CN}{|}}{C}}:^-$$

Propagation

$$B-CH_2-\underset{\underset{COOR}{|}}{\overset{\overset{CN}{|}}{C}}:^- \ + \ H_2C=\underset{\underset{COOR}{|}}{\overset{\overset{CN}{|}}{C}} \ \longrightarrow \ B-CH_2-\underset{\underset{COOR}{|}}{\overset{\overset{CN}{|}}{C}}-CH_2-\underset{\underset{COOR}{|}}{\overset{\overset{CN}{|}}{C}}:^- \ \text{etc.}$$

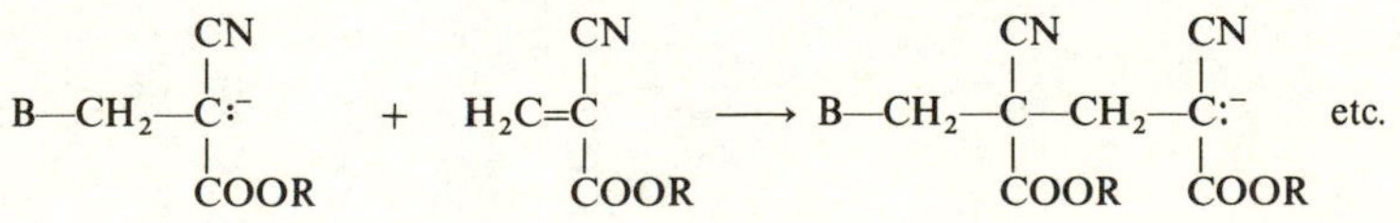

REFERENCES

1. Kuhn, W. *et al.* (1960) in *Size and Shape Changes of Contractile Polymers*, (ed. A. Wassermann), Pergamon Press, Oxford, p. 41.
2. Rehberg, C. E. and Fisher, C. H. (1948) *Ind. Eng. Chem.*, **40**, 1431.
3. Wakeman, R. L. (1947) *The Chemistry of Commercial Plastics*, Reinhold Publishing Corporation, New York, pp. 470, 471.
4. Grassie, N. (1960) *Trans. Plastics Institute*, **28**, 233.

BIBLIOGRAPHY

Riddle, E. H. (1954) *Monomeric Acrylic Esters*, Reinhold Publishing Corporation, New York.

Horn, M. B. (1960) *Acrylic Resins*, Reinhold Publishing Corporation, New York.

Volk, H. and Friedrich, R. E. (1980) in *Handbook of Water-Soluble Gums and Resins*, (ed. R. L. Davidson), McGraw Book Company, New York, ch. 16 (Polyacrylamide).

Greenwald, H. L. and Lyskin, L. S. (1980) in *Handbook of Water-Soluble Gums and Resins*, (ed. R. L. Davidson), McGraw Book Company, New York, ch. 17 (Poly(acrylic acid)).

FLUOROPOLYMERS

7.1 SCOPE

The commercial production of polymers containing fluorine is very small compared to the output of many other synthetic polymers. Nevertheless, several fluoropolymers are used in various important specialized applications. The principal commercial fluoropolymers at the present time are the homopolymers of tetrafluoroethylene (I), chlorotrifluoroethylene (II), vinyl fluoride (III) and vinylidene fluoride (IV) and various copolymers based on these monomers. These materials, together with a few other fluoropolymers of interest, form the contents of this chapter.

$$F_2C{=}CF_2 \qquad F_2C{=}\underset{\underset{Cl}{|}}{\overset{\overset{F}{|}}{C}} \qquad H_2C{=}CHF$$

(I) (II) (III)

$$H_2C{=}CF_2$$

(IV)

7.2 DEVELOPMENT

Polymerization of fluorine-containing olefins was first described in 1934 in a patent to I. G. Farbenindustrie (Germany) which related to polychlorotrifluoroethylene. However, this polymer was of limited value and it was not until 1938, when Plunkett in the United States prepared polytetrafluoroethylene, that commercial interest quickened. This polymer has found a great number of uses and probably accounts for at least 80% of the current output of fluoropolymers. The first pilot plant for polytetrafluoroethylene came into operation in 1943 and large scale production began in 1950 (E. I. du Pont de Nemours and Co.) (USA). In addition to polytetrafluoroethylene, many other fluoropolymers have been investigated and a few have reached commercial status during the period 1950–1980. In these investigations, attention has centered particularly on copolymers which are melt-processable and on those with rubbery characteristics.

Interest in fluoropolymers stems from the fact that the polymers have outstanding chemical and thermal resistance. These properties result from the great stability of the C–F bond, the effect of fluorine substitution in strengthening adjacent C–C bonds, and the screening effect of the fluorine atoms, which because of their small size may be tightly packed about the polymer backbone. Fluoropolymers have proved of great value in applications requiring good ageing properties, good resistance to chemicals and ability to withstand extremes of temperature, particularly elevated temperatures. The demand for such materials has increased considerably during recent years. It may be noted that fluoropolymers are invariably high-cost materials, a factor which generally limits their use to specialized fields.

7.3 RAW MATERIALS

7.3.1 Tetrafluoroethylene

Tetrafluoroethylene is currently prepared from chloroform by the following route:

$$CHCl_3 \xrightarrow{HF} CHClF_2 \xrightarrow{-HCl} F_2C{=}CF_2$$

chloroform   chlorodifluoromethane   tetrafluoroethylene

In the first step, chloroform is treated with hydrogen fluoride in the presence of antimony pentachloride as catalyst to give chlorodifluoromethane. This type of substitution reaction is very widely used for the preparation of organofluorine compounds (see later for further examples) and is generally carried out at 50–180°C and up to 3 MPa (30 atmospheres) pressure, depending on the products required. The chlorodifluoromethane is then pyrolysed; in one process this is achieved by mixing with steam at about 950°C. The exit gases are scrubbed to remove hydrogen chloride and then distilled to give tetrafluoroethylene. Tetrafluoroethylene is a colourless, odourless gas, b.p. −76°C.

7.3.2 Chlorotrifluoroethylene

Chlorotrifluoroethylene is prepared from ethylene by the following route:

$$H_2C{=}CH_2 \xrightarrow{Cl_2} Cl_3C{-}CCl_3 \xrightarrow{HF} ClF_2C{-}CCl_2F$$

ethylene   hexachloroethane   1,2,2-trichloro-1,1,2-
                                                 trifluoroethane

$$\xrightarrow{-Cl_2} F_2C{=}CClF$$

chlorotrifluoroethylene

Ethylene is treated with an excess of chlorine at 300–350°C in the presence of activated charcoal to give hexachloroethane. This product is then treated with hydrogen fluoride in the presence of antimony pentachloride to yield trichlorotrifluoroethane. Dechlorination in the liquid phase with zinc dust and ethanol results in the formation of chlorotrifluoroethylene which is washed with water to remove alcohol, dried and distilled under pressure. Chlorotrifluoroethylene is a gas, b.p. −27°C.

7.3.3 Vinyl fluoride

Vinyl fluoride may be obtained from acetylene by either of the two following routes:

$$HC \equiv CH \xrightarrow{\text{HF}} H_2C = CHF$$

acetylene $\qquad$ vinyl fluoride

$$H_3C - CHF_2$$
1,1-difluoroethane

In the first method, acetylene is heated with hydrogen fluoride in the presence of a catalyst of mercuric chloride on charcoal at about 40°C to yield vinyl fluoride directly. In the second method, acetylene is treated with an excess of hydrogen fluoride to form difluoroethane which is then pyrolysed at about 700°C in a platinum tube to give vinyl fluoride, which is separated by distillation under pressure. Vinyl fluoride is a gas, b.p. −72°C.

7.3.4 Vinylidene fluoride

Vinylidene fluoride is obtained from vinylidene chloride (section 4.3.1) by the following route:

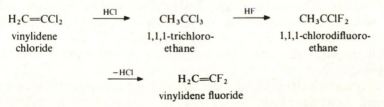

$$H_2C = CCl_2 \xrightarrow{\text{HCl}} CH_3CCl_3 \xrightarrow{\text{HF}} CH_3CClF_2$$

vinylidene $\qquad$ 1,1,1-trichloro- $\qquad$ 1,1,1-chlorodifluoro-
chloride $\qquad$ ethane $\qquad$ ethane

$$\xrightarrow{\text{−HCl}} H_2C = CF_2$$

vinylidene fluoride

In the first stage, vinylidene chloride undergoes addition with hydrogen chloride at about 30°C and atmospheric pressure in the presence of a Friedel-Crafts type catalyst. The resulting trichloroethane is then treated with hydrogen fluoride at about 180°C and 3 MPa (30 atmospheres) in the presence of antimony pentachloride to give chlorodifluoroethane. Pyrolysis of this product yields vinylidene fluoride. Vinylidene fluoride is a gas, b.p. −84°C.

7.3.5 Trifluoronitrosomethane

Trifluoronitrosomethane may be prepared by the reaction of bromotri-fluoromethane and nitric oxide in the presence of light and a mercury catalyst:

$$CBrF_3 \; + \; NO \longrightarrow CF_3NO \; + \; \tfrac{1}{2}Br_2$$

The nitroso compound is also obtained by the pyrolysis of trifluoroacetyl nitrite, which is prepared by reaction of trifluoroacetic anhydride and nitrogen trioxide:

$$(CF_3CO)_2O \; + \; N_2O_3 \longrightarrow 2CF_3COONO$$

$$CF_3COONO \longrightarrow CF_3NO \; + \; CO_2$$

Trifluoronitrosomethane is a deep blue gas, b.p. −84°C.

7.4   POLYTETRAFLUOROETHYLENE

7.4.1   Preparation

The preferred commercial method of preparation of polytetrafluoroethylene (PTFE) is by suspension polymerization. The details of the procedures employed have not been disclosed but it appears that two main processes are in use. In the first process, the conventional techniques of suspension polymerization are used to produce a granular product suitable for moulding and extrusion. The tetrafluoroethylene is polymerized under pressure in stainless steel autoclaves with a free radical initiator such as ammonium persulphate. The reaction is rapid and exothermic and requires careful control. The polymer granules are collected, washed and dried. In the second process, conditions are adjusted to give a dispersion of polymer of much finer particle size and lower molecular weight. The product may be stabilized and employed in latex form in such uses as film casting, coating and impregnation of fibres. Alternatively, the product may be coagulated to give a powder (often called 'dispersion polymer') used mainly for the extrusion of thin flexible sections.

Granular polymer of the type described above appears to have a very high average molecular weight, M_w being in the range 400 000–9 000 000.

7.4.2   Properties

Polytetrafluoroethylene is a white solid with a waxy appearance and feel. It is a tough flexible material of moderate tensile strength with a tendency to creep under compression. Comparative values for some properties of various fluoropolymers are given in Table 7.1. The electrical insulation properties are outstanding and are nearly as good as those of polyethylene. The coefficient of friction is unusually low and stated to be lower than that of any other solid; the non-stick properties are also excellent.

Polytetrafluoroethylene has exceptional resistance to chemical attack and is inert towards all types of reagents except molten alkali metals and fluorine. No solvent is known for the polymer, but it is swollen by some fluorocarbon oils at temperatures approaching the crystalline melting point; other organic materials do not even swell the polymer.

Polytetrafluoroethylene has high thermal stability and retains its properties over a wide temperature range. The polymer may be used up to about 300°C for long periods without loss of strength and thin sections remain flexible at temperatures below $-100°C$. There is some weight loss when polytetrafluoroethylene is heated above about 200°C but this very small up to about 350°C. The polymer also has good weather resistance.

Although polytetrafluoroethylene is a linear polymer with no significant amount of branching and is generally counted as a thermoplastic material it does not show normal melting behaviour in the sense of changing to a liquid or readily flowable melt. At the crystalline melting point there is an increase in volume of about 25% and the polymer changes to a weak, translucent material which has a very high viscosity and will not flow without fracture.

The high melting point and melt viscosity as well as the poor solubility characteristics of polytetrafluoroethylene make it impossible to fabricate by conventional means and special methods are required. Mouldings are made from granular polymer by a technique similar to that used in powder metallurgy. The polymer powder is preformed under pressure at room temperature, sintered at about 370°C when the particles fuse together, and then carefully cooled. Dispersion polymer, which gives products with improved tensile strength and flexural fatigue resistance, is not easily handled by this method. In this case, the polymer is usually mixed with about 20% of a lubricant (usually light petroleum, b.p. 100–120°C) and the paste is then extruded. The extrudate is heated to remove the lubricant and then sintered. Because of the need to evaporate the lubricant, only extrudates of thin section may be prepared by this technique.

The excellent electrical insulation properties and chemical inertness over a wide temperature range account for the two main fields of use of polytetrafluoroethylene, namely electrical and chemical applications. Electrical applications include wire coatings, holders and spacers for use in aggressive conditions whilst chemical applications include gaskets, pump parts and laboratory equipment. Other uses include non-stick coverings for kitchen utensils, low-friction linings for hoppers, and diaphragms for low temperature equipment.

Table 7.1 Typical values for various properties of some fluoropolymers

	PTFE	PCTFE	PVF	PVDF	FEP	EFTE	PFA	ECTFE
Specific gravity	2.1–2.3	2.1	1.4–1.6	1.8	2.2	1.7	2.15	1.2
Crystalline melting point (°C)	327	214	c. 200	171	c. 285	270	c. 300	c. 240
Tensile strength (MPa)	10–21	30–39	69–100	c. 48	19–21	45	29	49
(lbf/in^2)	1500–3000	4300–5700	10000–15000	c. 7000	2700–3100	6500	4200	7100
Elongation at break (%)	100–350	100–200	110–260	100–300	250–350	100–300	300	230

7.5   POLYCHLOROTRIFLUOROETHYLENE

Polychlorotrifluoroethylene (PCTFE) is more expensive than polytetrafluoroethylene and its use is comparatively limited.

The preferred commercial method of preparation of polychlorotrifluoroethylene is by suspension polymerization using procedures similar to those used for polytetrafluoroethylene. In order to obtain polymer of high molecular weight, reaction temperatures are generally kept in the range 0–40°C. Initiation is commonly by means of a redox system activated by a metal salt, e.g. persulphate – bisulphite with a ferrous salt. Average molecular weights of commercial polymers are within the range 50 000–500 000.

Polychlorotrifluoroethylene is a white solid. Compared to polytetrafluoroethylene, it has greater tensile strength, hardness and resistance to creep. Typical values for some properties are given in Table 7.1 The presence of a chlorine atom, which is larger than a fluorine atom, in the polymer does not permit such close chain packing as is possible with polytetrafluoroethylene. Thus polychlorotrifluoroethylene has a lower softening point and can be processed in the melt. Because of the lower tendency to crystallize, it is possible to produce thin transparent films by rapid quenching. The presence of chlorine in the polymer results in greater interchain attraction and accounts for the better mechanical strength of polychlorotrifluoroethylene. On the other hand, the unsymmetrical structure leads to poorer electrical insulation properties.

The chemical resistance of polychlorotrifluoroethylene is very good but not as good as that of polytetrafluoroethylene. Molten alkali metals and molten caustic alkalis attack the polymer and a few halogenated solvents swell or dissolve the polymer at elevated temperatures.

As mentioned previously, polychlorotrifluoroethylene may be melt processed. Standard techniques of extrusion and injection moulding may be used but careful temperature control is required since degradation occurs at processing temperatures, which are normally in the range 230–290°C. Degradation is accompanied by development of unsaturation and may involve the following chain splitting rearrangement [1]:

$$\sim CF_2 - CClF - CF_2 \overset{\vdots}{-} CClF - CF_2 - CClF \sim \longrightarrow$$

$$\sim CF_2 - CF = CF_2 \ + \ CCl_2F - CF_2 - CClF \sim$$

The temperature range of useful performance is from about $-100°C$ to 200°C.

Polychlorotrifluoroethylene finds limited use for such items as diaphragms, gaskets, electrical components and transparent windows for chemical equipment. Dispersions are used for the preparation of films and coatings.

7.6 POLY(VINYL FLUORIDE)

The polymerization of vinyl fluoride has been reported to be difficult but the polymer is available in film form. Details of the polymerization techniques used have not been disclosed. The polymer may be obtained by polymerization of vinyl fluoride in the presence of water, using benzoyl peroxide as initiator at 80°C and pressures up to 100 MPa (1000 atmospheres).

Poly(vinyl fluoride) (PVF) resembles poly(vinyl chloride) in chemical properties but its mechanical properties are generally superior because of its much greater ability to crystallize. The fluorine atom is sufficiently small to permit the molecules to pack in the planar zig-zag fashion of polyethylene. Some comparative values for various properties of poly(vinyl fluoride) are shown in Table 7.1. Thus films of poly(vinyl fluoride) have high tensile and impact strengths and outstanding resistance to flexural fatigue.

The polymer is insoluble below about 100°C but dissolves above this temperature in certain highly polar solvents such as dimethylformamide. The polymer is also highly resistant to hydrolysis.

Although poly(vinyl fluoride) is more heat resistant than poly(vinyl chloride), it has a tendency to eliminate hydrogen fluoride and instability at processing temperatures makes handling difficult. The weather resistance of poly(vinyl fluoride) is exceptionally good and the film is used particularly for outdoor applications such as glazing in solar energy collectors.

7.7 POLY(VINYLIDENE FLUORIDE)

Poly(vinylidene fluoride) (PVDF) is made by the polymerization of vinylidene fluoride in aqueous medium under pressure but details of the process have not been disclosed. Two grades of polymer are available with differing average molecular weight, namely 300 000 and 600 000.

Poly(vinylidene fluoride) has good tensile and impact strengths. Typical values for some properties are given in Table 7.1. The polymer has unusual piezoelectric and pyroelectric properties which have resulted in its use in solid state switching devices.

The resistance of poly(vinylidene fluoride) to solvents and chemicals is generally good but inferior to that of polytetrafluoroethylene and polychlorotrifluoroethylene. Some highly polar solvents such as dimethylacetamide dissolve the polymer at elevated temperature whilst organic amines cause discoloration and embrittlement. Fuming sulphuric acid leads to sulphonation.

Poly(vinylidene fluoride) has good heat resistance and may be used continuously at temperatures up to 150°C. The polymer may be melt processed by the standard techniques of injection moulding and extrusion. The polymer has very good weather resistance. Applications for poly(vinylidene fluoride) include moulded and lined tanks, valves, pumps and other equipment designed to handle corrosive fluids. The polymer is also used in coatings and wire covering.

7.8   TETRAFLUOROETHYLENE-HEXAFLUOROPROPYLENE COPOLYMERS

The essential feature of the commercial tetrafluoroethylene-hexafluoropropylene copolymers (sometimes called fluorinated ethylene-propylene (FEP) copolymers) is that the proportion of hexafluoropropylene (V) is such that they are truly thermoplastic whilst retaining other properties similar to those

$$F_2C=CF—CF_3$$
(V)

of polytetrafluoroethylene homopolymer. Thus, in contrast to the homopolymer, the copolymers can be melt processed by the standard techniques of injection moulding and extrusion.

The commercial copolymers have physical properties similar to those of polytetrafluoroethylene but with somewhat greater impact strength. Comparative values for some properties are given in Table 7.1. The copolymers also have similar excellent electrical insulation properties and chemical resistance. The maximum service temperature for the copolymers is about 60 deg C lower than that for the homopolymer under equivalent conditions. The temperature range of useful performance is from about $-80°C$ to $200°C$. Tetrafluoroethylene-hexafluoropropylene copolymers are used for various electrical and corrosion resistant mouldings, coatings and wire covering.

7.9   TETRAFLUOROETHYLENE-ETHYLENE COPOLYMERS

Tetrafluoroethylene-ethylene (ETFE) copolymers resemble the tetrafluoroethylene-hexafluoropropylene copolymers discussed in the preceding section in that they too can be melt processed by standard techniques.

The outstanding features of the tetrafluoroethylene-ethylene copolymers are very high impact strength and exceptional abrasion resistance for a fluoropolymer. Comparative values for some properties are given in Table 7.1. Resistance to chemicals and outdoor weathering approaches that of the fully fluorinated polymers. The temperature range of useful performance is from about $-80°C$ to $180°C$. Tetrafluoroethylene-ethylene copolymers find use in high performance electrical mouldings such as coil forms, connectors and sockets and cable insulation.

7.10   TETRAFLUOROETHYLENE-PERFLUOROALKOXY COPOLYMERS

A copolymer of tetrafluoroethylene and a small amount of perfluoro(propyl vinyl ether) (VI) (sometimes called PFA copolymer) is commonly regarded as the closest available melt-processable alternative to polytetrafluoroethylene.

$$F_2C=CF-O-CF_2-CF_2-CF_3 \qquad F_2C=CF-O-CF_3$$

$$\text{(VI)} \qquad\qquad\qquad\qquad \text{(VII)}$$

Thus the product resembles tetrafluoroethylene-hexafluoropropylene copolymers (section 7.8) but has better mechanical properties at elevated temperatures. Upper use temperatures are quoted as being the same as those for polytetrafluoroethylene. Comparative values for some properties are given in Table 7.1.

A terpolymer of tetrafluoroethylene (60 mole %), perfluoro(methyl vinyl ether) (VII) and a perfluorinated monomer with a reactive site is also commercially available. The material may be cross-linked to give an elastomer which has very low volume swell in a wide range of solvents and which withstands air oxidation up to 315°C.

7.11   CHLOROTRIFLUOROETHYLENE-ETHYLENE COPOLYMERS

Chlorotrifluoroethylene-ethylene copolymers (ECTFE) contain about 50 mole % chlorotrifluoroethylene. These copolymers are generally similar to the tetrafluoroethylene-ethylene copolymers (section 7.9) in that they are melt-processable and have high impact strength and good chemical resistance. Comparative values for some properties are given in Table 7.1. The temperature range of useful performance is from about −80°C to 170°C. Chlorotrifluoroethylene-ethylene copolymers find use in injection mouldings for chemical process equipment and cable insulation.

7.12   VINYLIDENE FLUORIDE-HEXAFLUOROPROPYLENE COPOLYMERS

As mentioned previously, interest in fluorine-containing copolymers has been particularly concerned with the development of rubbers. Many copolymers have been investigated but the bulk of the commercial production of fluoro-elastomers is accounted for by vinylidene fluoride copolymers, of which those with hexafluoropropylene are the most important.

Details of the processes used to produce vinylidene fluoride-hexafluoropropylene copolymers have not been disclosed but the copolymers may be prepared by emulsion polymerization under pressure using a persulphate-bisulphite initiator system. Highly fluorinated surfactants, such as ammonium perfluorooctoate, are most commonly used in order to avoid chain transfer reactions. The preferred vinylidene fluoride content for commercial copolymers is about 70% mole. The structure of such a copolymer might be represented as follows:

$$\sim CH_2-CF_2-CF_2-\overset{\overset{\displaystyle CF_3}{|}}{CF}-CH_2-CF_2-CH_2-CF_2 \sim$$

NMR studies confirm that head-to-tail arrangements predominate in the copolymers. It may be noted that under the conditions normally used to prepare commercial copolymers, hexafluoropropylene does not homopolymerize and so the molar proportion of this monomer in a copolymer cannot exceed 50% whatever the composition of the monomer feed.

Commercial copolymers typically have an average molecular weight (M_n) of about 70 000.

Since the copolymers are saturated, they cannot be vulcanized by conventional sulphur systems. However, they may be vulcanized by aliphatic amines and derivatives, aromatic dihydroxylic compounds and peroxides; all of these methods are practised commercially. Free diamines react too quickly, causing premature vulcanization or scorching during mixing. One method of reducing the tendency to scorch is to use diamines in the form of their inner carbamates; hexamethylenediamine carbamate (VIII) and ethylenediamine carbamate (IX) are employed commercially. Alternatively, the Schiff's bases of diamines function as delayed-action vulcanizing agents; examples of compounds of this type are N, N'-dicinnamylidene-1,6-hexanediamine (X) and N, N'-disalicylpropylenediamine (XI).

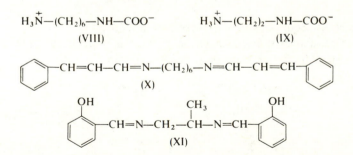

The usual commercial practice is to mix the rubber on a mill with the diamine derivative, together with filler (such as carbon black or silica) and a metal oxide. The compounded stock is then given a short press cure (typically, 0.5 hours at 150°C) and a long oven cure (typically, 24 hours at 200°C).

The reactions which occur in the vulcanization of vinylidene fluoride-hexafluoropropylene copolymers have been extensively studied [2] and a three-stage process is thought to be involved. The first step is postulated as the elimination of hydrogen fluoride from the polymer upon treatment with a base. Probably the tertiary fluorine atom of the hexafluoropropylene unit is the most readily removed. The initially formed double bonds then activate the elimination of hydrogen fluoride from neighbouring groups to give a conjugated system:

$$\sim CF_2-\underset{\underset{\displaystyle CF_3}{|}}{C}F-CH_2-CF_2-CH_2-CF_2\sim \xrightarrow{-HF} \sim CF_2-\underset{\underset{\displaystyle CF_3}{|}}{C}=CH-CF_2-CH_2-CF_2\sim$$

$$\xrightarrow{-HF} \sim CF_2-\underset{\underset{\displaystyle CF_3}{|}}{C}=CH-CF=CH-CF_2\sim \quad \text{etc.}$$

Elimination is thought to be catalysed by basic materials, such as magnesium oxide, which are commonly included in commercial formulations. Elimination takes place rapidly and probably occurs during milling when a diamine curing system is used. Evidence for the presence of unsaturation in amine-treated vinylidene fluoride-hexafluoropropylene copolymer comes from the infrared spectrum of the material.

The second step in the vulcanization process is regarded as involving addition of the curing agent at the sites of unsaturation. With a diamine $(H_2N-R-NH_2)$, cross-linking results:

$$
\begin{array}{ccc}
\sim CH{=}CF\sim & & \sim CH_2-CF\sim \\
| & & \quad\quad\quad| \\
NH_2 & & NH \\
| & \longrightarrow & \quad| \\
R & & R \\
| & & \quad| \\
NH_2 & & NH \\
| & & \quad\quad\quad| \\
\sim CH{=}CF\sim & & \sim CH_2-CF\sim
\end{array}
$$

This reaction occurs during the press cure operation.

The third step in the vulcanization process is considered to involve the elimination of hydrogen fluoride to form a diimine:

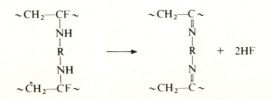

This reaction probably proceeds mainly during the oven cure. Besides permitting this reaction to go to completion, the long oven cure is thought to have the important function of ensuring the complete removal of water from the vulcanizate. Water is formed during the vulcanization process by reaction of the eliminated hydrogen fluoride with the metal oxide which is present:

$$MgO + 2HF \longrightarrow MgF_2 + H_2O$$

It is to be expected that the diimine cross-link shown above would be readily hydrolysed:

$$\sim CH_2-\underset{\substack{\| \\ N \\ | \\ R \\ | \\ N \\ \|}}{C}\sim \quad + \quad 2H_2O \rightleftharpoons \qquad \sim CH_2-\underset{\substack{\| \\ O}}{C}\sim \qquad + \quad H_2N-R-NH_2$$
$$\sim CH_2-\overset{}{C}\sim \qquad\qquad\qquad \sim CH_2-\overset{\substack{O \\ \|}}{C}\sim$$

If this is the case, it would be necessary to remove all the water present in order to force the equilibrium completely toward diimine formation and the establishment of permanent cross-links. Experimental evidence for this proposition is that a full state of cure is not attained if the long heating process is carried out in the mould, i.e. in a virtually closed system. Hydrolysis is also demonstrated by the fact that much of the diamine present in a vulcanizate can be extracted by treatment with water. It may be noted here that vinylidene fluoride-hexafluoropropylene copolymers which are peroxide-curable do not lead to water formation and so give vulcanizates with reduced porosity.

Vulcanized vinylidene fluoride-hexafluoropropylene copolymers show excellent resistance to oils and fuels, both at room temperature and at elevated temperature. Vulcanizates also have good resistance to most solvents, although polar solvents such as esters and ketones cause high swelling.

Vinylidene fluoride-hexafluoropropylene copolymer vulcanizates have excellent thermal stability, withstanding long periods at 250°C without serious deterioration. Low temperature performance is limited and the elastomers are not generally suitable for sub-zero use. Applications of the elastomers include seals and hose in contact with fuels and lubricants, pump components and tank linings.

A terpolymer of vinylidene fluoride, hexafluoropropylene and tetrafluoroethylene is also available and gives vulcanizates with improved resistance to heat ageing.

7.13   VINYLIDENE FLUORIDE-CHLOROTRIFLUORO-ETHYLENE COPOLYMERS

Vinylidene fluoride-chlorotrifluoroethylene copolymers constitute the second class of vinylidene fluoride copolymers which have achieved commercial importance as elastomers.

Details of the processes used to produce vinylidene fluoride-chlorotrifluoroethylene copolymers have not been disclosed but the copolymers may be prepared by techniques similar to those described in the previous section for the preparation of vinylidene fluoride-hexafluoropropylene copolymers. Copolymers with 50 and 70 mole % vinylidene fluoride are produced commercially. Although homopolymers of both vinylidene fluoride and chlorotrifluoroethylene are crystalline, copolymers containing more than about 25 mole % of either monomer are essentially amorphous.

Vinylidene fluoride-chlorotrifluoroethylene copolymers may be vulcanized by methods similar to those described in the previous section for the vulcanization of vinylidene fluoride-hexafluoropropylene copolymers. The chemistry of the reactions which occur has not been investigated so fully but it is likely that a similar three-stage process is involved. In the first stage, hydrogen chloride rather than hydrogen fluoride is eliminated:

$$\sim CF_2-CClF-CH_2-CF_2\sim \longrightarrow \sim CF_2-CF=CH-CF_2\sim +HCl$$

This may be inferred from the finding that treatment of vinylidene fluoride-chlorotrifluoroethylene copolymers with an amine produces the amine hydrochloride in high yield but very little amine hydrofluoride [3]. Since polychlorotrifluoroethylene is unreactive towards amines (RNH_2), the following substitution may be discounted:

$$\sim CF_2-CClF\sim \; + \; RNH_2 \longrightarrow \; \sim CF_2-\overset{\overset{\displaystyle RNH}{|}}{C}F\sim \; + \; HCl$$

The chemical resistance of vinylidene fluoride-chlorotrifluoroethylene copolymer vulcanizates is excellent, though not generally quite so good as that of the vinylidene fluoride-hexafluoropropylene copolymers. However, the vinylidene fluoride-chlorotrifluoroethylene copolymers do have better resistance to strong oxidizing acids.

Vulcanized vinylidene fluoride-chlorotrifluoroethylene copolymers also have good thermal stability, withstanding prolonged heating at 200°C without serious deterioration. Low temperature performance is limited; the glass transition temperatures of copolymers with 50 and 70 mole % vinylidene fluoride are 0°C and -15°C respectively. The principal application of the elastomers is for components of systems handling strong mineral acids.

7.14 FLUORONITROSOPOLYMERS

Nitrosopolymers have the general structure:

$$\left[-\overset{|}{N}-O-(-\overset{|}{\underset{|}{C}}-)_x- \right]_n$$

The nitrosopolymers which, so far, have been the most extensively studied have been highly fluorinated and thus these polymers are dealt with in this chapter. It is to be noted, however, that nitrosopolymers do not necessarily contain fluorine.

Fluoronitrosopolymers have potential primarily as low temperature, oil-resistant elastomers but although they have attracted a good deal of interest, these materials do not appear to have progressed beyond the pilot plant stage. The most thoroughly investigated nitrosopolymer is the copolymer of trifluoronitrosomethane and tetrafluoroethylene [4].

The preferred method of preparation of the copolymer is by suspension polymerization. In one method, polymerization is carried out in aqueous lithium bromide at $-25°C$. Magnesium carbonate is the suspending agent and no added initiator is required. A reaction time of approximately 20 hours is used. This is apparently a difficult polymerization to carry out; since no initiator is added, control and reproducibility are hard to achieve.

It is thought that copolymerization proceeds by a free radical mechanism since the rate of reaction is increased by exposure to ultraviolet light and decreased by the presence of hydroquinone. Reaction may be envisaged as proceeding according to the following scheme, in which trifluoronitroso-methane acts as initiator [5]:

Initiation

$$CF_3NO \rightleftharpoons F_3C—\dot{N}—O\cdot$$

$$F_3C—\dot{N}—O\cdot \; + \; F_2C{=}CF_2 \longrightarrow F_3C—\dot{N}—O—CF_2—\dot{C}F_2$$

Propagation

$$F_3C—\dot{N}—O—CF_2—\dot{C}F_2 \xrightarrow{CF_3NO} F_3C—\dot{N}—O—CF_2—CF_2—CF_3N—O\cdot$$

$$\xrightarrow{F_2C{=}CF_2} F_3C—\dot{N}—O—CF_2—CF_2—CF_3N—O—CF_2—\dot{C}F_2 \quad \text{etc.}$$

The reactivities of the monomers are such that whatever the composition of the monomer feed, a 1:1 copolymer is obtained. It is necessary to conduct the reaction at low temperature in order to obtain the copolymer; elevated temperatures favour the formation of the following cyclic oxazetidine, which is not a precursor for the polymer:

$$\begin{array}{c} F_3C—N—O \\ | \quad\; | \\ F_2C—CF_2 \end{array}$$

Trifluoronitrosomethane-tetrafluoroethylene copolymer may be vulcanized with a combination of triethylenetetramine and hexamethylenediamine carbamate; typically, vulcanization is effected by heating at 250°C for 1 hour in the press and then 210°C in an oven for 18 hours. The mechanism of the process is unknown. Vulcanization may be simplified by the insertion of reactive sites in the polymer. The inclusion of about 1 mole % 4-nitrosoperfluorobutyric acid (XII) in the polymerization formulation gives a terpolymer which can be cross-linked with metal salts, eg., chromium trifluoroacetate.

$$HOOC—CF_2—CF_2—CF_2—NO$$

(XII)

Trifluoronitrosomethane-tetrafluoroethylene copolymer vulcanizates have outstanding resistance to chemicals, being generally superior to vinylidene fluoride-hexafluoropropylene elastomers. Some fluorinated solvents cause appreciable swelling.

Trifluoronitrosomethane-tetrafluoroethylene copolymer has good thermal stability. The copolymer does not burn (even in pure oxygen) but decomposition into toxic gaseous products begins at about 200°C. The glass transition temperature of the copolymer is −54°C and vulcanizates remain serviceable down to about −40°C. A limitation of the vulcanizates is their low tensile strength, although silica effects some reinforcement. The elastomer has found very specialized use as containers for rocket propellants under arctic and space conditions.

7.15 OTHER FLUOROPOLYMERS

In addition to the polymers discussed above, a few other fluoropolymers have been produced on a commercial scale and are briefly described below.

7.15.1 Polyfluoroprene

The first commercial fluoroelastomer, introduced in 1948, was a polymer of fluoroprene (2-fluoro-1,3-butadiene) ($H_2C=CF–CH=CH_2$). In general, the elastomer has properties resembling those of neoprene and the advent of superior fluoroelastomers has led to the discontinuance of manufacture. Many copolymers of fluoroprene have been investigated but none have attained commercial status.

7.15.2 Fluoroacrylate polymers

Two fluoroacrylate elastomers have been produced commercially on a very limited scale, namely poly(1,1-dihydroperfluorobutyl acrylate) (XIII) and poly(3-perfluoromethoxy-1,1-dihydroperfluoropropyl acrylate) (XIV).

The two polymers have similar properties with the latter having better low temperature properties. In general, the polymers are inferior to the vinylidene fluoride copolymers described previously. However, they do show superior resistance to synthetic diester lubricants used in jet engines.

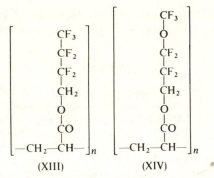

(XIII) (XIV)

7.15.3 Fluorosilicones

Fluorosilicone elastomers are commercially available. These polymers contain γ-trifluoropropylmethylsiloxane units:

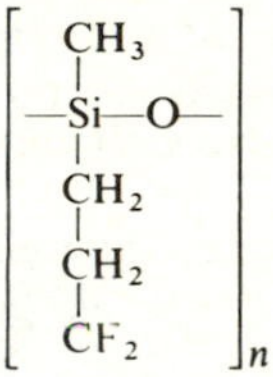

$$\left[\begin{array}{c} CH_3 \\ | \\ -Si-O- \\ | \\ CH_2 \\ | \\ CH_2 \\ | \\ CF_2 \end{array} \right]_n \quad (n = 500)$$

The vulcanizates have particularly good resistance to various fuels and remain serviceable from $-60°C$ to $200°C$. They are used as seals in aircraft fuel systems.

7.15.4 Hydropentafluoropropylene copolymers

Various copolymers of 2-hydropentafluoropropylene (XV) and vinylidene fluoride are commercially available.

$$F_2C=CH-CF_3$$

(XV)

These copolymers appear to have somewhat poorer solvent resistance than other fluoroelastomers but remain serviceable up to $185°C$.

7.15.5 Hexafluoroisobutene copolymers

A copolymer of hexafluoroisobutene (XVI) and vinylidene fluoride (50 mole %) is available.

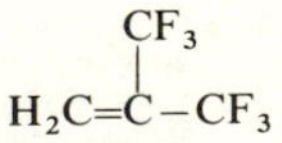

$$H_2C=\overset{\overset{\displaystyle CF_3}{|}}{C}-CF_3$$

(XVI)

This material is melt-processable and is comparable to polytetrafluoroethylene in its properties. Its impact strength, however, is somewhat lower.

REFERENCES

1. Iwasaki, M. *et al.* (1957) *J. Polymer Sci.*, **25**, 377.
2. Smith, J. F. (1959) *Proc. International Rubber Conference*, Washington D.C., 575; (1960) *Rubber World*, **142**, 102; Smith, J. F. and Perkins, G. T. (1961) *J. Applied Polymer Sci.*, **5**, No. 16, 460; (1961) *Rubber & Plastics Age*, **42**, 59.
3. Paciorek, K. L. *et al.* (1960) *J. Polymer Sci.*, **45**, 405.
4. Barr, D. A. and Haszeldine, R. N. (1955) *J. Chem. Soc.*, 1881.
5. Fetters, L. J. (1972) in *Fluoropolymers*, (ed. L. A. Wall), Wiley-Interscience, New York, p. 178.

BIBLIOGRAPHY

Rudner, M. A. (1958) *Fluorocarbons*, Reinhold Publishing Corporation, New York.
Montermoso, J. C. (1961) *Rubber Chem. Tech.*, **34**, 1521.
Cooper, J. R. (1968) in *Polymer Chemistry of Synthetic Elastomers*, Part I, (ed. J. P. Kennedy and E. G. M. Tornqvist), Interscience Publishers, New York, ch. 4.
Trappe, G. (1968) in *Vinyl and Allied Polymers*, Vol. I, (ed. P. D. Ritchie), Iliffe Books Ltd, London, ch. 12.
Wall, L. A. (ed.) (1972) *Fluoropolymers*, Wiley-Interscience, New York.

MISCELLANEOUS VINYL POLYMERS

8.1 SCOPE

In addition to the various vinyl polymers dealt with in the preceding chapters, many others have been described in the literature. A few have achieved some commercial significance. It is these miscellaneous vinyl polymers which are considered in this chapter.

8.2 COUMARONE-INDENE POLYMERS

Coumarone (benzofuran) (I) and indene (II) occur in coal tar naphtha (b.p. 150–200°C) together with various other compounds (mainly xylenes and

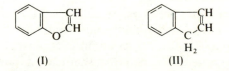

(I)   (II)

cumenes). In the preparation of polymers, no attempt is made to isolate these monomers and they are polymerized *in situ*. Indeed, resinification is used as a means of separating these substances from the other components of the cut which are subsequently distilled and sold as solvents. Polymerization of the coumarone-indene mixture is accomplished by adding concentrated sulphuric acid to the naphtha at 0°C. Reaction is stopped after 5–10 minutes by the addition of water and any sludge is removed. The solution is then neutralized and washed and residual napththa is distilled off to leave the resin. Depending on the relative proportions of coumarone and indene (the coumarone content is usually less than 10%) and the polymerization conditions, resins may vary from hard and brittle to soft and sticky materials. The resins have molecular weights of about 1000–3000 and are commonly assumed to be a random copolymer of the following type:

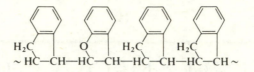

Coumarone-indene polymers are chemically inert and are used as binders in cheaper grades of floor tiles. They are also used as additives in various surface coating formulations.

8.3   POLY(VINYL CARBAZOLE)

N-Vinyl carbazole may be obtained from phenylhydrazine and cyclohexanone by the following route:

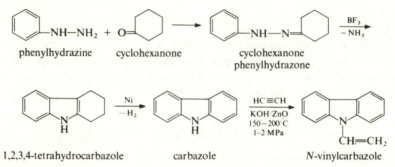

The second step in this synthesis is an example of the Fischer indole synthesis. Carbazole also occurs in coal tar.

Poly(vinyl carbazole) is produced by adiabatic bulk polymerization under nitrogen pressure using azobisisobutyronitrile and di-*tert*-butyl peroxide as initiators. Heating to 80–90°C causes an onset of polymerization and a rapid increase in temperature. After the maximum temperature has been reached the mass is cooled under pressure.

Poly(vinyl carbazole) is a very brittle material with a high softening point (about 195°C Vicat). It is insoluble in most organic solvents except aromatic and chlorinated hydrocarbons and tetrahydrofuran. The polymer is difficult to mould, requiring high process temperatures, namely about 300°C for injection moulding and 250°C for compression moulding. More usually the material is cast as thin film from solution. The most significant property of poly(vinyl carbazole) is its high photoconductivity and this has resulted in widespread use of the polymer in electrostatic dry-copying machines. When an electrostatic charge is applied to the polymer in the dark it discharges to an equilibrium value but when the polymer is illuminated almost complete discharging occurs. This phenomenon is used to create a latent electrostatic image which is then developed by transferring the charge to the toner. Poly(vinyl carbazole) also has good electrical insulation characteristics which are maintained over a wide range of temperature and frequency. On account of these characteristics, the polymer has been used as a capacitor dielectric, but this application is now of minor importance.

8.4 POLY(VINYL ETHER)S

Vinyl ethers are prepared by reaction of acetylene and alcohols in the presence of the potassium alkoxide as catalyst:

$$HC{\equiv}CH \ + \ ROH \ \xrightarrow[\substack{130–180°C \\ 0.5–2\,MPa}]{ROK} \ H_2C{=}CHOR$$

For the production of polymers, methyl, ethyl and isobutyl vinyl ethers are the most common monomers.

As indicated in section 1.4.2(a), vinyl ethers are susceptible only to cationic polymerization. In a typical process, methyl vinyl ether is agitated with boron trifluoride ($BF_3 . 2H_2O$). The temperature is kept at about 10°C by cooling for 3–4 hours; the reactor is then sealed and the temperature allowed to rise to 100°C. Reaction is complete in about 10 hours and the polymer is obtained as a viscous mass. It is interesting to note that poly(vinyl isobuty ether) may be obtained in crystalline form by conducting polymerization at −80° to −60°C in liquid propane using boron trifluoride etherate as initiator. This was the first stereoregular polymerization to be achieved[1]. Incidentally, isotactic polymer has also been prepared by the use of Ziegler-Natta type catalysts and this was the first indication that these catalysts could polymerize a suitable monomer by a cationic mechanism.

Poly(vinyl methyl ether) is water-soluble and is used in adhesive and textile sizes. The ethyl and isobutyl polymers find use in pressure-sensitive adhesives.

8.5 POLYVINYLPYRROLIDONE

N-Vinylpyrrolidone is prepared from 1,4-butanediol as follows:

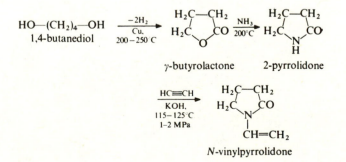

γ-butyrolactone   2-pyrrolidone

N-vinylpyrrolidone

N-Vinylpyrrolidone is water-soluble and is usually polymerized in aqueous solution at about 50°C with ammonia and hydrogen peroxide. The polymer is also water-soluble and is isolated by spray-drying. Commercial grades of polyvinylpyrrolidone (PVP) have average molecular weights (M_v) ranging from about 10 000 up to 360 000.

The largest use of polyvinylpyrrolidone is in cosmetic formulations, especially hair lacquers. In the latter applications, polyvinylpyrrolidone is the preferred film-former on account of good adhesion to hair, lustre of the film and ease of removal on washing. The polymer is also used as a binder in pharmaceutical tablets. Polyvinylpyrrolidone also finds use in the textile industry, particularly in colour stripping operations, where the great affinity of the polymer for dyestuffs is utilized. An interesting application of polyvinylpyrrolidone is in aqueous solution as a blood plasma substitute; such material was extensively used in Germany during the Second World War.

REFERENCES

1. Schildknecht, C. E. *et al.* (1948) *Ind. Eng. Chem.*, **40**, 2104.

BIBLIOGRAPHY

Blecher, L. *et al.* (1980) in *Handbook of Water-Soluble Gums and Resins*, (ed. R. L. Davidson), McGraw Book Company, New York, ch. 21, (Polyvinylpyrrolidone).
Pearson, J. M. and Stolka, M. (1981) *Poly(N-Vinylcarbazole)*, Gordon and Breach Science Publishers, New York.

9

ALIPHATIC POLYETHERS

9.1 SCOPE

For the purposes of this chapter, aliphatic polyethers are defined as aliphatic polymers which contain recurring ether groupings,

$$-\overset{\displaystyle |}{\underset{\displaystyle |}{C}}-O-\overset{\displaystyle |}{\underset{\displaystyle |}{C}}-,$$

as a part of the main polymer chain. This definition excludes poly(vinyl ether)s, in which the ether groupings are in the side-chains; these polymers are dealt with in Chapter 8. Aromatic polyethers are considered in Chapter 12. Also included in this chapter are furan polymers, which may be regarded as cyclic polyethers.

Aliphatic polyethers may be obtained from two different classes of monomer, namely carbonyl compounds and cyclic ethers. These two types of polyether are considered separately in this chapter.

9.2 POLYMERS OF CARBONYL COMPOUNDS

Polymers of several aldehydes and ketones and halogenated derivatives have been prepared. As yet, only those of formaldehyde have achieved commercial importance; these polymers are described in detail in the next section.

Polymers of aldehydes are commonly termed *polyacetals* (or *acetal resins*)* since their structure resembles that of polymers obtained by the acetal-forming reaction between aldehydes and diols:

$$nR{-}CHO \;+\; nHO{-}R'{-}OH \longrightarrow [{-}O{-}RCH{-}O{-}R'{-}]_n \;+\; nH_2O$$

It may be noted that the reaction shown above can actually be used for the preparation of polymers but it does not represent a commercially used

---

* Since the most important commercial polymers of aldehydes are based on formaldehyde, the term polyacetal (or acetal resin) is commonly used as applying solely to polymers of formaldehyde.

method. Similarly, polymers of ketones are sometimes termed *polyketals*. The aforementioned polymers should not be confused with the poly(vinyl acetal)s and poly(vinyl ketal)s discussed in Chapter 5.

As mentioned earlier, polymers of formaldehyde are described in the next section. Polymers of higher aliphatic aldehydes and ketones have been extensively investigated but, in general, they do not have the stability necessary for commercial development. For example, acetaldehyde may be polymerized using organometallic initiators at low temperatures, e.g. triethyl-aluminium at $-78°C$; in this case crystalline isotactic polymer is obtained. Also, the polymerization of acetaldehyde may be effected with cationic initiators at low temperatures (e.g. aluminium chloride at $-65°C$) or with metal oxides at low temperatures (e.g. alumina at $-70°C$); in these cases amorphous atactic polymer is obtained. The tacticity of polyacetaldehyde arises because the polymer comprises structural units which contain an asymmetric carbon atom:

$$\left[-\overset{\displaystyle CH_3}{\underset{\displaystyle H}{\overset{*}{C}}} -O- \right]_n$$

Polyacetone is reported to be formed when a mixture of magnesium vapour and acetone vapour is condensed on a cold finger at $-196°C$ [1] or when solid acetone is irradiated [2]. The unstable, rubbery product has been assigned the polyketal structure:

$$\left[-\overset{\displaystyle CH_3}{\underset{\displaystyle CH_3}{C}} -O- \right]_n$$

However, subsequent attempts to repeat earlier work have failed and there is some doubt as to the identity of this product.

9.3 POLYFORMALDEHYDE

9.3.1 Development

Formaldehyde will polymerize in a number of ways and although commercially useful polymers have been prepared only relatively recently, several different types of formaldehyde polymers have been recognized for over a century. For example, distillation of aqueous solutions of formaldehyde containing 2% of sulphuric acid yields the trimer, *trioxan*:

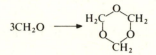

$$3CH_2O \longrightarrow$$

The trioxan distils as an azeotrope with water and may be recovered by solvent extraction or crystallization. Trioxan is a white crystalline solid, m.p. 60–62°C. In the presence of weak alkalis such as aqueous calcium hydroxide, formaldehyde undergoes condensation to give a mixture, known as *formose*, which contains a number of hexoses, e.g.

$$6CH_2O \longrightarrow HOCH_2—(CHOH)_4—CHO$$

When aqueous solutions of formaldehyde are evaporated to dryness, a white crystalline solid (m.p. 121–123°C) known as *paraformaldehyde* is obtained. This comprises polyoxymethylene glycols formed by condensation of methylene glycol (hydrated formaldehyde; see section 9.3.2):

$$HO—CH_2—OH \xrightarrow[-H_2O]{HO—CH_2—OH} HO—CH_2—O—CH_2—OH \xrightarrow[-H_2O]{HO—CH_2—OH}$$

$$HO—CH_2—O—CH_2—O—CH_2—OH \xrightarrow[-H_2O]{HO—CH_2—OH} HO—[CH_2—O]_n—H$$

The value of n generally lies in the range 6–50. By carrying out the reaction in the presence of strong acids and bases the degree of polymerization may be increased to approximately 200. On heating, paraformaldehyde readily decomposes to give formaldehyde; in fact, paraformaldehyde is a convenient source of anhydrous formaldehyde gas.

During the 1920s Staudinger prepared somewhat higher molecular weight polymer by carrying out the polymerization in the absence of water and at lower temperatures; various initiators were used, e.g. boron trifluoride. Solution polymerization in ether at 0°C gave a brittle, friable and thermally unstable product whilst polymerization in bulk at −80°C gave a material which although thermally unstable, possessed some degree of toughness. Besides thermal instability, the polymers showed insolubility in all common solvents and were therefore not readily processable and commercial development was inhibited.

In about 1947 an intensive research programme on the polymerization of formaldehyde was initiated by E. I. du Pont de Nemours and Co. (USA). As a result of this work tough, melt-processable homopolymers of formaldehyde were developed and commercial production began in 1959. The du Pont monopoly was unusually short-lived, as a formaldehyde copolymer (made from trioxan and a small amount of a cyclic ether) with properties similar to those of the homopolymer was introduced in 1960 by Celanese Corporation (UK); full-scale production of the copolymer began in 1962. A patent suit brought by du Pont against Celanese was dropped in 1963.

9.3.2 Raw materials

The most common route for the preparation of formaldehyde is the following:

$$CO \xrightarrow{H_2} CH_3OH \xrightarrow[-H_2O]{O_2} CH_2O$$

carbon monoxide methanol formaldehyde

Methanol is prepared by the interaction of carbon monoxide and hydrogen. In older plants in which a promoted zinc oxide catalyst is utilized, reaction conditions are 300–400°C and about 30 MPa (300 atmospheres). In newer plants a copper-based catalyst is employed; this allows the use of milder conditions, namely 200–300°C and 5–10 MPa (50–100 atmospheres). The methanol is condensed out and unreacted gases, with fresh make-up gas, recycled to the converters. In the second stage, methanol is oxidized to formaldehyde. In one process a mixture of methanol vapour and air is passed over a catalyst of molybdenum oxide promoted with iron at 350–450°C. The exit gases are scrubbed with water and the formaldehyde is isolated as an aqueous solution.

Formaldehyde is a colourless gas with a pungent odour, b.p. −19°C. Formaldehyde is commercially most commonly available as an aqueous solution, known as *formalin*. Generally, formalin contains 40% w/v (37% w/w) formaldehyde and about 8% methanol which acts as a stabilizer, retarding polymerization and preventing precipitation of insoluble polymers. In aqueous solution, monomeric formaldehyde is mainly in the form of methylene glycol rather than in the free state:

$$CH_2O + H_2O \rightleftharpoons HO—CH_2—OH$$

The preparation of high molecular weight polyformaldehyde requires extremely pure formaldehyde. Formaldehyde obtained directly from formalin often contains impurities such as water, methanol and formic acid and is not suitable. Thus, trioxan and paraformaldehyde (see section 9.3.1.), which can be obtained in a state of high purity, are convenient sources of formaldehyde in commercial polyacetal processes. It may be noted that the production of polyformaldehyde consumes a very minor proportion of the total output of formaldehyde.

9.3.3 Preparation

As mentioned previously two types of formaldehyde polymer have become commercially available; these are homopolymers and copolymers.

(a) Homopolymers

In the preparation of high molecular weight polyformaldehyde the initial operation consists of the production of pure formaldehyde, free from low molecular weight polymers and other hydroxy compounds which cause chain transfer. In a typical process potassium hydroxide-precipitated paraformaldehyde (degree of polymerization approximately 200) is carefully washed with water and dried for several hours *in vacuo* at 80°C. The dried polymer is then decomposed in nitrogen at 150–160°C; the product is passed through several traps at −15°C to remove water, glycols, and other impurities. The resulting formaldehyde has a water content (free and combined) of less than 0.1%.

The formaldehyde is then introduced into a reactor where it passes over the surface of a rapidly stirred solution of initiator (either a Lewis acid or base; triphenylphosphine appears to be favoured) in a carefully dried inert medium such as heptane at about 40°C. The process is designed to give a very low concentration of formaldehyde to minimize transfer from polymer to monomer. To the initiator solution may be added a polymer stabilizer (e.g. diphenylamine) and transfer agents (e.g. traces of water or methanol). Polymerization is continued until the concentration of polymer in the slurry is about 20% and then the polymer is collected by filtration.

In the final stage the polymer is subjected to an esterification reaction to improve its thermal stability. The esterification may be effected with a number of anhydrides, but acetic anhydride is generally preferred. Typically, the polyformaldehyde is heated under slight pressure to about 160°C with acetic anhydride and a small amount of sodium acetate (catalyst). The polymer is soluble in acetic anhydride at this temperature but is precipitated when the solution is cooled. The acetylated polymer is collected by filtration, washed with water (to remove the anhydride and catalyst) and then acetone (containing di-β-naphthyl-p-phenylenediamine as antioxidant), and dried *in vacuo* at 70°C. The product is then extruded and chopped into granules. The average molecular weight (M_n) of the polyformaldehyde produced by these methods is generally in the range 30 000–100 000.

The polymerization of formaldehyde by Lewis bases such as triaryl amines (R_3N), arsines, and phosphines proceeds by the following anionic mechanism:

Initiation

$$R_3N \; + \; H_2O \longrightarrow [R_3NH]^+OH^-$$

$$[R_3NH]^+OH^- \; + \; H_2C{=}O \longrightarrow HO{-}CH_2{-}O^-[R_3NH]^+$$

Propagation

$$HO{-}CH_2{-}O^-[R_3NH]^+ \; + \; H_2C{=}O \rightleftharpoons HO{-}CH_2{-}O{-}CH_2{-}O^-[R_3NH]^+ \; \text{etc.}$$

Chain transfer

$$HO{-}[CH_2{-}O]_n{-}CH_2{-}O^-[R_3NH]^+ \; + \; H_2O \longrightarrow$$

$$HO{-}[CH_2{-}O]_n{-}CH_2OH \; + \; [R_3NH]^+OH^-$$

The polymerization of formaldehyde by Lewis acids such as boron trifluoride proceeds according to following cationic mechanism:

Initiation

$$BF_3 \; + \; H_2O \longrightarrow [BF_3OH]^- H^+$$

$$[BF_3OH]^- H^+ \; + \; O{=}CH_2 \longrightarrow HO{-}\overset{+}{C}H_2[BF_3OH]^-$$

Propagation

$$HO{-}\overset{+}{C}H_2[BF_3OH]^- \; + \; O{=}CH_2 \rightleftharpoons HO{-}CH_2{-}O{-}\overset{+}{C}H_2[BF_3OH]^- \; etc.$$

Chain transfer

$$HO{-}\!\!\left[CH_2{-}O\right]_{\!n}\!\!\overset{+}{C}H_2[BF_3OH]^- \; + \; H_2O \longrightarrow$$

$$HO{-}\!\!\left[CH_2{-}O\right]_{\!n}\!\!CH_2OH \; + \; [BF_3OH]^- H^+$$

The hydroxy-terminated polymers have poor thermal stability. Loss of a proton, possibly to an initiator residue, from a chain end gives an anion capable of decomposing to formaldehyde by the reverse of the propagation process. The stability of the polymer is therefore improved if the hydroxy end-groups are removed by esterification:

$$HO{-}\!\!\left[CH_2{-}O\right]_{\!n}\!\!CH_2{-}OH \xrightarrow{\;(CH_3CO)_2O\;}$$

$$CH_3COO{-}\!\!\left[CH_2{-}O\right]_{\!n}\!\!CH_2{-}OOCCH_3$$

It may be noted here that the polymerization of formaldehyde cannot be effected with free radical initiators.

(b)   Copolymers

Details of the procedures used in the preparation of commercial formaldehyde copolymers have not been fully disclosed. The principal monomer is trioxan and the second monomer is a cyclic ether such as ethylene oxide, 1,3-dioxolane or an oxetane; ethylene oxide appears to be the preferred comonomer and is used at a level of about 2%. Boron trifluoride (or its etherate) is apparently the most satisfactory initiator, although many cationic initiators are effective; anionic and free radical initiators are not effective. The reaction is carried out in bulk. The rapid solidification of the polymer requires a reactor fitted with a powerful stirrer to reduce particle size and permit adequate temperature control. The copolymer is then heated at 100°C with aqueous ammonia; in this step, chain-ends are depolymerized to the copolymer units to give a thermally-stable product. The polymer is filtered off and dried prior to stabilizer incorporation, extrusion and granulation.

The mechanism of polymerization of trioxan has not been completely elucidated. A possible scheme, in which boron trifluoride-water is the initiator is as follows:

Initiation

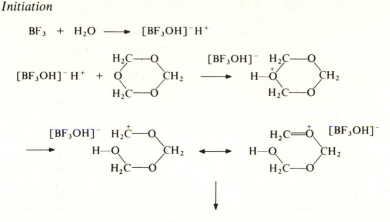

$$BF_3 + H_2O \longrightarrow [BF_3OH]^- H^+$$

$$[BF_3OH]^- H^+ + 3H_2C{=}O$$

Propagation

$$HO{-}CH_2{-}O{-}CH_2{-}O{-}\overset{+}{C}H_2[BF_3OH]^- + O{=}CH_2 \longrightarrow$$

$$HO{-}CH_2{-}O{-}CH_2{-}O{-}CH_2{-}O{-}\overset{+}{C}H_2[BF_3OH]^- \text{ etc.}$$

In the first step, trioxan is protonated by the complex protic acid formed by interaction of boron trifluoride and water. (It has been shown that no reaction occurs in the complete absence of water [3]). The resulting oxonium ion undergoes ring-opening to give a resonance-stabilized species. This then depolymerizes to build up an equilibrium concentration of formaldehyde, which remains constant during the polymerization. The actual propagation step then involves the addition of formaldehyde rather than trioxan. This scheme accounts for the observation that the polymerization of pure trioxan involves an induction period which may be reduced or even eliminated by the addition of formaldehyde.

9.3.4 Properties

The homopolymers and copolymers of formaldehyde, prepared as described above, are rigid materials with broadly similar properties. They are particularly noted for their stiffness, fatigue resistance and creep resistance and are counted as one of the 'engineering plastics'. They find application principally in injection moulded mechanical parts such as gears, cams and plumbing components. Comparative values for some properties of typical commercial products are given in Table 9.1. The copolymers are somewhat less crystalline and therefore have lower density, melting point, hardness, tensile strength and flexural modulus. The main advantage claimed for the copolymers is improved processability, with less degradation at processing temperatures.

Table 9.1 Typical values for various properties of formaldehyde homopolymer and copolymer

	Homopolymer	Copolymer
Specific gravity	1.425	1.410
Crystalline melting point (°C)	175	163
Tensile strength (MPa)	69	59
(lbf/in^2)	10 000	8 500
Flexural modulus (MPa)	2 800	2 500
(lbf/in^2)	410 000	360 000
Elongation at break (%)	15	60
Impact strength, Izod (J/m)	75	64
(ft lbf/in^2)	1.4	1.2
Hardness, Rockwell M	94	80

As is characteristic of crystalline polymers which do not interact with any
liquids, there are no effective solvents at room temperature for the commercial formaldehyde polymers. At temperatures above 70°C, solution occurs in
a few solvents such as the chlorophenols. The resistance of the polymers to
inorganic reagents is not, however, outstanding. Strong acids, strong alkalis
and oxidizing agents cause a deterioration in mechanical properties. (The
copolymers are significantly superior to the homopolymers in alkali resistance.)

Oxidation of polyformaldehyde occurs in air on prolonged exposure to
ultraviolet light and/or elevated temperature. Antioxidants are therefore
commonly added to the polymers.

9.4   POLYMERS OF CYCLIC ETHERS

Several polymers have been prepared by ring-opening polymerization of
cyclic ethers of general formula

$$(CH_2)_x \; O$$

and their derivatives. Polymers of commercial interest have been obtained from the epoxides (oxiranes)($x = 2$), ethylene oxide, propylene oxide
and epichlorhydrin; the oxacyclobutane (oxetane)($x = 3$), 3,3-bis(chloromethyl)oxacyclobutane; and the furan ($x = 4$), tetrahydrofuran. These polymers make up the contents of this section. It has not been found possible to
polymerize pyrans ($x = 5$), the 6-membered ring being strain free. (It may be
noted, however, that the 6-membered ring compound, trioxan, can be polymerized; for convenience, this polymer is dealt with in section 9.3.) The
polymerization of cyclic formals containing 7-, 8- and 11-membered rings has

also been reported but there has been no commercial development of the resulting polymers.

9.4.1 Raw materials

(a) Ethylene oxide

The dominant commercial route to ethylene oxide is by direct oxidation of ethylene:

$$H_2C{=}CH_2 \ + \ \tfrac{1}{2}O_2 \longrightarrow \overset{O}{\overbrace{CH_2{-}CH_2}}$$

A mixture of ethylene, oxygen and a diluent (e.g. carbon dioxide or nitrogen) is passed over a silver catalyst at about 250°C and 2MPa (20 atmospheres). Although other materials have been mentioned, silver appears to be the only catalyst used in commercial-scale ethylene oxide production. Commercial catalysts are prepared by depositing metallic silver on supports such as alumina and silicon carbide. Such catalysts give a selectivity of about 70%, i.e. 70% of the ethylene consumed is converted to ethylene oxide and 30% to carbon dioxide:

$$H_2C{=}CH_2 \ + \ 3O_2 \longrightarrow 2CO_2 \ + \ 2H_2O$$

The last reaction is highly exothermic and reactors must be designed to ensure the rapid removal of the large amount of heat evolved. The exit gases from the reactor are cooled and compressed and then scrubbed with water, which extracts the ethylene oxide. The aqueous solution is distilled to recover high-purity ethylene oxide.

In an older process, ethylene oxide is obtained from ethylene by the following route:

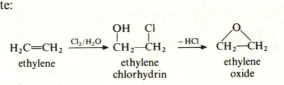

Compared to direct oxidation, the chlorhydrin route gives higher yields of ethylene oxide (75–80%); however, the additional cost of the chlorine makes the route uneconomical.

Ethylene oxide is a colourless gas, b.p. 11°C.

(b) Propylene oxide

In contrast to ethylene oxide, the bulk of propylene oxide is produced by the chlorhydrin route as follows:

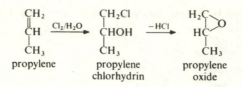

In the first stage a mixture of chlorine, water and excess of propylene is fed into a packed tower at 35–50°C and a dilute solution of propylene chlorhydrin is formed. It was previously thought that this reaction consisted of the addition of preformed hypochlorous acid to the olefin but it now appears that electrophilic attack by chlorine is the first step, followed by reaction of the carbenium ion with water:

$$H_3C-CH=CH_2 \ + \ Cl_2 \ \longrightarrow \ H_2C-\overset{+}{C}H-CH_2Cl \ + \ Cl^-$$

$$H_3C-\overset{+}{C}H-CH_2Cl \ + \ H_2O \ \longrightarrow \ H_3C-\overset{\overset{\displaystyle OH}{|}}{C}H-CH_2Cl \ + \ H^+$$

The chlorhydrin solution is then treated with calcium hydroxide at about 100°C to give propylene oxide, which is flashed off and purified by distillation.

Although many efforts have been made to develop a direct oxidation route to propylene oxide, none has been successful. The low yields of propylene oxide obtained render the route uneconomical; the methyl group is readily oxidized so that substantial amounts of acrolein are also formed. However, hydroperoxidation processes have been developed and are becoming of increasing importance. In fact, it seems likely that these processes eventually will displace the chlorohydrin method. Two variations of the hydroperoxidation method are operated:

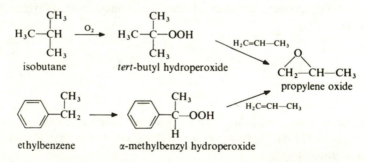

In the first process, isobutane is oxidized by air at 120–150°C and about 3 MPa (30 atmospheres) to give *tert*-butyl hydroperoxide. This is then treated with propylene in the liquid phase at 120–140°C and about 3.5 MPa (35 atmospheres) in the presence of a molybdenum catalyst to give propylene oxide. *tert*-Butyl alcohol is obtained as a by-product. In the second process, ethylbenzene is utilized under similar conditions and 1-phenylethanol (which can be dehydrated to styrene) is the by-product.

Propylene oxide is a colourless liquid, b.p. 34°C.

(c) Epichlorhydrin

The preparation of epichlorhydrin,

$$CH_2—CH—CH_2Cl,$$

is outlined in section 11.3.2(a). Epichlorhydrin is a colourless liquid, b.p. 115°C.

(d) 3,3-Bis(chloromethyl)oxacyclobutane

3,3,-Bis(chloromethyl)oxacyclobutane (3,3-bis(chloromethyl)oxetane) may be prepared from pentaerythritol by the following route:

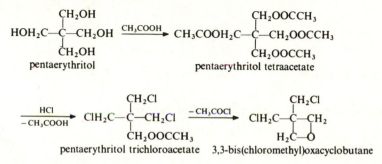

pentaerythritol trichloroacetate   3,3-bis(chloromethyl)oxacyclobutane

Pentaerythritol is esterified to the tetraacetate which is then treated with dry hydrogen chloride at 200°C in the presence of zinc chloride as catalyst. Only three acetoxy groups can be replaced and the product is the trichloroacetate. The crude trichloroacetate is then treated with sodium hydroxide and the resulting oxetane is purified by distillation under reduced pressure. 3,3-Bis(chloromethyl)oxacyclobutane is a solid, m.p. 19°C.

(e) Tetrahydrofuran

The standard, and until recently the only, industrial route to tetrahydrofuran is as follows:

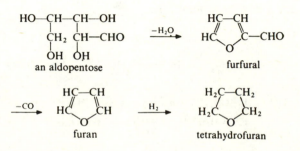

Agricultural waste products such as corn cobs and oat hulls, which are rich in pentosans (polysaccharides containing pentose residues), are heated with dilute sulphuric acid to effect hydrolysis and cyclization. Furfural is isolated from the mixture by steam distillation and then pyrolysed at about 400°C over a chromium-containing catalyst to give furan. The furan is hydrogenated over nickel at about 125°C and 10 MPa (100 atmospheres) to give tetrahydrofuran.

The furfural route is still extensively used in the USA but in other parts of the world it has been largely replaced by routes based on 1,4-butanediol and maleic anhydride (section 11.2.2(b)):

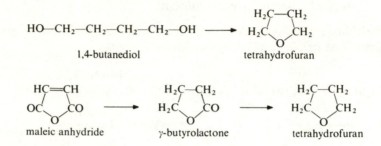

Tetrahydrofuran is a colourless liquid, b.p. 65°C.

9.4.2   Poly(ethylene oxide)

The ring-opening polymerization of ethylene oxide is readily effected by a variety of ionic reagents and several types of polymer have been prepared. For commercial purposes, poly(ethylene oxide)s of low molecular weight and of very high molecular weight are of interest.

(a)   Low molecular weight polymers

Poly(ethylene oxide)s of low molecular weight, i.e. below about 3000, are generally prepared by passing ethylene oxide into ethylene glycol at 120–150°C and about 0.3 MPa (3 atmospheres) pressure, using an alkaline initiator such as sodium hydroxide. Anionic polymerization proceeds according to the following scheme:

Initiation

$$HO-CH_2-CH_2-OH \ + \ NaOH \ \rightleftharpoons \ HO-CH_2-CH_2-O^-Na^+ \ + \ H_2O$$

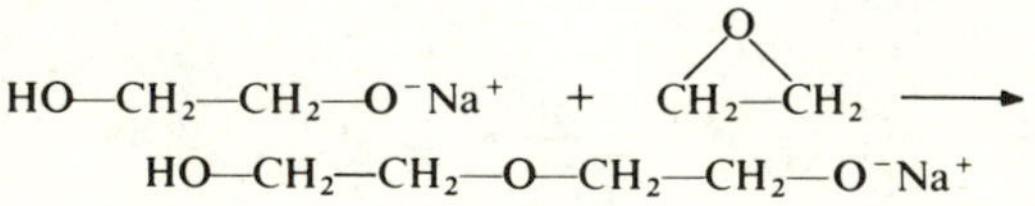

$$HO-CH_2-CH_2-O^-Na^+ \ + \ CH_2-CH_2 \longrightarrow$$

$$HO-CH_2-CH_2-O-CH_2-CH_2-O^-Na^+$$

and $\overset{\displaystyle O}{\overset{\displaystyle /\!\!\backslash}{CH_2\!\!-\!\!CH_2}}$ + NaOH $\longrightarrow$ HO—CH$_2$—CH$_2$—O$^-$Na$^+$

Propagation

$$HO\!-\!CH_2\!-\!CH_2\!-\!O^-Na^+ \; + \; \overset{\displaystyle O}{\overset{\displaystyle /\!\!\backslash}{CH_2\!\!-\!\!CH_2}} \longrightarrow$$

HO—CH$_2$—CH$_2$—O—CH$_2$—CH$_2$—O$^-$Na$^+$ etc.

Transfer

HO—[CH$_2$—CH$_2$—O]$_n$—CH$_2$—CH$_2$—O$^-$Na$^+$ + HO—CH$_2$—CH$_2$—OH $\rightleftharpoons$

HO—[CH$_2$—CH$_2$—O]$_n$—CH$_2$—CH$_2$—OH + HO—CH$_2$—CH$_2$—O$^-$Na$^+$

HO—[CH$_2$—CH$_2$—O]$_n$—CH$_2$—CH$_2$—O$^-$Na$^+$ + H$_2$O $\rightleftharpoons$

HO—[CH$_2$—CH$_2$—O]$_n$—CH$_2$—CH$_2$—OH + NaOH

The polymers produced by these methods are thus terminated mainly by hydroxy groups (a few unsaturated end-groups are also formed) and are often referred to as poly(ethylene glycol)s. Poly(ethylene glycol)s with molecular weights in the range 200–600 are viscous liquids which find use as surfactants in inks and paints and as humectants. At molecular weights above about 600, poly(ethylene glycol)s are low-melting waxy solids, uses of which include pharmaceutical and cosmetic bases, lubricants and mould release agents.

It may be noted that homogeneous cationic polymerization of ethylene oxide also generally leads to low molecular weight products; typical initiators include aluminium chloride, boron trifluoride and titanium tetrachloride. Systems of this type are not utilized on a commercial scale.

(b) *High molecular weight polymers*

Poly(ethylene oxide)s of molecular weight ranging from about 100 000 to 5×10^6 and above are available. Details of the techniques used to manufacture these polymers have not been disclosed, but the essential feature is the use of (generally) heterogeneous initiator systems. Effective initiators are mainly of two types, namely alkaline earth compounds (e.g. carbonates and oxides of calcium, barium and strontium) and organometallic compounds (e.g. aluminium and zinc alkyls and alkoxides, commonly with added co-initiators).

The precise modes of action of these initiators have not, as yet, been fully resolved. However, it is now generally thought that polymerization occurs through a co-ordinated anionic mechanism, in which the ethylene oxide is co-ordinated to the initiator through an unshared electron pair on the oxirane oxygen atom:

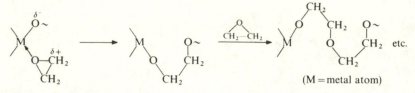

(M = metal atom)

Unlike the low molecular weight poly(ethylene oxide)s, the high molecular weight polymers are tough and extensible. They are highly crystalline, with a melting point of 66°C. Unlike most water-soluble polymers, the high molecular weight poly(ethylene oxide)s may be melt processed; they may be injection moulded, extruded and calendered without difficulty.

Poly(ethylene oxide)s are soluble in an unusually broad range of solvents, which includes water; chlorinated hydrocarbons such as carbon tetrachloride and methylene dichloride; aromatic hydrocarbons such as benzene and toluene; ketones such as acetone and methyl ethyl ketone; and alcohols such as methanol and isopropanol. There is an upper temperature limit of solubility in water for the high molecular weight poly(ethylene oxide)s; this varies with concentration and molecular weight but is usually between 90 and 100°C. Water-solubility is due to the ability of the polyether to form hydrogen bonds with water; these bonds are broken when the temperature is raised, restoring the anhydrous polymer which is precipitated from the solution.

High molecular weight poly(ethylene oxide)s find use as water-soluble packaging films and capsules for such products as laundry powders, colour concentrates, tablets and seeds. In solution, the polymers are used as thickeners in pharmaceutical and cosmetic preparations, textile sizes and latex stabilizers.

9.4.3 Poly(propylene oxide)

Propylene oxide may be polymerized by methods similar to those described in the preceding section for ethylene oxide. Similarly, polymers of low molecular weight and of high molecular weight are of commercial interest.

(a) Low molecular weight polymers

Poly(propylene oxide)s of low molecular weight, i.e. in the range 500–3500, are important commercial materials principally on account of their extensive use in the production of both flexible and rigid polyurethane foams (Chapter 16).

At first, the most common polyether used in flexible polyurethane foams was a linear poly(propylene glycol) with a molecular weight of about 2000. This is prepared by polymerizing the oxide at about 160°C in the presence of propylene glycol and sodium hydroxide. Polymerization takes place according to the anionic mechanism given in section 9.4.2(a). The resulting polymer has the following general form:

$$HO-CH(CH_3)-CH_2-O-[CH_2-CH(CH_3)-O]_n-CH_2-CH(CH_3)-OH$$

The majority of the hydroxyl groups in the polymer are secondary groups and are rather unreactive in the urethane reaction. Initially, this limitation was overcome by the preparation of pre-polymers (see Chapter 16) and by the use of block copolymers with ethylene oxide. The latter products are 'tipped' with poly(ethylene oxide) and are terminated with primary hydroxyl groups of enhanced reactivity:

$$HO-[CH_2-CH_2-O]_x-[CH_2-CH(CH_3)-O]_y-[CH_2-CH_2-O]_z-CH_2-CH_2-OH$$

(It may be noted that straight poly(ethylene glycol) is not satisfactory for foam production owing to its water sensitivity and tendency to crystallize.)

The advent of more effective catalyst systems, however, now makes it possible for poly(propylene oxide)s to be used in the preparation of flexible polyurethane foams without recourse to the above mentioned procedures. Also, it is now common practice to use polyethers which are triols rather than diols; these lead to slightly cross-linked flexible foams with improved load bearing characteristics. The triols are produced by polymerizing propylene oxide in the presence of a trihydroxy compound such as glycerol, 1,1,1-trimethylolpropane or 1,2,6-hexane triol; the use of, for example, trimethylolpropane leads to the following polyether triol:

$$H_3C-CH_2-C\begin{cases} CH_2-O-[CH_2-CH(CH_3)-O]_x-CH_2-CH(CH_3)-OH \\ CH_2-O-[CH_2-CH(CH_3)-O]_y-CH_2-CH(CH_3)-OH \\ CH_2-O-[CH_2-CH(CH_3)-O]_z-CH_2-CH(CH_3)-OH \end{cases}$$

Polyethers of molecular weights in the range 3000–3500 are normally used.

For the production of rigid polyurethane foams, polyether triols of lower molecular weight (about 500) are used so that the degree of cross-linking is increased. Alternatively, polyethers of higher functionality may be used; these are prepared by polymerizing propylene oxide in the presence of hydroxy compounds such as pentaerythritol and sorbitol.

(b)   High molecular weight polymers

Poly(propylene oxide)s of high molecular weight, i.e., greater than 100 000 have been prepared by the use of initiators similar to those employed in the preparation of high molecular weight poly(ethylene oxide)s (section 9.4.2(b)). The most extensively investigated initiators have been organoaluminium and organozinc compounds, generally with added co-initiators. As has been noted previously (section 1.6.4), the structural unit of poly(propylene oxide) contains an asymmetric carbon atom and the polymer can exhibit tacticity. Both atactic and isotactic poly(propylene oxide) have been prepared. As normally obtained, i.e. from D-L-propylene oxide, the isotactic polymer is optically inactive but optically active isotactic polymer has been produced from L-propylene oxide. Except for their optical activity, both forms of isotactic polymer are very similar in properties; they are both crystalline and have a melting point of 74°C. Isotactic poly(propylene oxide)s have not yet found commercial application. Atactic poly(propylene oxide) has been investigated as a rubber but does not appear to have been produced in any quantity.

9.4.4   Epichlorhydrin polymers

It might be anticipated that epichlorhydrin would give rise to a range of polymers similar to those obtained from propylene oxide:

$$n CH_2\text{---}CH\text{---}CH_2Cl \longrightarrow \left[\text{---}CH_2\text{---}\overset{\displaystyle CH_2Cl}{\underset{|}{CH}}\text{---}O\text{---} \right]_n$$

Indeed, several polymers have been prepared, ranging from liquids and rubbery amorphous materials to a crystalline polymer of melting point 119°C. Of these products, epichlorhydrin elastomers are available commercially. These materials are prepared by solution polymerization in an aromatic solvent such as benzene or toluene at 40–130°C with an alkylaluminium-water initiator. The solution is treated with steam to give the solid product. Two main types of material are produced. The first is a homopolymer of epichlorhydrin and the second is a 1 : 1 copolymer of epichlorhydrin and ethylene oxide. These polymers are saturated but are readily vulcanized through the chloromethyl group by a variety of reagents, e.g. ethylene thiourea. A terpolymer containing a small amount of allyl glycidyl ether besides epichlorhydrin and ethylene oxide is also available and this may be vulcanized by more conventional reagents such as sulphur systems and peroxides. The vulcanizates have a good combination of heat and ozone resistance, fuel and oil resistance, low temperature flexibility and low permeability to air. The homopolymer shows better flame resistance and lower permeability to air than the copolymer, whereas the latter is superior in low temperature flexibility. The largest use of the elastomers is in the automotive field as fuel, lubricating fluid, air and vacuum hose.

It may be noted here that epichlorhydrin is also used for the preparation of epoxies and phenoxies. Strictly, these polymers may be regarded as polyethers but they are dealt with elsewhere (Chapters 12 and 18) since they arise by chemical processes which are somewhat different to those described in this chapter.

9.4.5 Poly-3,3-bis(chloromethyl)oxacyclobutane

Oxacyclobutane may be polymerized to a high molecular weight, crystalline polymer:

$$n \quad \begin{matrix} CH_2-CH_2 \\ | \quad\quad | \\ H_2C-\!\!-\!\!-O \end{matrix} \quad \xrightarrow{BF_3} \quad [-CH_2-CH_2-CH_2-O-]_n$$

The polymer has a low melting point (35°C) and is not of commercial interest. However, a derivative of oxacyclobutane, 3,3-bis(chloromethyl)oxacyclobutane, does lead to a polymer with more useful properties:

$$n ClCH_2-\overset{\overset{\displaystyle CH_2Cl}{|}}{\underset{\underset{\displaystyle H_2C-O}{|}}{C}}-CH_2 \quad \longrightarrow \quad \left[-CH_2-\overset{\overset{\displaystyle CH_2Cl}{|}}{\underset{\underset{\displaystyle CH_2Cl}{|}}{C}}-CH_2-O- \right]_n$$

Polymerization of 3,3-bis(chloromethyl)oxacyclobutane may be effected by Friedel-Crafts reagents (e.g. aluminium chloride and boron trifluoride) at low temperatures (about −50°C).

The polymer has a high melting point (α-form 188°C, β-form 180°C) and shows good heat stability and chemical resistance. The material was available commercially for some years but is no longer manufactured.

9.4.6 Polytetrahydrofuran

Low molecular weight polymers of tetrahydrofuran were introduced commercially in 1955. The polymers were intended for use in the preparation of flexible polyurethane foams, being terminated by hydroxyl groups:

$$n \quad \begin{matrix} H_2C-CH_2 \\ / \quad\quad \backslash \\ H_2C \quad\quad CH_2 \\ \backslash \quad\quad / \\ O \end{matrix} \quad \longrightarrow \quad OH-\!\!\left[(CH_2)_4-O\right]_{n-1}\!\!-(CH_2)_4-OH$$

Although these polyethers produced good foams they were rather expensive and were soon displaced by the cheaper propylene oxide-based polyethers. However, these polymers of tetrahydrofuran (commonly called poly(oxytetramethylene)glycol) now find use in the preparation of polyamide, polyester and polyurethane thermoplastic elastomers (see sections 10.3, 11.8 and 16.6.3).

Details of the manufacture of polytetrahydrofuran have not been disclosed. Polymerization of tetrahydrofuran can be effected only by cationic initiators. Combinations of metal halides with water are not effective but protic acids (e.g. $HClO_4$) and oxonium salts (e.g. $(C_2H_5)_3O^+BF_4^-$) are efficient initiators. Polymerization occurs by a ring-opening mechanism involving a tertiary oxonium ion, e.g.

Initiation

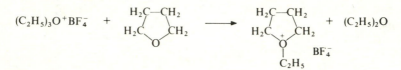

Propagation

9.5   FURAN POLYMERS

Furfural and furfuryl alcohol are used to make a variety of polymers which are known collectively as 'furan' polymers.

The preparation of furfural is described in section 9.4.1(e). Furfuryl alcohol is obtained by the hydrogenation of furfural in the presence of a copper chromite catalyst:

9.5.1   Polymers from furfural

Furfural is converted to a dark cross-linked polymeric material on heating with a strong acid or Friedel-Crafts catalyst; the mechanism of the reaction is uncertain. Polymers of this kind are brittle and are of no commercial significance.

More important polymers of furfural are based on its ability to react with phenols to form thermosetting resins. The most common resins of this kind are of the novolak type made by reaction of a slight molar excess of phenol with furfural in the presence of an alkaline catalyst (cf. Chapter 14). Usually, blends of phenol-furfural novolaks and phenol-formaldehyde novolaks are used in the production of moulding powders with enhanced flow properties.

9.5.2 Polymers from furfuryl alcohol

Furfuryl alcohol may also be converted into polymeric materials in the presence of acids and such products have found some commercial use. Typically, furfuryl alcohol is heated at about 100°C with an acid catalyst such as phosphoric acid. The reaction is extremely exothermic and efficient cooling is necessary to prevent premature cross-linking. When the required degree of reaction is reached the system is neutralized and dehydrated under reduced pressure. The product is a dark free-flowing liquid. Often urea and formaldehyde are also included in resin formulations. Final cure of the resin is effected *in situ* by addition of an acid just prior to application. Weak acids (e.g. phthalic anhydride and phosphoric acid) give mixtures with long pot life which cure at 100–200°C whilst strong acids (e.g. *p*-toluenesulphonic and sulphuric acids) are effective at room temperature.

The first product of furfuryl alcohol polymerization is thought to be a linear polymer, formed as follows:

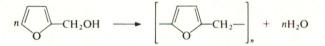

The mechanism whereby this product is converted into a cross-linked polymer is uncertain. Loss of unsaturation during the process suggests the following reaction:

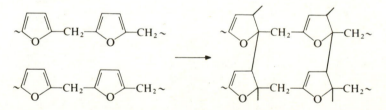

Cured furfuryl alcohol polymers have outstanding heat and chemical resistance and find use in cements for chemical plant, adhesives and foundry binders.

REFERENCES

1. Kargin, V. A. *et al.* (1960) *Dokl. Akad. Nauk SSSR*, **134**, 1098.
2. Okamura, S. *et al.* (1961) *Doitai To Hoshasen*, **4**, 70; (1963) Japan Pat. 16, 341.
3. Collins, G. L. *et al.* (1979) *J. Polym. Sci. Polym. Lett. Ed.*, **17**, 667.

BIBLIOGRAPHY

Gaylord, N. G. (ed.) (1961) *Polyethers, Part I: Polyalkylene Oxides and other Polyethers*, Interscience Publishers, New York.

Akin, R. B. (1962) *Acetal Resins*, Reinhold Publishing Corporation, New York.

Furukawa, J. and Saegusa, T. (1963) *Polymerization of Aldehydes and Oxides*, Interscience Publishers, New York.

Sittig, M. (1963) *Polyacetal Resins*, Gulf Publishing Co., Houston.

Boardman, H. (1964) in *Manufacture of Plastics*, (ed. W. M. Smith), Reinhold Publishing Corporation, New York, ch. 15, (Poly-3,3-bis(chloromethyl)-oxacyclobutane).

Vogl, O. (ed.) (1967) *Polyaldehydes*, Edward Arnold (Publishers) Ltd, London.

Barker, S. J. and Price, M. B. (1970), *Polyacetals*, Iliffe Books, London.

Bailey, F. E. and Koleske, J. V. (1976) *Poly(ethylene oxide)*, Academic Press, New York.

Dreyfuss, P. (1982) *Poly(tetrahydrofuran)*, Gordon and Breach Science Publishers, New York.

POLYAMIDES AND POLYIMIDES

10.1 SCOPE

For the purposes of this chapter, polyamides are defined as polymers which contain recurring amide groups (–CO–NH–) in the main polymer chain. Various types of polyamides fall within this definition.

The most important commercial polymers in this group are synthetic linear aliphatic polyamides which are capable of fibre formation; these polymers are very commonly termed *nylons*. It may be noted that by far the greater part of the total output of nylons is used for fibre production but the materials have also become of some importance in non-fibrous applications, particularly engineering applications.

There are also available more complex synthetic aliphatic polyamides which are not fibre-forming but which have found use in such applications as adhesives and coatings. These products are known as *fatty polyamides* and are not generally regarded as nylons.

More recently, synthetic aromatic polyamides which are capable of fibre formation have been introduced; these polymers are commonly called *aramids*.

Finally, a very important class of polyamides consists of the naturally-occurring *proteins*. The importance of proteins extends beyond their various fundamental roles in living matter to several technological applications such as adhesives and fibres. These applications are reviewed briefly in this chapter.

Polyimides are polymers in which the chain contains recurring imide groups:

In polymers of commercial interest, the imide groups generally form part of cyclic structures which lead to materials of high thermal stability.

10.2 NYLONS

10.2.1 Development

As mentioned in Chapter 1, the commercial development of nylons was an outcome of the fundamental researches into polymerization which were begun in 1929 by Carothers of E. I. du Pont de Nemours and Co. Several polyamides and polyesters were investigated; nylon 6,6 (see later for nomenclature) was first synthesized in 1935 and was selected as the most promising fibre-forming material. Commercial production was started in 1938; nylon stockings were put on trial sale in 1939 and became generally available in the USA in 1940. The first nylon mouldings were produced in 1941 but the polymer did not become well known in this form until about 1950.

The immediate spectacular success of nylon 6,6 led to an intensive search throughout the world for competitive fibre-forming polymers not covered by the du Pont patents. As a result of such work, I. G. Farbenindustrie (Germany) developed nylon 6; production was started in 1940. After the Second World War, the manufacture of nylon 6 was undertaken in many countries since it was free from patent restrictions. At the present time nylons 6,6 and 6 account for nearly all of the total output of polyamides for fibre production. The two materials have very similar properties and are virtually interchangeable in their applications. The relative importance of the products shows marked geographical variation. In the USA and the UK nylon 6,6 predominates, whilst in most other regions nylon 6 is the predominant material.

Besides nylons 6,6 and 6, several other nylons have achieved commercial status. It should be noted, however, that since nylons 6,6 and 6 are produced in very large quantities, they are substantially cheaper than other nylons, which do not have the same benefit of economy of scale. Hence the other nylons are generally restricted to applications for which nylons 6,6 and 6 are unsuitable.

The principal nylons which are currently commercially available are shown below. For the preparation of these nylons, three general methods have been developed, namely reaction of a diamine and a dicarboxylic acid; self-condensation of an ω-amino acid; and ring-opening polymerization of a lactam. The method usually applied for each nylon is also shown below. The various nylons are distinguished from one another by a numbering system based on the number of carbon atoms in the starting materials.

Nylon 6,6

$$n\mathrm{H_2N-(CH_2)_6-NH_2} \quad + \quad n\mathrm{HOOC-(CH_2)_4-COOH} \longrightarrow$$

hexamethylenediamine   adipic acid
(6 carbon atoms)   (6 carbon atoms)

$$[\mathrm{-HN-(CH_2)_6-NH-OC-(CH_2)_4-CO-]}_n \quad + \quad 2n\mathrm{H_2O}$$

Nylon 6,9

$$nH_2N-(CH_2)_6-NH_2 \;+\; nHOOC-(CH_2)_7-COOH \longrightarrow$$

hexamethylene diamine   azelaic acid
(6 carbon atoms)   (9 carbon atoms)

$$[-HN-(CH_2)_6-NH-OC-(CH_2)_7-CO-]_n \;+\; 2nH_2O$$

Nylon 6,10

$$nH_2N-(CH_2)_6-NH_2 \;+\; nHOOC-(CH_2)_8-COOH \longrightarrow$$

hexamethylenediamine   sebacic acid
(6 carbon atoms)   (10 carbon atoms)

$$[-HN-(CH_2)_6-NH-OC-(CH_2)_8-CO-]_n \;+\; 2nH_2O$$

Nylon 6,12

$$nH_2N-(CH_2)_6-NH_2 \;+\; nHOOC-(CH_2)_{10}-COOH \longrightarrow$$

hexamethylene diamine   dodecanedioic acid
(6 carbon atoms)   (12 carbon atoms)

$$[-HN-(CH_2)_6-NH-OC-(CH_2)_{10}-CO-]_n \;+\; 2nH_2O$$

Nylon 11

$$nH_2N-(CH_2)_{10}-COOH \longrightarrow [HN-(CH_2)_{10}-CO-]_n \;+\; nH_2O$$

ω-aminoundecanoic acid
(11 carbon atoms)

Nylon 6

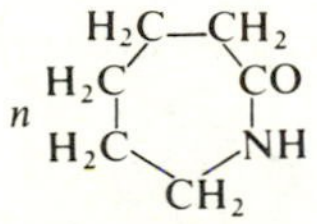

caprolactam
(6 carbon atoms)

$$\longrightarrow [-(CH_2)_5-CO-NH-]_n$$

Nylon 12

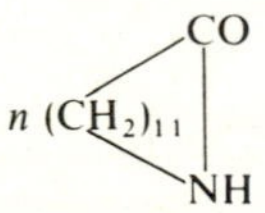

dodecyl lactam
(12 carbon atoms)

$$\longrightarrow [-(CH_2)_{11}-CO-NH-]_n$$

10.2.2   Raw materials

(a)   *Adipic acid*

The principal commercial route to adipic acid is from benzene as follows:

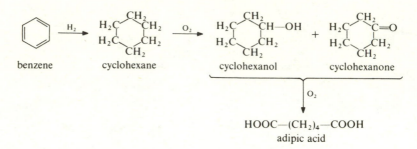

$$HOOC—(CH_2)_4—COOH$$
adipic acid

Typically, benzene is hydrogenated in the liquid phase at 150–200°C and about 2.5 MPa (25 atmospheres) using a Raney nickel catalyst. Cyclohexane may also be obtained by fractionation of natural gas and petroleum streams. It is possible to oxidize pure cyclohexane to adipic acid in one step but normal commercial practice is to use a two-step process. In the first step, cyclohexane is oxidized with air at about 160°C and 1 MPa (10 atmospheres) using cobalt naphthenate as catalyst. Unchanged cyclohexane is removed by distillation to give a mixture consisting of cyclohexanol and cyclohexanone (in approximately equal amounts). The second oxidation step is carried out by treating the cyclohexanol/cyclohexanone mixture with 50% aqueous nitric acid in the presence of an ammonium vanadate-copper catalyst. Reaction conditions are generally about 80°C and 0.3 MPa (3 atmospheres). The adipic acid is crystallized from the reaction mass, separated by centrifuging and dried.

Some older processes, now of minor importance, are based on phenol rather than cyclohexane. Phenol is hydrogenated to cyclohexanol which is then oxidized to adipic acid with nitric acid.

Adipic acid is a white crystalline solid, m.p. 152°C.

(b) Hexamethylenediamine

The standard commercial route to hexamethylenediamine is from adipic acid as follows:

$$HOOC—(CH_2)_4—COOH \xrightarrow{\text{NH}_3} NC—(CH_2)_4—CN$$
adipic acid   adiponitrile

$$\xrightarrow{\text{H}_2} H_2N—(CH_2)_6—NH_2$$
hexamethylenediamine

In a typical process, adipic acid is converted to the nitrile in the vapour phase by treatment with ammonia at 350–450°C using boron phosphate as catalyst. The adiponitrile is separated by distillation and then hydrogenated at about 130°C and 30 MPa (300 atmospheres). Hydrogenation is carried out in the presence of a supported cobalt catalyst and ammonia (which minimizes cyclization to hexamethyleneimine). Pure hexamethylenediamine is isolated by distillation under reduced pressure.

Two alternative routes to adiponitrile have become significant relatively recently. The first of these routes is based on butadiene:

$$H_2C=CH—CH=CH_2 \xrightarrow{\text{HCN}} H_2C=CH—CH_2—CH_2—CN$$

butadiene various cyanobutenes

$$\xrightarrow{\text{HCN}} NC—(CH_2)_4—CN$$

adiponitrile

In the first step, butadiene is heated with hydrogen cyanide to give a mixture of various cyanobutenes. The cyanobutenes are then separated, isomerized and treated with hydrogen cyanide to yield adiponitrile. All stages are carried out in the liquid phase at about 100°C.

The second alternative route to adiponitrile involves the electrolytic hydro-dimerization of acrylonitrile:

$$2H_2C=CH—CN + 2H^+ + 2e^- \longrightarrow NC—(CH_2)_4—CN$$

Hexamethylenediamine is a white crystalline solid, m.p. 41°C.

(c) Azelaic acid

Azelaic acid is made by the ozonolysis of oleic acid:

$$H_3C—(CH_2)_7—CH=CH—(CH_2)_7—COOH \xrightarrow{\text{O}_3/\text{H}_2\text{O}} H_3C—(CH_2)_7—COOH$$

$$+ \quad HOOC—(CH_2)_7—COOH$$

Azelaic acid is a crystalline solid, m.p. 107–108°C.

(d) Sebacic acid

Sebacic acid is normally made from castor oil, which is essentially glycerol triricinoleate. The castor oil is heated with sodium hydroxide at about 250°C. This treatment results in saponification of the castor oil to ricinoleic acid which is then cleaved to give 2-octanol and sebacic acid:

$$\overset{\displaystyle OH}{H_3C—(CH_2)_5—CH—CH_2—CH=CH—(CH_2)_7—COOH} \longrightarrow$$

$$\overset{\displaystyle OH}{H_3C—(CH_2)_5—CH—CH_3} + HOOC—(CH_2)_8—COOH$$

This process results in low yields of sebacic acid (about 50% based on the castor oil) but, nevertheless, other routes have not proved competitive. Sebacic acid is a colourless crystalline solid, m.p. 134.5°C.

(e)   Dodecanedioic acid

Dodecanedioic acid is prepared from cyclododecene (obtained from butadiene) by methods which are entirely analogous to those used to prepare
adipic acid from benzene (section 10.2.2(a)). The cyclododecene is reduced to
cyclododecane, which is oxidized firstly to a mixture of cyclododecanol and
cyclododecanone and then to dodecanedioic acid. Dodecanedioic acid is a
colourless crystalline solid, m.p. 129°C.

(f)   ω-Aminoundecanoic acid

ω-Aminoundecanoic acid is normally prepared from castor oil (glycerol
triricinoleate). Firstly the castor oil is subjected to methanolysis to yield the
methyl ester of ricinoleic acid and then the following route is used:

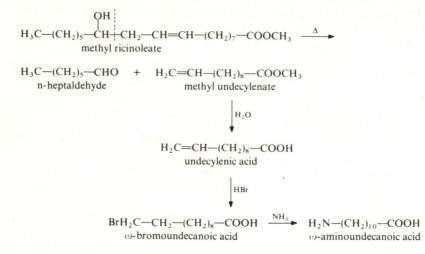

In the first step, the pyrolysis of methyl ricinoleate is carried out at about
500°C; the principal products are n-heptaldehyde and methyl undecylenate.
The latter is hydrolysed to give undecylenic acid which is treated with
hydrogen bromide in a non-polar solvent in the presence of a peroxide.
Under these conditions, reverse Markownikoff addition occurs and the main
product is ω-bromoundecanoic acid. This product is then treated with
ammonia to give ω-aminoundecanoic acid, which is a crystalline solid, m.p.
189°C.

(g)   Caprolactam

The major commercial route to caprolactam is from cyclohexane (section
10.2.2(a)) as follows:

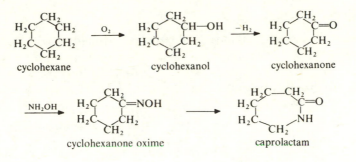

The first step consists of the air oxidation of cyclohexane to a mixture of cyclohexanol and cyclohexanone as described in section 10.2.2(a). The mixture is passed over a zinc oxide catalyst at about 400°C, when most of the cyclohexanol is dehydrogenated to cyclohexanone. The cyclohexanone is separated by distillation and is then treated with aqueous hydroxylamine sulphate at about 100°C to form the oxime. The reaction mixture is neutralized with aqueous ammonia or sodium hydroxide and the crude oxime separated as an oily layer. This is stirred with oleum (sulphur trioxide in sulphuric acid) at 120°C and the oxime undergoes the Beckmann rearrangement to give caprolactam. In one process, the solution containing the lactam is continuously withdrawn from the reactor and rapidly cooled to below 75°C to minimize hydrolysis. The solution is then further cooled and neutralized with aqueous ammonia. Crude caprolactam separates as an oil and is purified by distillation under reduced pressure or by recrystallization.

The profitability of the above route to caprolactam is intimately bound up with the value of the substantial quantities of ammonium sulphate which are produced as a by-product and several other routes have been investigated. A particularly attractive reaction is the photochemical combination of nitrosyl chloride and cyclohexane which gives cyclohexanone oxime hydrochloride directly:

$$H_2C \begin{matrix} CH_2 \\ CH_2 \\ CH_2 \end{matrix} CH_2 \quad + \quad NOCl \quad \xrightarrow{h\nu} \quad H_2C \begin{matrix} CH_2 \\ C=NOH.\ HCl \\ CH_2 \end{matrix} CH_2$$

In one commercial process, the reaction is carried out by irradiating a mixture of cyclohexane, nitrosyl chloride and hydrochloric acid at −10° to 16°C with light from a mercury vapour lamp. The raw material costs in this process are lower than those in the conventional process and the ammonium sulphate production is about half.

Caprolactam is a white crystalline solid, m.p. 70°C.

(h)   Dodecyl lactam

One route to dodecyl lactam is from butadiene as follows:

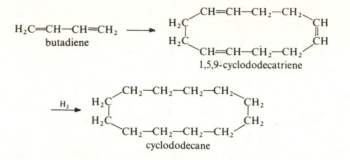

$H_2C=CH-CH=CH_2$
butadiene

1,5,9-cyclododecatriene

cyclododecane

Butadiene is trimerized with a Ziegler-Natta catalyst to give 1,5,9-cyclo-dodecatriene which is then hydrogenated to cyclododecane. Dodecyl lactam is then obtained by a series of reactions similar to those used to prepare caprolactam from cyclohexane (section 10.2.2(g)).

10.2.3   Preparation

(a)   Nylon 6,6

For the production of polyamide fibres, it is essential that the polymer is of high molecular weight, i.e. 10 000 or more. In order to achieve such molecular weight when the polyamide is prepared by reaction of a diamine and a dibasic acid, it is necessary to have exact equivalence of the reactants. It is common practice to ensure such equivalence by the preparation of a 'nylon salt' prior to polymerization.

Thus in the production of nylon 6,6, the first step consists of the neutralization of an aqueous solution of adipic acid by addition of hexamethylenediamine; the exact end-point is determined electrometrically. The aqueous salt solution is then concentrated to 60–75% solids content before transfer to the polymerization reactor. In an alternative procedure, approximately equivalent quantities of adipic acid and hexamethylenediamine are mixed in boiling methanol, from which the 1:1 salt is precipitated. The salt is isolated by filtration and then dissolved in water and the solution fed to the polymerization reactor. The nylon salt, hexamethylenediammonium adipate may be represented as follows:

$$\begin{bmatrix} {}^-OCC-(CH_2)_4-COO^- \\ {}^+H_3N-(CH_2)_6-NH_3{}^+ \end{bmatrix}$$

On heating, the nylon salt dissociates and then polymerization occurs. Conversion of the salt to polymer may be carried out either batch-wise or continuously. In a typical batch process, the salt solution is fed into an autoclave together with a small amount of acetic acid to limit the molecular weight to the desired level (generally 10 000–15 000). The reactor is sealed and the temperature raised to about 220°C; a pressure of about 2 MPa (20 atmospheres) develops. After 1–2 hours the temperature is increased to 270–280°C and steam is bled off to maintain the pressure at 2 MPa. Heating is then continued for a further 2 hours, during which period the pressure is allowed to fall to atmospheric. In some processes, the reactor is evacuated at this stage in order to obtain polymer of high molecular weight. The molten polymer is then extruded from the reactor by pumping in nitrogen. (The melt must be protected from oxygen which causes discoloration.) The extrudate is fed on to a water-cooled drum to form a ribbon which is then disintegrated. The product has a moisture content of about 1% at this stage.

In the continuous production of nylon 6,6, similar reaction conditions are used but the reaction mixture moves slowly through various zones of a reactor. One type of reactor in use consists of three separate tubes. In the first tube polymerization is begun with no removal of water; in the second tube steam is removed as polymerization continues; and in the third tube polymerization is completed. The product is then generally directly melt spun into fibre.

(b)   Nylons 6,9, 6,10 and 6,12

Nylons 6,9, 6,10 and 6,12 are prepared from the appropriate nylon salt by methods exactly similar to those described above for nylon 6,6. The nylon salts are prepared from hexamethylenediamine and azelaic, sebacic and dodecanedioic acids respectively.

(c)   Nylon 11

Nylon 11 is produced continuously by heating ω-aminoundecanoic acid at 200–220°C with continuous removal of water. The latter stages of the reaction are conducted under reduced pressure in order to produce polymer of high molecular weight. In addition to the formation of polymer, about 0.5% of the 12-membered ring lactam is produced by intramolecular condensation but this is not normally removed since its presence has little effect on the properties of the polymer.

(d)   Nylon 6

Both batch and continuous processes are used for the production of nylon 6. In a typical batch process, a mixture of caprolactam, water (5–10% by weight) and acetic acid (about 0.1%) is fed into a reactor which has been purged with nitrogen (to prevent discoloration by oxygen). The mixture is heated at about 250°C for 12 hours; a pressure of about 1.5 MPa (15 atmospheres) is maintained by venting off steam. The product is then extruded as ribbon, quenched in water and chopped into chips. At this point the material consists of high molecular weight polymer (about 90%) and low molecular weight compounds (about 10%); the latter are caprolactam (mainly), higher lactams and amino acids. In order to obtain the best physical properties in the final product, the low molecular weight compounds are removed either by leaching with water at about 85°C or by heating at about 180°C at 5 kPa (0.05 atmosphere).

In the continuous production of nylon 6, similar reaction conditions are used. In one process, a mixture of molten caprolactam, water and acetic acid is pumped continuously to a reactor operating at about 260°C. The feed slowly traverses through the reactor whilst steam is bled off so as to maintain atmospheric pressure. Residence time is 18–20 hours. The product is stripped of low molecular weight material by heating *in vacuo* and then directly spun into fibre

The water-initiated polymerization of caprolactam is believed to proceed according to the following scheme:

Initiation

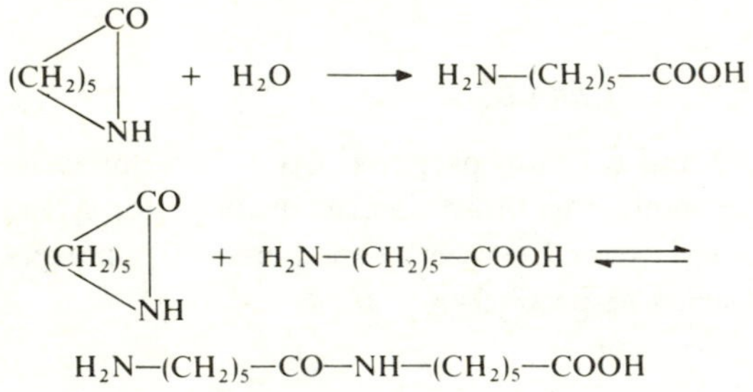

$$(CH_2)_5 \begin{array}{c} CO \\ | \\ NH \end{array} + H_2O \longrightarrow H_2N-(CH_2)_5-COOH$$

$$(CH_2)_5 \begin{array}{c} CO \\ | \\ NH \end{array} + H_2N-(CH_2)_5-COOH \rightleftharpoons$$

$$H_2N-(CH_2)_5-CO-NH-(CH_2)_5-COOH$$

Propagation

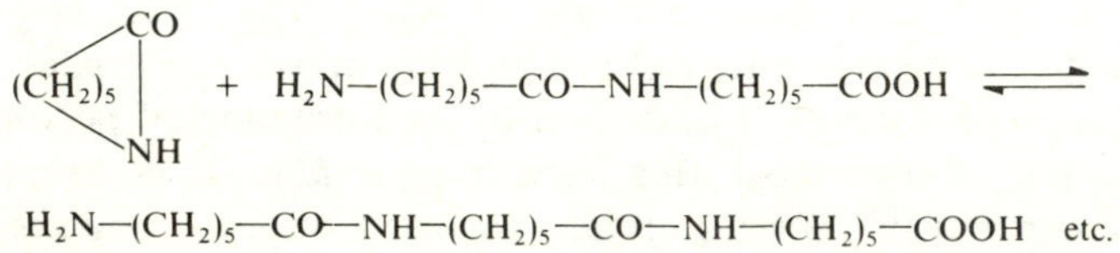

$$(CH_2)_5 \begin{array}{c} CO \\ | \\ NH \end{array} + H_2N-(CH_2)_5-CO-NH-(CH_2)_5-COOH \rightleftharpoons$$

$$H_2N-(CH_2)_5-CO-NH-(CH_2)_5-CO-NH-(CH_2)_5-COOH \quad \text{etc.}$$

Besides the amide interchange reaction shown above, propagation may also involve reaction of amino and carboxyl end-groups:

$$\sim CO-(CH_2)_5-NH_2 + HOOC-(CH_2)_5-NH \sim \rightleftharpoons$$

$$\sim CO-(CH_2)_5-NH-CO-(CH_2)_5-NH \sim + H_2O$$

The same mechanism applies to initiation with either nylon 6,6 salt or a preformed amino acid, both of which have been used extensively for this purpose.

A rather different technique for the preparation of nylon 6 consists of polmerization *in situ* in a mould. In this process, rapid polymerization is achieved by the use of anionic initiators. Anionic polymerization of caprolactam is effected by strong bases $(B^- M^+)$ such as metal amides, metal hydrides and alkali metals according to the following scheme:

Initiation

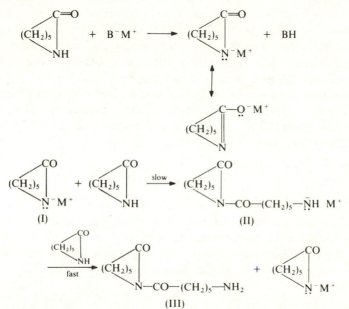

Propagation

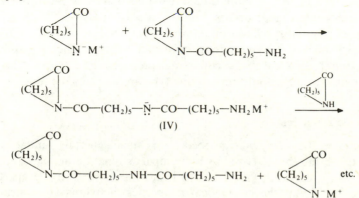

The second step in the initiation sequence is energetically unfavourable and very slow because the lactam anion (I) is stabilized by conjugation whereas the amine anion (II) which is produced is not so stabilized. The slow reaction in the initiation sequence is responsible for the induction period which occurs in the polymerization of lactams by strong bases alone. Once the active primary amine anion (II) is formed it rapidly undergoes proton-exchange to reform the stable lactam anion and an acyllactam (III). Compounds of the latter type readily participate in the propagation reaction since the amide anion (IV) which is produced is stabilized by conjugation. The induction period mentioned above may therefore be eliminated by the addition of preformed N-acyllactam or related compound at the start of the reaction. In the polymerization casting of caprolactam, a typical initiating system consists of a mixture of the sodium salt of caprolactam and N-acetylcaprolactam. The reaction temperature is initially about 150°C but during polymerization it rises to about 200°C.

Anionic polymerization of caprolactam is utilized in casting processes, by which means it is possible to produce very large mouldings and in reaction injection moulding (RIM), in which liquid components are mixed by high-pressure impingement and then injected into a mould.

(e) Nylon 12

Nylon 12 is produced by heating dodecyl lactam at about 300°C in the presence of aqueous phosphoric acid. The polymerization of dodecyl lactam does not involve an equilibrium reaction (unlike the polymerization of caprolactam) and hence an almost quantitative yield of polymer is obtained and the removal of low molecular weight material is unnecessary.

(f) Nylon copolymers

When a blend of two or more different nylons is heated above the melting point, amide interchange occurs. Initially, block copolymers are formed but prolonged reaction results in random copolymers. For example, a blend of nylons 6,6 and 6,10 gives a random copolymer after being heated for 2 hours; the product is identical to the copolymer prepared directly from the mixed monomers. Various copolymers of this type are available commercially.

10.2.4 Properties

The various types of nylon described above have generally similar physical properties, being characterized by high impact strength, toughness, flexibility and abrasion resistance. (See Table 10.1 for some comparative properties.) It may be noted that the mechanical properties of nylons are considerably

Table 10.1 Comparative properties of typical commercial grades of nylons

	6,6	6	6,9	6,10	6,12	11	12	6,6–6,10 (35:65)	6,6–6,10–6 (40:30:30)
Specific gravity	1.14	1.13	1.09	1.09	1.07	1.04	1.02	1.08	1.09
Crystalline melting point (°C)	264	215	205	215	210	185	175	195	160
Tensile strength (MPa)	79	76	59	59	59	38	46	38	52
(1bf/in^2)	11 500	11 000	8500	8500	8500	5500	6600	5500	7500
Elongation at break (%)	80–100	100–200	–	100–150	150–300	280–300	200	200	300
Impact strength, Izod (J/m)	27–53	32–53	59	85–107	53–101	96	101	107	–
(ft 1bf/in)	0.5–1.0	0.6–1.0	1.1	1.6–2.0	1.0–1.9	1.8	1.9	2.0	–
Heat, distortion temperature at 1.86 MPa (264 1bf/in^2) (°C)	75	60	60	55	65	55	50	–	30
Water absorption at saturation (%)	8.0	9.0	–	2.4	3	2.0	1.8	6.5	10.7

affected by the amount of crystallization in the test piece, temperature and humidity and it is necessary to control carefully these factors in the determination of comparative properties. The principal structural difference between the various types of nylon is in the length of the aliphatic chain segments separating adjacent amide groups. The amide groups lead to intermolecular hydrogen bonding so that crystallization readily occurs to give materials of high melting point and tensile strength. As the length of the aliphatic segment increases there is a reduction in melting point (which facilitates processing), tensile strength, heat distortion temperature and water absorption and an increase in elongation and impact strength. This effect is illustrated in Table 10.1 where a general progression in the values of the properties given may be observed. Copolymerization results in polymer chains of irregular structure with reduced ability to crystallize; the copolymers thus have lower melting points and tensile strengths than the corresponding homopolymers. The selection of a particular nylon for a given application generally involves consideration of the relative importance of mechanical properties, water resistance and ease of processing. For example, textile fibres are usually prepared from nylons 6,6 and 6 since these polymers have the highest tensile strengths. Monofilaments for use in such applications as brushes, sports equipment and surgical sutures are normally of nylons 6,10 and 11 since these polymers have greater flexibility and water resistance. Extruded products such as cable sheathing and tubing are often of nylons 11 and 12 and nylon copolymers because of ease of processing. All the nylons may be injection moulded and mouldings have found use in engineering applications such as bearings, cams and gears. The nylons have reasonably good electrical insulation properties at low frequencies and in dry conditions. Because of their polar structure nylons are not good insulators at high frequencies and since water is absorbed they are generally unsatisfactory in humid conditions.

Because of their crystallinity, the nylons are soluble at room temperature only in liquids capable of interaction with the polymer. Thus nylons dissolve in strong proton donors such as acetic acid, formic acid and phenols. Liquids of similar high solubility parameter, e.g. alcohols, have some swelling action and even readily dissolve some copolymers. Nitrobenzene, benzyl alcohol and glycols are effective solvents at elevated temperatures; towards most other organic solvents, fuels and oils nylons show outstanding resistance.

Concentrated mineral acids attack nylons rapidly at room temperature but dilute acids have a less marked effect. Resistance to alkalis is very good at room temperature but is generally reduced at elevated temperatures. Oxidizing agents such as chlorine and hydrogen peroxide attack nylons but few other inorganic reagents have any effect. Oxidation of nylons also occurs in air on exposure to ultraviolet light and/or temperatures above about 70°C; discoloration and reduction in mechanical properties occur. The following scheme may be involved in the photo-oxidation of nylons:

Initiation

$$\sim CO-NH\sim \xrightarrow{\ h\nu\ } \sim CO\cdot\ +\ \cdot NH\sim$$

$$\sim CO\cdot\ +\ \sim CO-NH-CH_2\sim \longrightarrow \sim CHO\ +\ \sim CO-NH-\overset{\cdot}{C}H\sim$$

$$\sim \overset{\cdot}{N}H\ +\ \sim CO-NH-CH_2\sim \longrightarrow \sim NH_2\ +\ \sim CO-NH-\overset{\cdot}{C}H\sim$$

Propagation

$$\sim CO-NH-\overset{\cdot}{C}H\sim\ +\ O_2 \longrightarrow \sim CO-NH-\overset{\overset{\displaystyle O-O\cdot}{|}}{C}H\sim \xrightarrow{\ \sim CO-NH-CH_2\sim\ }$$

$$\sim CO-NH-\overset{\cdot}{C}H\sim\ +\ \sim CO-NH-\overset{\overset{\displaystyle OOH}{|}}{C}H\sim$$

$$\downarrow$$

$$\sim CO-NH-\overset{\overset{\displaystyle O\cdot}{|}}{C}H\sim\ +\ HO\cdot$$

$$\sim CO-NH-CH_2\sim\ +\ \sim CO-NH-\overset{\overset{\displaystyle O\cdot}{|}}{C}H\sim \longrightarrow$$

$$\sim CO-NH-\overset{\cdot}{C}H\sim\ +\ \sim CO-NH-\overset{\overset{\displaystyle OH}{|}}{C}H\sim$$

$$\sim CO-NH-CH_2\sim\ +\ HO\cdot \longrightarrow \sim CO-NH-\overset{\cdot}{C}H\sim\ +\ H_2O$$

Termination

$$\begin{array}{l} -CO-NH-\overset{\cdot}{C}H\sim \\ \sim CO-NH-\overset{\cdot}{C}H\sim \end{array} \longrightarrow \begin{array}{l} \sim CO-NH-\overset{|}{C}H\sim \\ \sim CO-NH-\overset{|}{C}H\sim \end{array}$$

Subsequent reactions

$$\sim CO-NH-\overset{\overset{\displaystyle OH}{|}}{C}H\sim \longrightarrow \sim CO-NH_2\ +\ \overset{\overset{\displaystyle O}{\|}}{C}H\sim$$

$$\sim CHO\ +\ \tfrac{1}{2}O_2 \longrightarrow \sim CO_2H$$

Evidence for this scheme comes largely from studies of the products obtained by degradation of model amides such as *N*-pentylhexanamide [1].

It is possible to bring about chemical modification of nylons by reaction of the amide group. A useful reaction of this type is methoxymethylation, which may be accomplished by dissolving the nylon in a solvent such as formic acid (90%) and treating the solution with formaldehyde and methanol in the presence of an acid catalyst, e.g. phosphoric acid:

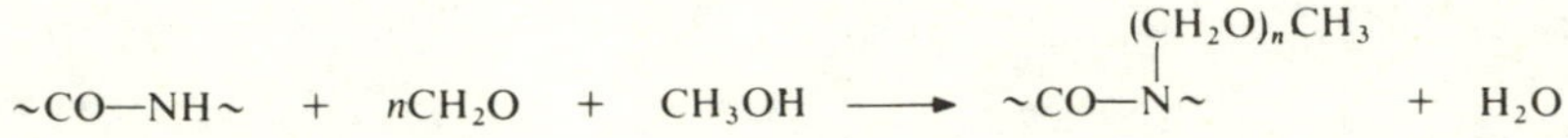

$$\sim CO-NH\sim \quad + \quad nCH_2O \quad + \quad CH_3OH \quad \longrightarrow \quad \overset{\overset{\displaystyle (CH_2O)_nCH_3}{|}}{\sim CO-N\sim} \quad + \quad H_2O$$

where $n = 1$ or 2

Methylmethoxy nylons in which about 33% of the amide groups have been substituted are commercially available. Such polymers are soluble in alcohols and may be cross-linked by heating at about 120°C in the presence of an acid catalyst, e.g. citric acid:

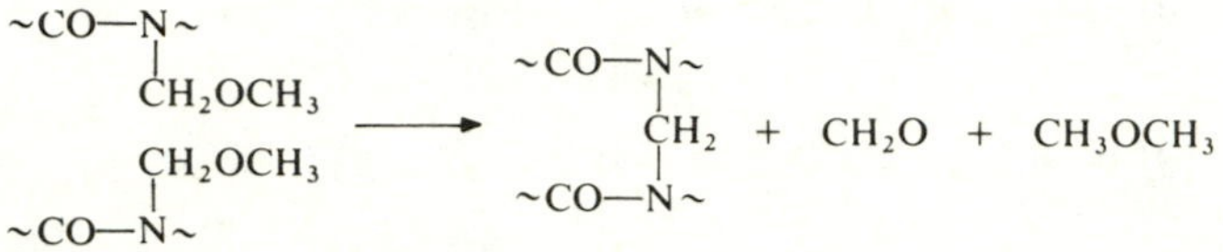

$$\begin{array}{c} \sim CO-N\sim \\ \quad | \\ \quad CH_2OCH_3 \\ \\ \quad CH_2OCH_3 \\ \quad | \\ \sim CO-N\sim \end{array} \quad \longrightarrow \quad \begin{array}{c} \sim CO-N\sim \\ \quad | \\ \quad CH_2 \\ \quad | \\ \sim CO-N\sim \end{array} \quad + \quad CH_2O \quad + \quad CH_3OCH_3$$

These materials find limited application in coatings where good abrasion and flexing resistance are required. It may be noted that nylons may be *N*-methylolated by treatment with formaldehyde but the products are too unstable to be of commercial use.

10.3   POLYAMIDE THERMOPLASTIC ELASTOMERS

As discussed previously, thermoplastic elastomers are materials which have the functional properties of conventional vulcanized rubbers but which may be processed as normal thermoplastics (see section 2.9).

Polyamide thermoplastic elastomers are block copolymers made up of alternating hard and soft segments. The hard segments are polyamide blocks and the soft segments are either aliphatic polyether or polyester blocks. A typical product has the following structure:

$$-\left[-NH-(CH_2)_{10}-CO- \right]_x \left[-O-\left[(CH_2)_4-O- \right]_n \right]_y -$$

A copolymer of this kind is synthesized in two steps. In the first step, a low molecular weight polyamide ($M_n = 300-15\,000$) with terminal carboxyl groups is prepared. (In the above example, nylon 11 is shown as the polyamide block but any nylon may be used.) In the second step, the carboxyl-terminated polyamide is treated with a polymeric diol ($M_n = 100-6000$). (In the above example, poly(oxytetramethylene) glycol (section 9.4.6) is used to obtain a polyether block but a wide variety of diols may be used.)

The properties of polyamide thermoplastic elastomers depend on the type and length of the two constituent blocks. By appropriate selection of the segments, mechanical, thermal and chemical properties can be regulated within a broad range. Generally, polyamide thermoplastic elastomers are high-cost materials characterized by high tensile strength, good recovery and flex fatigue properties, good abrasion resistance and ability to withstand service temperatures up to 150°C. Applications include hoses, gaskets and seals and high-performance athletics goods such as ski-boots and footballs.

10.4 FATTY POLYAMIDES

Fatty polyamides are products obtained by the reaction of di- and polyfunctional amines with polybasic acids prepared from unsaturated vegetable oil acids. Suitable amines include ethylenediamine and diethylenetriamine whilst the acids are prepared as follows. The mixture of fatty acids obtained by the saponification of unsaturated vegetable oils such as linseed, soyabean and tung oils is heated for several hours at about 300°C (or at lower temperatures in the presence of a catalyst); the resulting mixture is then heated under reduced pressure to remove the more volatile components (mainly monobasic acids). The residue, which is termed *dimer acid*, is used for the production of fatty polyamides. This material is a complex mixture consisting principally of dimerized fatty acids (60–75%) together with lesser amounts of trimerized acids and higher polymers. The structures of all the components in such mixtures have not been resolved but it is believed that the main constituents are substituted cyclohexanes, which arise through Diels-Alder reactions. The following dimerization of linoleic acid illustrates the kind of process which is probably involved:

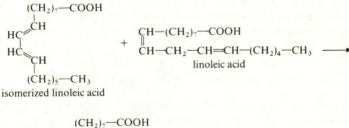

The formation of a fatty polyamide from the above dimer is illustrated by the following reaction involving ethylenediamine:

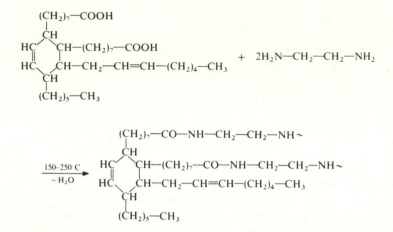

The fatty polyamides prepared from dimer acids are of two main types, namely solid polymers and liquid polymers. The solid polymers are largely linear products obtained by reaction of dimer acids and diamines and have molecular weights in the range 2000–15 000. The liquid polymers are highly branched products of lower molecular weight which result from the interaction of dimer acids and polyamines containing three or more amino groups; a stoichiometric excess of amine is used so that the polymers contain free amino groups. Solid fatty polyamides find use in such applications as coatings for flexible substrates (e.g. paper, plastic films and aluminium foil); the coatings are heat-sealable and relatively impermeable to water vapour. The principal use of liquid fatty polyamides is in conjunction with epoxy resins; the interaction of the free amino groups of the polyamides with epoxy groups forms the basis of this use (section 18.3.3(b)).

10.5   AROMATIC POLYAMIDES

Although aromatic polyamides are characterized by great thermal stability and strength, commercial development has been relatively slow largely because of production and processing difficulties. The former are due to such factors as slow reaction rates of aromatic amines and tendency to discolour during polymerization whilst the latter arise because the extensive decomposition which generally accompanies fusion of the polymers precludes melt-processing. Most interest has centred on fibre-forming aromatic polyamides (aramids) and commercial eminence has been achieved by two such polymers, namely poly(m-phenylene isophthalamide) and poly(p-phenylene terephthalamide). Both of these products were introduced by du Pont, the first in the late 1960s and the second in 1972.

10.5.1 Poly(*m*-phenylene isophthalamide)

Poly(*m*-phenylene isophthalamide is produced by reaction of isophthaloyl chloride and *m*-phenylenediamine:

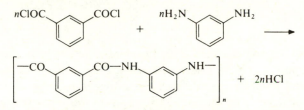

This reaction occurs rapidly in the presence of an acid acceptor under mild conditions. The conventional melt polymerization techniques, as used for the preparation of nylons, cannot be applied to aromatic polyamides since the melting points of the polymers are too high. Polymerization is therefore conducted either in solution (e.g. in methylene chloride) or in suspension. In the latter case, the diamine is dissolved in water, together with an acid acceptor (e.g. sodium carbonate) and the diacid chloride is dissolved in a solvent which is immiscible with water (e.g. carbon tetrachloride or cyclohexanone). The two solutions are then subjected to intensive mixing. Rapid reaction occurs at the liquid interface or just inside the solvent boundary and this technique is therefore commonly termed *interfacial polymerization*.

Poly(*m*-phenylene isophthalamide) has an extremely high melting point (380–390°C) and cannot be melt processed by the usual means. Commercial material is supplied as fibre and as a paper and is used directly in these forms. Fibre is prepared by extruding a solution of the polymer in a mixture of dimethylformamide and lithium chloride into hot air. The aromatic polyamide papers are produced from a combination of chopped fibres and chopped film (prepared continuously by interfacial polymerization).

At ordinary temperatures, poly(*m*-phenylene isophthalamide) has mechanical properties comparable to those of the aliphatic nylons but at elevated temperatures the aromatic polymer is greatly superior. Broadly, the mechanical properties of poly(*m*-phenylene isophthalamide) show little change up to about 200°C. The aromatic polymer also shows little change in electrical insulation properties over a similar temperature range. The polymer resists ignition and is free of after-glow. The chemical resistance of poly(*m*-phenylene isophthalamide) appears to be generally similar to that of the aliphatic nylons except that resistance to mineral acids is rather better. The fibre is used for heat and flame protective clothing. The paper has found use in electrical insulating applications where resistance to elevated temperatures is required.

10.5.2 Poly(*p*-phenylene terephthalamide)

Poly(*p*-phenylene terephthalamide) is produced by reaction of terephthaloyl chloride and *p*-phenylenediamine:

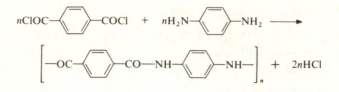

Full details of the manufacture of the polymer have not been disclosed. The reaction is reported to be carried out in a solvent mixture of hexamethylphosphoramide and N-methylpyrrolidone. The polymer is supplied in fibre form.

Poly(p-phenylene terephthalamide) is an extraordinarily strong material. The specific tensile strength (tensile strength/density) of the fibre is higher than that of any other continuous fibre commercially available. The polymer does not melt but begins to carbonize at about 425°C. The material retains its mechanical properties over a wide temperature range for prolonged periods. There is little change up to about 180°C and even at 250°C approximately half of the initial tensile strength is retained after 100 hours. The material is attacked by mineral acids and alkalis and is also sensitive to ultraviolet light.

Poly(p-phenylene terephthalamide) fibres currently have three major applications, namely as reinforcement in radial tyres and mechanical rubber goods, in ballistic protective fabrics and ropes, and as reinforcement in polymer (particularly epoxy) composites for aircraft and aerospace components.

10.6   PROTEINS

Proteins are naturally occurring polyamides. They contain recurring amide groups and are polymers of amino acids. They differ from the synthetic polyamides described previously in this chapter in that they have a much more complicated molecular architecture. Whereas the structures of synthetic polyamide polymers such as nylons 6 and 11 are each built up of the residues of only one amino acid, protein structures are each composed of several different amino acid residues. Commonly, a protein structure contains the residues of about 15 different amino acids, with 3 or 4 making up the bulk of the polymer. Hydrolysis of proteins yields the free amino acids upon which the polymer is based:

$$\sim NH{-}R{-}CO{-}NH{-}R'{-}CO{-}NH{-}R''{-}CO \sim \xrightarrow{H_2O} H_2N{-}R{-}COOH \; +$$

$$H_2N{-}R'{-}COOH \; + \; H_2N{-}R''{-}COOH \; \text{etc}$$

A total of some 25 amino acids has been obtained by hydrolysis of proteins. All of these amino acids, except two, are α-amino acids; the exceptions are proline and hydroxyproline which are imino acids. The more important of these amino acids are shown in Table 10.2. It may be noted that amino acids with an equal number of amino and carboxyl groups are called *neutral* amino

Table 10.2 Amino acids from proteins

Neutral amino acids

Aliphatic

$$\underset{\text{glycine}}{\overset{\overset{\displaystyle NH_2}{|}}{CH_2-CO_2H}} \qquad \underset{\text{alanine}}{\overset{\overset{\displaystyle NH_2}{|}}{H_3C-CH-CO_2H}} \qquad \underset{\text{valine}}{\overset{\overset{\displaystyle NH_2}{|}}{(CH_3)_2CH-CH-CO_2H}}$$

$$\underset{\text{leucine}}{\overset{\overset{\displaystyle NH_2}{|}}{(CH_3)_2CH-CH_2-CH-CO_2H}} \qquad \underset{\text{isoleucine}}{\overset{\overset{\displaystyle CH_3\ NH_2}{|\ \ \ |}}{H_3C-CH_2-CH-CH-CO_2H}}$$

$$\underset{\text{serine}}{\overset{\overset{\displaystyle OH\ \ NH_2}{|\ \ \ \ |}}{CH_2-CH-CO_2H}} \qquad \underset{\text{threonine}}{\overset{\overset{\displaystyle OH\ \ NH_2}{|\ \ \ \ |}}{H_3C-CH-CH-CO_2H}}$$

$$\underset{\text{cysteine}}{\overset{\overset{\displaystyle SH\ \ \ NH_2}{|\ \ \ \ |}}{CH_2-CH-CO_2H}} \qquad \underset{\text{methionine}}{\overset{\overset{\displaystyle CH_3S\ \ \ \ \ NH_2}{|\ \ \ \ \ \ \ \ \ |}}{CH_2-CH_2-CH-CO_2H}}$$

$$\begin{array}{c} \overset{\displaystyle NH_2}{|} \\ CH_2-CH-CO_2H \\ | \\ S \\ | \\ S\ \ \ \ \overset{\displaystyle NH_2}{|} \\ CH_2-CH-CO_2H \\ \text{cystine} \end{array}$$

Aromatic

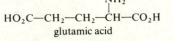

$$\underset{\text{phenylalanine}}{\text{C}_6\text{H}_5-CH_2-\overset{\overset{\displaystyle NH_2}{|}}{CH}-CO_2H} \qquad \underset{\text{tyrosine}}{HO-C_6H_4-CH_2-\overset{\overset{\displaystyle NH_2}{|}}{CH}-CO_2H}$$

Heterocyclic

proline   hydroxyproline   tryptophan

Acidic amino acids

Aliphatic

$$\underset{\text{aspartic acid}}{HO_2C-CH_2-\overset{\overset{\displaystyle NH_2}{|}}{CH}-CO_2H} \qquad \underset{\text{asparagine}}{H_2N-CO-CH_2-\overset{\overset{\displaystyle NH_2}{|}}{CH}-CO_2H}$$

$$\underset{\text{glutamic acid}}{HO_2C-CH_2-CH_2-\overset{\overset{\displaystyle NH_2}{|}}{CH}-CO_2H} \qquad \underset{\text{glutamine}}{H_2N-CO-CH_2-CH_2-\overset{\overset{\displaystyle NH_2}{|}}{CH}-CO_2H}$$

Table 10.2   (Continued)

Basic amino acids

Aliphatic

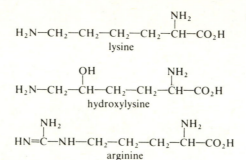

$$\underset{\text{lysine}}{H_2N-CH_2-CH_2-CH_2-CH_2-\overset{\displaystyle NH_2}{\overset{|}{CH}}-CO_2H}$$

$$\underset{\text{hydroxylysine}}{H_2N-CH_2-\overset{\displaystyle OH}{\overset{|}{CH}}-CH_2-CH_2-\overset{\displaystyle NH_2}{\overset{|}{CH}}-CO_2H}$$

$$\underset{\text{arginine}}{\overset{\displaystyle NH_2}{\overset{|}{HN=C}}-NH-CH_2-CH_2-CH_2-\overset{\displaystyle NH_2}{\overset{|}{CH}}-CO_2H}$$

Heterocyclic

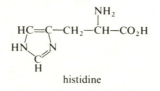

histidine

NOTE: α-Aminoacids normally exist as dipolar ions; the structures given in this table are used for convenience.

acids, whereas those with a greater number of carboxyl (or amide) groups are called *acidic* and those with a greater number of amino (or imino) groups are called *basic*. Proteins are highly stereospecific and the amino acids derived from them always have the L-configuration (unless, of course, the conditions of hydrolysis are such as to cause racemization). Quantitative analysis of the amino acid mixture given by a protein on hydrolysis indicates the relative frequency of the various residues in the protein. If the molecular weight of the protein also is known, the actual numbers of each type of amino acid residue involved can be computed. However, these measurements give no indication of the sequence of the amino acid residues along the protein chain. Since the molecular weights of proteins range from several thousand to several million, it will be appreciated that the number of possible arrangements of amino acid residues in a protein is very large. For example, a mere 10 different residues can be arranged in 3 628 800 different sequences. Nevertheless, the sequence of amino acid residues has been determined for several proteins.

Proteins are commonly classified into two main classes, viz. *fibrous* and *globular proteins*. The former are the principal structural components of such materials as feathers, hair, nails and silk; they consist of essentially linear polymers which are not dispersible in water. Globular proteins occur in materials such as egg-white, blood and milk; they are relatively compact cross-linked polymers and are dispersible in water.

A few proteins have found technological use as adhesives, fibres and plastics and are described briefly in the following sections.

10.6.1   Animal glues

Animal glues are most commonly manufactured from the bone or hide of cattle and sheep. In the preparation of bone glue, the bones are firstly degreased by exposure to the vapour of boiling benzene. The bones are then transferred to autoclaves and treated alternately with steam under pressure and hot water, in several cycles. As a result of this treatment the protein in the bone, collagen, is degraded and passes into solution. The aqueous solution is concentrated in a vacuum evaporator and allowed to form a gel which is then dried and pulverized. Hide glue is prepared in a similar fashion. Animal glues are applied as aqueous solutions and 'set' by loss of water. The glues are much used in furniture and other wood assemblies; glues containing glycerol are flexible and are used, for example, in book binding. It may be noted that if bone and hide are subjected to rather milder hydrolysis than that used for the preparation of glue, a higher molecular weight product is obtained. This product, gelatin, is used for edible and photographic puposes. The amino acids found in the hydrolysate of collagen are given in Table 10.3.

10.6.2   Blood albumin glue

In the preparation of blood albumin glue, blood serum is evaporated under reduced pressure and low temperature to give a powder which is water-soluble. It is essential to avoid coagulation during this process since coagulated albumin is insoluble and has little bonding power. The soluble blood powder is then dissolved in aqueous alkali to give the glue, which is extensively used in the hot pressing of interior grades of plywood.

10.6.3   Casein

Casein occurs in several animal and vegetable materials but the only source of commercial importance is cow's milk, in which the casein content is about 3%. In commercial practice, casein is isolated from skimmed milk by either acid coagulation or by rennet coagulation. In acid coagulation, dilute sulphuric acid is added to the milk at about 35°C, with stirring. Coagulation

Table 10.3　Amino acids found in the hydrolysates of some proteins [2]

	Casein	Collagen	Fibroin	Wool keratin
Glycine	30	363	581	87
Alanine	43	107	334	46
Valine	54	29	31	40
Leucine	60	28	7	86
Isoleucine	49	15	8	–
Serine	60	32	154	95
Threonine	41	19	13	54
Cystine	2	0	–	49
Methionine	17	5	–	5
Phenylalanine	28	15	20	22
Tyrosine	45	5	71	26
Proline	65	131	6	83
Hydroxyproline	0	107	0	0
Aspartic acid + asparagine	63	47	21	54
Glutamic acid + glutamine	153	77	15	96
Lysine	61	31	5	19
Hydroxylysine	0	7	0	0
Arginine	25	49	6	60
Tryptophan	8	0	0	9
Histidine	19	5	2	7

Figures given are amino acid residues per molecular weight of 100 000 i.e. mole/10^5 g.

occurs at pH 4.6, the isoelectric point; excess of acid is avoided in order to limit hydrolysis of the protein. The mixture is then heated to about 50°C to give a more granular solid, which is filtered off, washed and dried at 60°C. Rennet coagulation is carried out in a similar manner. About 0.03% by weight of rennet (an enzyme obtained from calf's stomach lining) is added to milk at pH 6 and about 40°C; after coagulation, the temperature is raised to about 65°C and then the solid is filtered off, washed and dried.

Acid casein and rennet casein differ principally in their inorganic content. This non-protein material consists mainly of calcium and phosphorus compounds. Acid casein has an ash content of 1–2% whilst the value for rennet casein is 7–9%. The presence of these inorganic materials markedly affects the technological properties of casein, as is discussed later. The amino acids found in the hydrolysate of casein are given in Table 10.3.

Acid casein finds use in a variety of applications, including adhesives (mainly for paper and wood) and paper coatings. In these latter uses, the casein is applied as a dispersion in aqueous alkali. Casein fibre may be produced by forcing an alkaline dispersion through a spinneret into an acid coagulating bath; the fibre (commonly termed 'casein wool') was produced commercially during World War II in Italy and the USA but it is now of little importance. Acid casein is used rather than rennet casein in these various applications since it is more readily dispersed in aqueous alkali.

Rennet casein is preferred to acid casein for the production of solid articles on the grounds of easier processability. The casein is firstly mixed with about 20% water together with colourants and then passed through an extruder. Under the influence of heat and pressure the mixture is converted into a rubbery material which is extruded as rod. Most commonly, the rod is chopped into discs (for use as button blanks) but sheet may be made by laying rods in a mould and pressing. At this stage the material is soft, very water-sensitive and liable to putrefy; it is converted into a hard, stable substance which is less sensitive to water by treatment with formaldehyde. Formoliz-ation is carried out by immersing the casein in formalin (4–5%) at about 60°C for a period ranging from 2 days to several months, depending on the thickness of the section. During this period the casein becomes cross-linked, possibly through amide groups present in the main polymer chain and pendant amine groups of lysine residues:

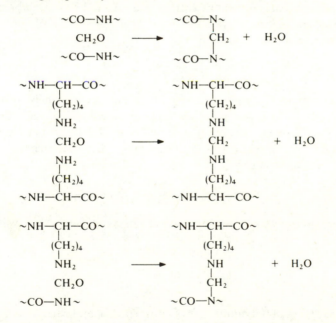

The formolization reaction must be carried out slowly to avoid rapid hardening of the surface of the section which would reduce permeability and make it difficult to cure the centre of the material. The hardened material contains large quantities of water which must be removed slowly to prevent cracking and warping; drying is thus carried out in humid air at about 45°C over a period of several days. The formolized material is then machined and polished to give the finished article. The use of casein for the preparation of solid articles has now virtually disappeared in the face of competition from synthetic polymers.

10.6.4 Soyabean glues

Soyabean glues are based on the residue (flour) remaining after the oil has been extracted from the bean. This residue contains 45–50% protein, the remainder being mainly carbohydrates and cellulose. The flour may be dispersed in aqueous alkali to give a glue used in the production of interior grades of plywood. Alternatively, the protein content of the flour may be isolated by alkaline extraction and use for paper coating.

10.6.5 Silk

Silk is produced by the larvae of the silk moth *Bombyx mori*. The fibrous material of the cocoon is composed of at least two proteins, sericin and fibroin. The sericin is present as a sticky coating on the fibroin fibres (silk) and is removed by treatment with hot water. The composition of the hydrolysate of fibroin is given in Table 10.3.

10.6.6 Wool

Wool is composed largely of the protein keratin, which is also found in feathers, hair, horn, nails and scales. The composition of the hydrolysate of wool keratin is given in Table 10.3. It will be seen that this protein has a high content of cystine; wool thus contains many disulphide cross-links which render the fibre insoluble.

10.7 POLYIMIDES

As indicated previously, polyimides are characterized by a repeating imide group:

Two basic types of polyimides are of commercial significance. In the first type, which are unmodified polyimides, imide is the principal functional group present. In the second type, which are modified polyimides, other functional groups are also present; modified polyimides include poly(amide-imide)s, polyesterimides, polyetherimides and polybismaleimides. The development of this large variety of modified polyimides has been stimulated by the desire to find materials which are more readily processable than the unmodified polyimides whilst retaining the same superior properties.

10.7.1 Unmodified polyimides

The classic method of polyimide synthesis is by reaction of a dianhydride and a diamine. The first commercial polyimide (du Pont, 1961) was based on pyromellitic dianhydride and 4,4'-diaminodiphenyl ether:

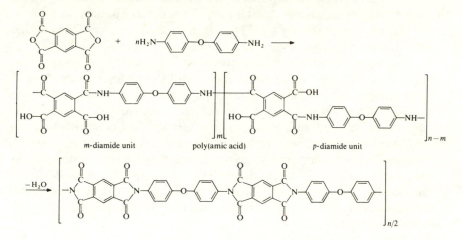

m-diamide unit   poly(amic acid)   p-diamide unit

The formation of poly(amic acid), which is the first step in this synthesis, is a fast exothermic reaction which is generally carried out at room temperature or slightly above in a solvent for the polymer. Under these conditions the extent of imidization is negligible. In a typical process, the diamine is dissolved in dimethylformamide, dimethylacetamide or dimethyl sulphoxide and the dianhydride is added portion-wise with cooling. A viscous solution of the poly(amic acid) is formed. The poly(amic acid) contains a mixture of m- and p- diamide units as shown above. The second step in the preparation of the polyimide is cyclization of the poly(amic acid). In addition to intramolecular condensation leading to the linear polymer shown above, some intermolecular condensation occurs to give a cross-linked structure. Thus the final product is insoluble and infusible and it is thus necessary to convert the poly(amic acid) to the polyimide in that physical form in which the final polymer is desired. For example, in the preparation of film the poly(amic acid) solution is cast and then heated firstly to about 150°C and finally to 300°C. During imidization, water is formed and may cause scission of the polymer chains, particularly in the casting of thick film. The addition of acetic anhydride and pyridine (catalyst) to the poly(amic acid) solution results in the effective removal of water from the reaction system and prevents molecular weight reduction.

The outstanding property of aromatic polyimides is their thermal stability. Films of the polyimide prepared from pyromellitic dianhydride and 4,4'-diaminophenyl ether are stable in air up to about 420°C; above this temperature volatilization occurs. Mechanical and electrical insulation properties are maintained after prolonged exposure to elevated temperatures; for example, after 1000 hours in air at 300°C the polymer retains 90% of its initial tensile strength and 90% of its initial dielectric strength. At room temperature the physical and electrical properties of the polyimide film are comparable to those of poly(ethylene terephthalate) film. The resistance of the polyimide to most organic solvents is very good but the polymer is somewhat hydrophilic and is degraded by aqueous acids and alkalis.

The principal drawbacks of this polyimide are the impossibility of processing by conventional methods and the difficulty of removing water formed during imidization and residual solvent; these by-products cause voids in the finished material unless completely removed. Nevertheless, the polyimide finds use in such applications as coatings for electrical components, insulating film for electric motors and cables and moulded engineering parts.

One method of circumventing the somewhat cumbersome two-step process described above is to replace the diamine by a diisocyanate, and a polyimide based on benzophenone tetraacid dianhydride and 4,4′-diaminodiphenylmethane diisocyanate is commercially available. The polymer has the following structure:

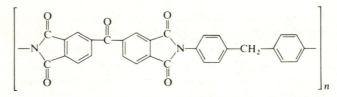

This type of polymer does not contain crosslinks and can be processed by compression moulding. The polymer is soluble and films and fibres can be prepared from solutions.

10.7.2   Poly(amide-imide)s

Poly(amide-imide)s are formed when a triacid derivative is treated with a diamine. The resulting polymer contains both amide and imide links. One such product is prepared from trimellitic anhydride (as the acid chloride) and 4,4′-diaminodiphenylmethane as shown on page 219.

Poly(amide-imide)s may be processed by conventional means such as injection moulding and extrusion. The polymers are somewhat less heat resistant than straight polyimides but useful mechanical properties are maintained up to about 250°C. Poly(amide-imide)s have found extensive use in the aerospace industry, affording weight reduction by replacing metals in such components as jet engine parts and electrical devices. The polymers are also used for automotive parts.

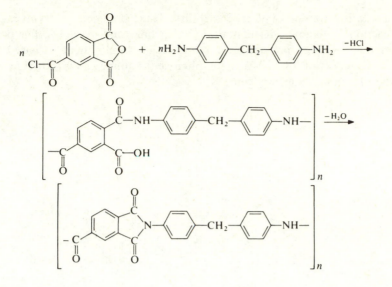

10.7.3 Polyesterimides

Polyesterimides are synthesized by reaction of dianhydrides containing ester links and diamines. A typical polyesterimide is prepared by treating trimellitic acid anhydride with bis(2-hydroxyethyl) terephthalate to give an intermediate product with terminal anhydride groups:

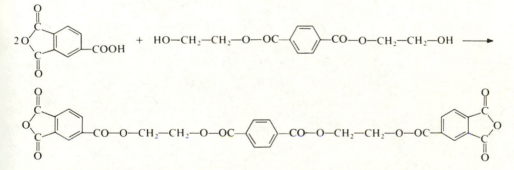

The intermediate dianhydride is then treated with a diamine such as 4,4′-diaminodiphenylmethane to give a poly(amic acid) which is imidized as described in section 10.7.1. Polyesterimides are used mainly for heat resistant wire coatings.

10.7.4 Polyetherimides

Polyetherimides are the most recently introduced polyimides (General Electric Co., 1982) and probably represent the most successful combination of polyimide performance and processability. These polymers are synthesized by reaction of dianhydrides containing ether links and diamines. The poly-etherimide which is preferred for commercial development is based on bisphenol A (section 18.3.1 (a)) and *m*-phenylenediamine and is thought to be produced by the following route:

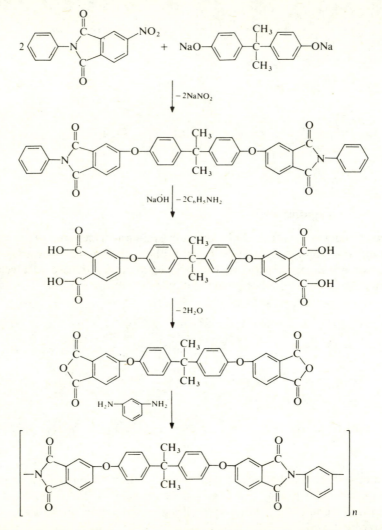

This polyetherimide is characterized by high strength, rigidity and dimen-sional stability and good electrical insulation properties. These properties are largely maintained after prolonged exposure to temperatures up to about 180°C. The polymer is resistant to a wide range of solvents including alcohols and fully halogenated compounds; partially halogenated compounds can be good solvents. The polymer is resistant to mineral acids but is attacked by alkalis. The polyetherimide is readily processed by standard methods such as injection moulding, extrusion and thermoforming. Applications include electrical/electronic components, jet engine parts and computer components.

10.7.5 Polybismaleimides

Polybismaleimides utilize an approach to polyimide synthesis which is different from that used for the polyimides described above. In the latter cases, imidization takes place after polymerization has occurred. In polybismaleimides, preformed imide groups are present in low molecular weight intermediates which are subsequently polymerized.

Bismaleimides are obtained from maleic anhydride (section 11.2.2 (b)) and a diamine such as 4,4'-diaminodiphenylmethane:

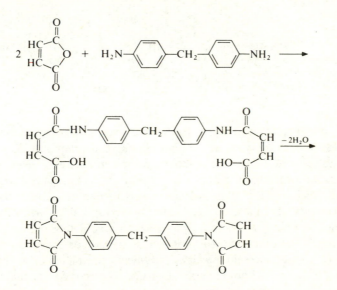

Bismaleimides can be polymerized directly through the double bonds at each end of the molecule but homopolymers prepared in this way tend to be very brittle. More useful products are obtained if chain extension involves a co-reactant. A variety of compounds will add to bismaleimides but the most commonly used are diamines, e.g. 4,4'-diaminodiphenylmethane. Usual practice is to heat the bismaleimide with a molar deficiency of diamine at about 230°C; under these conditions both addition of the amino group to the maleic double bond and direct interaction of double bonds occur simultaneously. This type of procedure is mostly used for the production of compression moulded printed circuit boards.

10.8 POLYBENZIMIDAZOLES

Polybenzimidazoles may be regarded as derivatives of aromatic polyamides and are thus conveniently considered in this chapter. They are prepared by reaction of aromatic tetraamines and dicarboxylic compounds.

The polymer which has been preferred for commercial development is that based on 3,3'-diaminobenzidine and diphenyl isophthalate and the reaction between these compounds illustrates the general method of preparation of polybenzimidazoles:

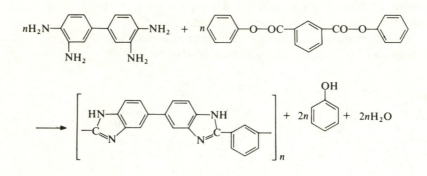

The diphenyl ester is preferred to the free acid in this reaction because the acid tends to decarboxylate at the high temperatures required for polymerization; the acid chloride is too reactive and the methyl ester leads to partial methylation of the amine groups. Polymerization is normally carried out in two stages because the high molecular weight material is intractable. In a typical process, the reactants are heated at about 300°C for 2 hours under nitrogen and then for 0.5 hours *in vacuo* to give a low molecular weight prepolymer. The prepolymer is applied as a melt or in solution and the conversion to high molecular weight polymer is completed *in situ* by heating at about 370°C for 1–3 hours under pressure (to give void-free material). The following mechanism has been proposed for this reaction [3] (see page 223).

Polybenzimidazoles are of interest chiefly because of the maintenance of physical properties at elevated temperatures. Prolonged heating in air at temperatures up to about 250°C leads to little change in properties; above 250°C, oxidative degradation results in gradual deterioration. Polybenzimidazoles have been employed in laminates and adhesives for use at high temperatures, mainly in the aerospace field.

More recently, polybenzimidazole fibre has become available commercially. The fibre is obtained by spinning from a solution of the polymer in dimethylacetamide. Suggested applications include protective clothing and dust filters for hot flue gases.

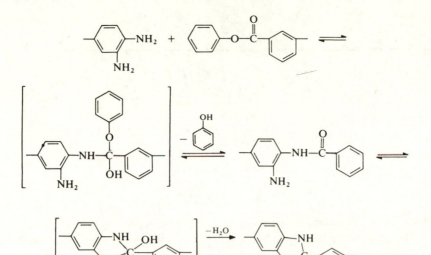

REFERENCES

1. Sharkey, W. H. and Mochel, W. E. (1959) *J. Am. Chem. Soc.*, **81**, 300.
2. Bryan, W. P. and Hein, G. E. (1969) in *Encyclopedia of Polymer Science and Technology*, Vol. 11, Interscience Publishers, New York, p. 621.
3. Gray, D. N. and Shulman, G. P. (1965) A.C.S. Symposium, Atlantic City, N.J.

BIBLIOGRAPHY

Collins, J. H. (1952) *Casein Plastics and Allied Materials*, 2nd edn, The Plastics Institute, London.
Floyd, D. E. (1962) *Polyamide Resins*, 2nd edn, Reinhold Publishing Corporation, New York.
Kohan, M. I. (ed.) (1973) *Nylon Plastics*, John Wiley & Sons, Inc., New York.
Nelson, W. E. (1976) *Nylon Plastics Technology*, Newnes-Butterworths, London.
Mittal, K. L. (ed.) (1984) *Polyimides*, Vol. 1 and Vol. 2, Plenum Press, New York.
Elias, H-G. and Vohwinkel, F. (1986) *New Commercial Polymers 2*, Gordon and Breach Science Publishers, New York, ch. 11, (Polyimides).

POLYESTERS

11.1 SCOPE

An ester is a compound whose structure may be derived by the replacement of the replaceable hydrogen of an acid by an alkyl, aryl, alicyclic or heterocyclic group. The most important esters are those derived from carboxylic acids; the structural formula of these esters is R–CO–O–R′. For the purposes of this chapter, polyesters are defined as polymers containing recurring –CO–O– groups in the main chain. It may be noted that this definition excludes polymers of esters such as vinyl acetate and methyl methacrylate since in these polymers the ester groups occur in side-chains and not in the main chain.

A large number of polyesters is commercially available. These polymers are conveniently classified into the following types
Unsaturated polyesters
Alkyds
Poly(allyl ester)s
Poly(ethylene terephthalate)
Poly(butylene terephthalate)
Cyclohexylenedimethylene terephthalate polymers
Polyester thermoplastic elastomers
Polyarylates
Poly(*p*-hydroxybenzoate)s
Polyester plasticizers

11.2 UNSATURATED POLYESTERS

11.2.1 Development

The polymers making up this first group of polyesters are linear polyesters containing aliphatic unsaturation which provides sites for subsequent cross-linking. A polymer of this type first became available in the USA in 1946; the polymer was prepared from diethylene glycol and maleic anhydride and could be cross-linked by reaction with styrene. The polymer was of interest for the preparation of glass-fibre laminates by techniques which did not involve high pressures. This application stemmed from the discovery of the reinforcing properties of glass fibres by the United States Rubber Co. (USA) in 1942. The commercial production of glass-fibre reinforced polyester laminates was firmly established by about 1949 and such laminates are widely used for large structures such as boat hulls, sports car bodies and roofing panels. During the 1950s bulk moulding compounds (BMC) were developed and in the late 1960s sheet moulding compounds (SMC) were introduced. These moulding compositions, which are essentially mixtures of unsaturated polyester, glass-fibre and filler, can be processed by the standard techniques used for thermosetting materials. During the past few years the use of polyester moulding compositions has grown significantly, particularly in the transportation field.

11.2.2 Raw materials

Linear unsaturated polyesters are prepared commercially by the reaction of a saturated diol with a mixture of an unsaturated dibasic acid and a 'modifying' dibasic acid (or corresponding anhydrides). In principle, unsaturation desired in a polyester can be derived from either an unsaturated diol or an unsaturated acid; for economic reasons the latter is invariably preferred. As mentioned previously, the unsaturated acid provides sites for subsequent cross-linking; the function of the modifying acid is to reduce the number of reactive unsaturated sites along the polymer and hence to reduce the cross-link intensity and brittleness of the final product. Some acids and anhydrides which are used to modify polyesters are, in fact, unsaturated but the double bonds are not sufficiently reactive to represent sites for subsequent cross-linking.

(a) Diols

Propylene glycol.   Propylene glycol is the diol most widely used for the manufacture of linear unsaturated polyesters; it is prepared by the hydration of propylene oxide (section 9.4.1(b)):

$$H_3C-CH-CH_2 \quad + \quad H_2O \quad \longrightarrow \quad HO-CH-CH_2-OH$$

Commonly the reaction is carried out without a catalyst at about 200°C and 2 MPa (20 atmospheres). Propylene glycol is isolated by distillation under reduced pressure; it is a colourless liquid, b.p. 189°C. Propylene glycol is the preferred diol because it forms polyesters which are compatible with styrene

(for the function of which, see later) and which show little tendency to
crystallize; it is also readily available at low cost.

Other diols.   Diols other than propylene glycol are utilized to a lesser extent,
being used to impart special properties to the polymer. For example, diethylene glycol (I) leads to greater flexibility although water-sensitivity is increased
and neopentylene glycol (2,2-dimethyl-1,3-propanediol) (II) gives polymers
with improved resistance to thermal degradation. Polyesters with good alkali
resistance are prepared from the diether of propylene glycol and bisphenol A
(2,2-bis(4'-hydroxyphenyl)propane) (III). The simplest diol, ethylene glycol,
gives polyesters with a tendency to crystallize but is sometimes used in
admixture with other diols.

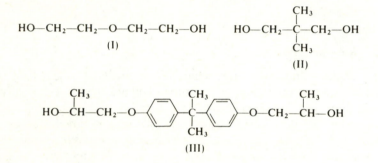

$$HO{-}CH_2{-}CH_2{-}O{-}CH_2{-}CH_2{-}OH$$
(I)

$$HO{-}CH_2{-}\underset{CH_3}{\overset{CH_3}{C}}{-}CH_2{-}OH$$
(II)

(III)

(b)   Unsaturated acids and anhydrides

Maleic anhydride.   Maleic anhydride is the most important unsaturated
component used in the manufacture of linear unsaturated polyesters. It is
mostly obtained by the oxidation of benzene:

$$\text{benzene} + \tfrac{9}{2}O_2 \longrightarrow \text{maleic anhydride} + 2H_2O + 2CO_2$$

The reaction is carried out in the vapour phase by passing a mixture of
benzene and excess of air over a vanadium pentoxide catalyst at 350–450°C.
The effluent is cooled and most of the maleic anhydride is condensed. The
non-condensed material passes to a scrubber where the remaining anhydride
is hydrolysed to maleic acid. The maleic acid solution then passes to evapor-
ators for concentration and dehydration. High-purity anhydride is obtained
by distillation under reduced pressure.

Because of a rise in price of benzene, increasing amounts of maleic
anhydride are being obtained from cheaper *n*-butane:

$$C_4H_{10} + \frac{7}{2}O_2 \longrightarrow \begin{array}{c} HC{-}CO \\ \| \quad \rangle O \\ HC{-}CO \end{array} + 4H_2O$$

Reaction conditions are similar to those used for the oxidation of benzene.

Maleic anhydride is a white crystalline solid, m.p. 52–53°C. Maleic anhydride is preferred to maleic acid since it is more reactive and gives rise to less water on esterification.

Other acids.   Fumaric acid (IV), the *trans*-isomer of maleic acid is sometimes preferred to maleic anhydride as it is less corrosive and gives lighter-coloured products with slightly improved heat resistance. Chloromaleic acid (V) and chlorofumaric acid may be used in the production of self-extinguishing resins.

$$\begin{array}{cc} CH{-}COOH & CCl{-}COOH \\ \| & \| \\ HOOC{-}CH & CH{-}COOH \\ (IV) & (V) \end{array}$$

(c)   Modifying acids and anhydrides

Phthalic anhydride.   The most important modifying component used in the manufacture of linear unsaturated polyesters is phthalic anhydride. The anhydride is generally obtained by the oxidation of *o*-xylene:

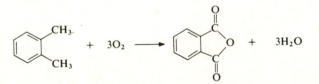

The reaction is carried out in the vapour phase by passing a mixture of *o*-xylene and air over a catalyst such as vanadium pentoxide supported on silica and promoted with titanium dioxide at about 400°C. The exit gases are cooled and the phthalic anhydride is collected and purified by distillation under reduced pressure.

Older processes based on the air-oxidation of naphthalene (obtained from coal) are now of minor importance.

Phthalic anhydride is a white crystalline solid, m.p. 131°C. Phthalic anhydride gives polyesters which are compatible with styrene and the cross-linked products are hard and rigid.

Other acids and anhydrides.   Modifying components other than phthalic anhydride are frequently used in the preparation of unsaturated polyesters in order to impart special properties to the final product. Adipic and sebacic acids are employed to give flexible materials and isophthalic acid (VI) is used for tough products with higher heat distortion temperatures. The use of *endo*-methylenetetrahydrophthalic anhydride ('nadic anhydride') (VII), the Diels-Alder reaction product of cyclopentadiene and maleic anhydride, leads to a substantial improvement in heat resistance. Flame resistant materials are obtained by the use of chlorinated acids and anhydrides, e.g. tetrachlorophthalic anhydride (VIII) and 'chlorendic acid' (HET acid) (IX). The latter is prepared by the Diels-Alder reaction of hexachlorocyclopentadiene and maleic anhydride; the initial product is 'chlorendic anhydride' but this rapidly absorbs water from the air to give the acid.

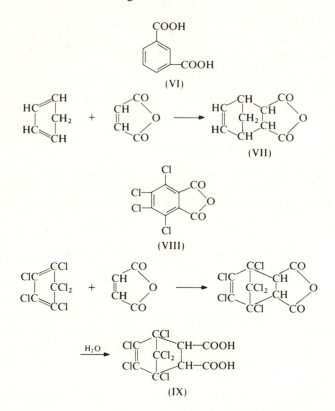

(d)   Cross-linking monomers

It is possible to cross-link unsaturated linear polyester chains directly one to another; however, reaction is slow and a low degree of cross-linking is achieved. These limitations are overcome by the introduction of a material which forms bridges between the chains. The materials most commonly used to cross-link unsaturated linear polyesters in this way are vinyl monomers. The addition of a liquid vinyl monomer to the polymer also leads to a reduction in viscosity and this facilitates the impregnation of glass-fibre in the preparation of laminates.

Styrene is the most widely used cross-linking monomer, being preferred because of its compatibility, low viscosity, ease of use and low price. Other materials are sometimes employed when special properties are required. For example, methyl methacrylate is used, often in conjunction with styrene, for the preparation of translucent sheeting. Diallyl phthalate (X) and triallyl cyanurate (XI) are used for heat resistant products. Partially polymerized diallyl phthalate (solid) is used as the cross-linking agent in moulding powders (the so-called 'alkyd polyester' moulding powders) based on linear unsaturated polyesters.

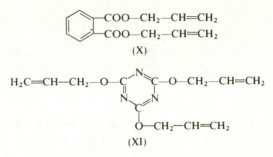

$$\text{(X)}$$

$$\text{(XI)}$$

11.2.3   Preparation

Linear unsaturated polyesters are prepared batch-wise by heating a mixture of the appropriate acidic and hydroxy components in a reactor jacketed for heating and cooling and fitted for distillation. A typical formulation for a general-purpose material might be as follows:

Propylene glycol	100 parts by weight
Maleic anhydride	72
Phthalic anhydride	54

The molar ratio of the ingredients shown above is $1.2 : 0.67 : 0.33$; the excess of glycol is to allow for loss during the reaction and to restrict the molecular weight of the polymer. The mixture is heated at 150–200°C for 6–16 hours and water is continuously distilled from the reactor. Sometimes xylene is added to the reaction mixture to assist in the removal of water by azeotropic distillation and sometimes a catalyst such as p-toluenesulphonic acid is added to reduce the reaction time. In order to prevent discoloration, the reaction is carried out in an inert atmosphere of either carbon dioxide or nitrogen. Heating is continued until the average molecular weight of the polyester reaches about 1000–2000. The polymer is then cooled to about 90°C and pumped into a blending tank containing vinyl monomer to which has been added an inhibitor such as hydroquinone. In a general purpose material, the weight of styrene used is about half that of the polymer. The blend (which is commonly referred to as 'polyester resin') is then allowed to cool to room temperature.

The reaction between a hydroxy-compound and an anhydride proceeds in two distinct steps. In the first step, esterification of the anhydride occurs to form a free acid group which is then esterified in the second step, e.g.

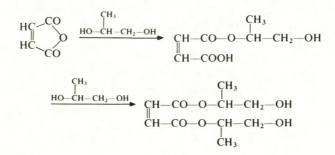

The first step proceeds more rapidly than the second since the anhydride group is more reactive than the free acid group. Clearly, a diol and an anhydride may interact through a sequence of reactions of the foregoing type to yield a linear polyester. Thus a segment of the polyester obtained from propylene glycol, maleic anhydride and phthalic anhydride might have the following structure:

$$\sim OC-CH=CH-CO-O-\underset{\underset{CH_3}{|}}{CH}-CH_2-O-OC \quad CO-O-\underset{\underset{CH_3}{|}}{CH}-CH_2-O-OC-CH=CH-CO-O-\underset{\underset{CH_3}{|}}{CH}-CH_2-O\sim$$

Appreciable *cis-trans* isomerization generally occurs during the polyesterification of unsaturated dibasic acids and anhydrides. Such isomerization is particularly marked with maleic anhydride, which becomes incorporated into the polymer chains mostly as fumarate groups. The extent of isomerization is governed by several factors including the structure of the diol, the reaction conditions, the catalyst used (if any) and the molecular weight of the polymer. The polymerization of maleic anhydride with propylene glycol gives almost entirely poly(propylene fumarate) [1] but in most other cases the final polymer contains 70–90% fumarate groups and 10–30% maleic groups. It is fortunate that maleate-fumarate isomerization does occur because the fumarate group shows much greater reactivity towards vinyl monomers than the maleate group; hence subsequent cross-linking proceeds more readily.

11.2.4   Cross-linking

As has been noted previously, the cross-linking of unsaturated linear polyesters involves the reaction of the unsaturated sites in the polymer chain with a vinyl-type monomer. This reaction is analogous to conventional vinyl copolymerization and proceeds by an essentially similar mechanism. As carried out in commercial practice, cross-linking of unsaturated polyesters is invariably a free radical reaction. Two types of initiating systems are commonly employed for this reaction, namely those effective at elevated temperatures and those effective at room temperature.

The most important initiators used at elevated temperatures are peroxides, which liberate free radicals as a result of thermal decomposition. A peroxide which is widely used in this way is benzoyl peroxide (XII), the decomposition of which is discussed in section 1.4.2(a). Other peroxides which find application in the cross-linking of unsaturated polyesters are 2,4-dichlorobenzoyl peroxide (XIII), di-*tert*-butyl peroxide (XIV) and lauroyl peroxide (XV). Mixtures of polyester resin and this type of peroxide are comparatively stable at room temperature but rapidly cross-link at temperatures ranging from about 70 to 150°C, depending on the choice of peroxide. Such peroxides are used principally in processes employing moulding compositions, when short curing times are required.

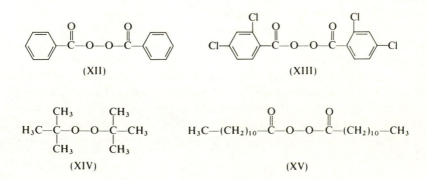

Initiating systems which are effective at room temperature normally consist of mixtures of a peroxy compound and an activator ('accelerator'). In the presence of the accelerator, the peroxy compound rapidly decomposes without the application of heat into free radicals. The two most important peroxy materials now used for the 'cold' curing of polyester resins are methyl ethyl ketone peroxide (MEKP) and cyclohexanone peroxide. These names are rather misleading in that neither of these materials is a single compound and both have a variable composition depending on their method of manufacture. The main components of commercial methyl ethyl ketone peroxide are the following [2]:

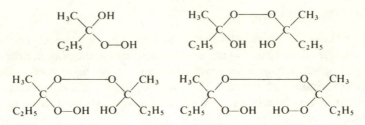

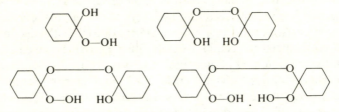

The principal components of commercial cyclohexanone peroxide are the following [3]:

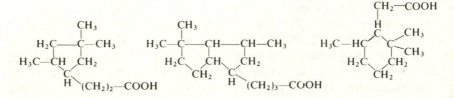

Thus it is seen that both methyl ethyl ketone peroxide and cyclohexanone peroxide are essentially hydroperoxides.

The most common accelerators for methyl ethyl ketone peroxide and cyclohexanone peroxide are salts of metals which exhibit more than one valency. The most widely used metal of this kind is cobalt, although salts of cerium, iron, manganese, tin and vanadium also find some application. In order to be effective as an accelerator a metal salt must be soluble in the polyester resin. The most commonly used salts are naphthenates, which are readily soluble; octoates also may be used. (Naphthenic acid is extracted from the gas oil and kerosene fractions of petroleum and consists of a complex mixture of carboxylic acids of substituted cyclopentanes and cyclohexanes*.) The decomposition of a hydroperoxide (ROOH) by a metal salt such as cobalt naphthenate to give free radicals proceeds according to the following chain reaction:

$$ROOH + Co^{2+} \longrightarrow RO\cdot + OH^- + Co^{3+}$$
$$ROOH + Co^{3+} \longrightarrow ROO\cdot + H^+ + Co^{2+}$$

*Examples of carboxylic acids which have been isolated from naphthenic acid are the following:

This cycle is repeated until all the hydroperoxide has been decomposed. Cobalt naphthenate-methyl ethyl ketone peroxide or -cyclohexanone peroxide systems are very extensively used in the production of large glass-fibre laminates made by hand lay-up and cured at room temperature.

It may be noted that the foregoing metal-based accelerators, which are highly reactive towards hydroperoxides, have little influence on the decomposition of peroxides such as those used for curing polyester resins at elevated temperatures. However, peroxides do decompose rapidly at room temperature into free radicals in the presence of tertiary amines. Amines such as dimethylaniline, diethylaniline and dimethyl-*p*-toluidine react violently with benzoyl peroxide and initiating systems based on these materials have found some use, principally on account of the long pot-life of resin containing either benzoyl peroxide or tertiary amine. On the other hand, polyester resins cured with benzoyl peroxide-tertiary amine tend to discolour and craze on ageing. The reaction between benzoyl peroxide and a tertiary amine is thought to proceed via a one-electron transfer from nitrogen, as shown in the following example involving dimethylaniline [4]:

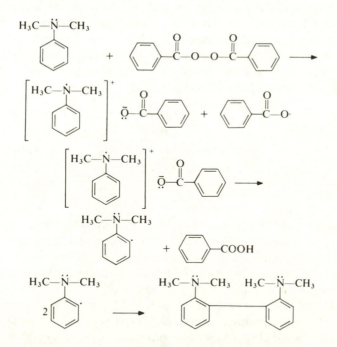

The cross-linking of an unsaturated linear polyester by means of a vinyl monomer such as styrene may be represented as shown overleaf:

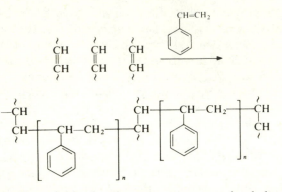

The average value of n in the above structure may be deduced from both spectroscopic and degradation studies [1]. In the first method, the reduction in fumarate unsaturation which occurs when various polyester resins of known styrene content are cured is determined from the infrared spectra. If it is assumed that none of the styrene originally present homopolymerizes but all enters into the cross-linking reaction, the average length of the cross-links can be estimated. This assumption may be justified on the grounds that polystyrene homopolymer can be extracted by toluene only from cured polyester resins in which the original styrene content is greater than about 50% by weight. (Commercial resins generally contain about 35% styrene by weight.) Hydrolytic degradation of the cross-linked structure shown above results in a polymer of the following type:

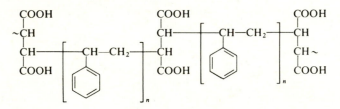

Determination of the equivalent weight of the polymer permits calculation of the average value of n. Both the spectroscopic and degradation studies indicate that in a typical cured general purpose polyester resin the cross-links contain 1–3 styrene units. Determination of the molecular weight of the polymer shown above (which may be regarded as a copolymer of fumaric acid and styrene) enables estimation of the average number of chains linked together by a continuous cross-link. Typically such copolymers have $M_n \simeq 2500$ which indicates that each continuous styrene-fumarate copolymer chain passes through 6–10 cross-link sites in the undegraded cured polyester.

It will be noted that the cross-linking of linear unsaturated polyesters by vinyl monomers does not involve the elimination of any volatile by-products. Hence it is possible to cure the resins without the application of pressure.

Since it is also possible to cure the resins without the application of heat, they are very useful in the manufacture of large structures such as boat hulls.

11.2.5 Properties of cross-linked polymers

When polyester resins have been cross-linked they are rigid, infusible and insoluble. There are so many varieties of polyester resins now commercially available that it is difficult to give typical values for physical properties of cured materials. Furthermore, polyester resins are mostly used in conjunction with glass fibre and the physical properties of the final products greatly depend on the type and quantity of glass fibre incorporated. This point is illustrated by Table 11.1, in which are given comparative values of some properties of cured polyester, both unfilled and reinforced. Cross-linked polyesters have good heat stability, showing little weight loss up to about 200°C. The mechanical strength of general purpose materials begins to decline at about 100°C and the maximum service temperature of glass-fibre laminates and mouldings is about 150°C; for heat-resistant grades of resins these temperature are of the order of 160°C and 200°C respectively. The electrical insulating properties of cured polyesters are satisfactory for many purposes but the polar nature of the ester group results in a relatively high power factor and dielectric constant and so the use of the resins in high frequency applications is limited.

Cross-linked polyesters are resistant to a wide range of organic solvents but they are attacked by chlorinated hydrocarbons (e.g. chloroform, ethylene dichloride and trichloroethylene), esters (e.g. ethyl acetate) and ketones (e.g. acetone and methyl ethyl ketone). The ester groups in the polymer provide sites for hydrolytic attack and strong alkalis cause appreciable degradation. The polymer is, however, resistant to most inorganic and organic acids, with the exception of strong oxidizing acids.

11.3   ALKYDS

11.3.1   Development

Alkyds are network polymers which find extensive use as surface coatings. The most important polyesters of this type are derived from phthalic anhydride and glycerol. The esterification of glycerol with phthalic anhydride was first investigated by Smith in 1901, but since the ultimate product was always a glassy, brittle material there was no commerical interest. During the period 1910–1916, the General Electric Co. (USA) investigated the products of this reaction in some detail and developed resins which found limited use for bonding mica flake into sheet for electrical applications. In 1924, Kienle (also of the General Electric Co.) began an important series of investigations which

Table 11.1 Typical values for various properties of cured polyester, unfilled and reinforced

	Unfilled casting	Glass chopped strand mat laminate (Hand lay-up)	Glass woven cloth laminate (Hand lay-up)	Glass roving rod (Extruded)	Bulk moulding compound	Sheet moulding compound
Specific gravity	1.2	1.6	1.7	1.9	1.8	1.8
Tensile strength (MPa)	62	140	340	830	41	83
(lbf/in^2)	9 000	20 000	50 000	120 000	6 000	12 000
Flexural strength (MPa)	120	210	410	1 000	110	190
(lbf/in^2)	18 000	30 000	60 000	150 000	16 000	27 000
Compressive strength (MPa)	140	140	240	480	140	190
(lbf/in^2)	20 000	20 000	35 000	70 000	20 000	27 000
Impact strength (unnotched) (J/m)	110	1 100	1 300	3 700	370	800
(ft lbf/in)	2	20	25	70	7	15
Glass content (% weight)	0	30	55	70	30	30

culminated in 1933 in a patent covering the modification of polyesters with drying oils. Such modification has three significant effects, namely the resins become soluble in aliphatic solvents; atmospheric oxygen brings about rapid cross-linking of resin films; and the cross-linked films are flexible and durable. This development led to a speedy acceptance of oil-modified polyesters as surface coatings. Oil-modified polyesters, which are commonly known as *alkyds* (which term is derived from *al*cohol and *acid*), now account for approximately half of all the resins consumed in the protective coatings field.

11.3.2 Raw materials

The principal raw materials involved in the preparation of alkyd resins are polyhydric alcohols (polyols) and dibasic acids (or corresponding anhydrides) together with the modifying oils (or corresponding acids). A great variety of reactants is used in the manufacture of commercial alkyd resins; only the most common are discussed in this section.

(a) Polyhydric alcohols

Glycerol. Glycerol is the polyhydric alcohol most widely used for the preparation of alkyd resins and is obtained both synthetically and as a by-product in the manufacture of soap. Most synthetic glycerol is obtained from propylene via allyl chloride:

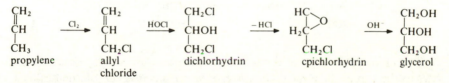

The first step is the 'hot' chlorination of propylene. A mixture of propylene and chlorine (4:1 molar) is heated at about 500°C and 0.2 MPa (2 atmospheres). Under these conditions a free radical substitution reaction occurs rather than addition at the double bond and allyl chloride is the main product. This is treated with pre-formed hypochlorous acid (formed in a separate reactor by passing chlorine into water) at about 30°C to give the addition product, dichlorhydrin. The reaction mixture separates into two layers. The aqueous layer is removed to leave dichlorhydrin which is then stirred with a lime slurry to give epichlorhydrin. The epichlorhydrin is then hydrolysed to glycerol by treatment with aqueous sodium hydroxide at 150°C. The glycerol is produced as a dilute solution containing sodium chloride. Most of the water is evaporated off, during which operation the salt crystallizes out and is removed from the bottom of the evaporator to leave crude glycerol. The crude product is purified by distillation under reduced

pressure. It may be noted that the intermediate epichlorhydrin has significance in its own right for the manufacture of epoxy resins (Chapter 18).

A disadvantage of the process outlined above is that chlorine is required
and is converted into valueless calcium and sodium chlorides. For this
reason, newer glycerol plants utilize the following route, which does not
involve chlorine:

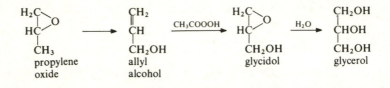

Propylene oxide (section 9.4.1(b)) is isomerized to allyl alcohol by heating at
200–250°C in the presence of a trilithium phosphate catalyst. The allyl
alcohol is then epoxidized to glycidol with peracetic acid in ethyl acetate. The
glycidol is then hydrolysed to glycerol.

In the preparation of soap, oils and fats (which are essentially glyceryl
esters of fatty acids) are saponified by heating with aqueous sodium hydroxide:

$$
\begin{array}{l}
CH_2\!-\!OOCR \\
CH\!-\!OOCR \quad + \quad 3NaOH \quad \longrightarrow \\
CH_2\!-\!OOCR
\end{array}
\quad
\begin{array}{l}
CH_2OH \\
CHOH \quad + \quad 3RCOONa \\
CH_2OH
\end{array}
$$

Sodium chloride is added to the reaction mixture to 'salt out' the soap which
rises to the top of the liquid. The lower aqueous layer is run off, neutralized
and evaporated. The salt is precipitated and is filtered off to leave crude
glycerol which is purified by distillation under reduced pressure.

Glycerol is a colourless, viscous liquid, b.p. 290°C. Glycerol is the preferred
polyol for the preparation of alkyd resins because of its low cost and high
boiling point (which enables reactions to be carried out at high temperatures);
also glycerol-based alkyd resins have good solubility and compatibility
characteristics and good film properties.

Other polyols. Polyols other than glycerol are utilized to a lesser extent in
the manufacture of alkyd resins, being used to impart special properties. Such
polyols include pentaerythritol (XVI), trimethylolpropane (XVII) and
sorbitol (XVIII). Sometimes a proportion of diols such as ethylene glycol and
propylene glycol is included in alkyd resin formulations to reduce cross-link
intensity.

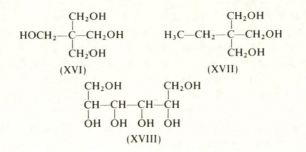

(XVI)

(XVII)

(XVIII)

(b)   Dibasic acids and anhydrides

Phthalic anhydride is the difunctional acidic component most widely used for the preparation of alkyd resins. Manufacture of the material is dealt with in Section 11.2.2(c). Phthalic anhydride is preferred in this application on account of its low cost and because it gives rise to resins with good solubility and compatibility characteristics and good film properties. Other dibasic acids and anhydrides, which are used to a lesser extent, include maleic anhydride (section 11.2.2(b)), isopthalic acid (VI), adipic acid and sebacic acid (section 10.2.2).

(c)   Modifying oils and acids

As indicated earlier, oils are esters of glycerol and fatty acids and have the following structure:

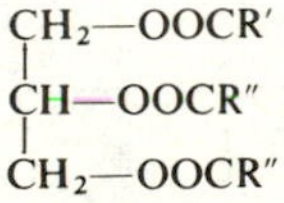

$$CH_2\text{---}OOCR'$$
$$CH\text{---}OOCR''$$
$$CH_2\text{---}OOCR''$$

Oils which find commercial utilization are of both animal and vegetable origin, but for the manufacture of alkyd resins only oils from plant sources are important. Glycerides are said to be 'simple' when only one acid residue is present and 'mixed' when there is more than one acid residue. Oils are invariably mixed glycerides. The acid residues present in oils are usually straight chains containing an even number of carbon atoms and may be saturated or unsaturated. The acids obtained by the hydrolysis of some oils commonly used for the manufacture of alkyd resins are shown in Table 11.2. Oils are commonly designated as 'drying', 'non-drying' and 'semi-drying' according to the effect of atmospheric oxygen on thin films of the oils. Generally, films of drying oils become dry and insoluble after exposure for 2–6 days whereas films of non-drying oils are still fluid after 20 days; films of semi-drying oils become tacky after about 7 days. It will be noticed from

Table 11.2 that drying oils contain a high proportion of unsaturated acid residues whereas non-drying oils are based chiefly on saturated acids.

In addition to the above-mentioned oils there are also several naturally-occurring and synthetic acids which are used for modifying purposes in the manufacture of alkyd resins. These include rosin (the main constituent of which is abietic acid (XIX)) and the adduct of rosin and maleic anhydride (XXI); in the formation of this adduct it is supposed that abietic acid isomerizes to the conjugated diene, levopimaric acid (XX), before addition occurs. The synthetic fatty acids pelargonic acid (XXII) and isooctanoic acid (XXIII) are also sometimes used as modifying acids in the preparation of alkyd resins.

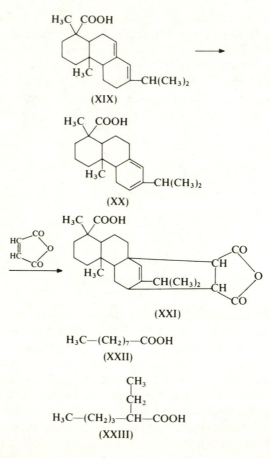

$$H_3C-(CH_2)_7-COOH$$
(XXII)

$$H_3C-(CH_2)_3-\underset{\underset{\displaystyle CH_3}{\overset{\displaystyle |}{\underset{\displaystyle CH_2}{\overset{\displaystyle |}{}}}}{CH}-COOH$$
(XXIII)

11.3.3   Types of alkyd resins

The properties of alkyd resins depend to a large extent on the nature and amount of modifying oil incorporated. It is therefore convenient to classify

Table 11.2 Acids found in the hydrolysates of some common oils (parts by weight)

Acids			Drying oils			Semi-drying oils			Non-drying oils		
Name	Formula	I.U.P.A.C Nomenclature	Linseed	Perilla	Tung	Dehydrated castor	Safflower	Soybean	Castor	Coconut	Cottonseed
Caprylic	$C_8H_{16}O_2$	Octanoic								6	
Capric	$C_{10}H_{20}O_2$	Decanoic								6	
Lauric	$C_{12}H_{24}O_2$	Dodecanoic								44	
Myristic	$C_{14}H_{28}O_2$	Tetradecanoic								18	1
Palmitic	$C_{16}H_{32}O_2$	Hexadecanoic	6	7	4	2	8	11	2	11	29
Palmitoleic	$C_{16}H_{30}O_2$	cis-9-Hexa-decenoic									2
Stearic	$C_{18}H_{36}O_2$	Octadecanoic	4	2	1	1	3	4	1	6	4
Oleic	$C_{18}H_{34}O_2$	cis-9-Octa-decenoic	22	13	8	7	13	25	7	7	24
Ricinoleic	$C_{18}H_{34}O_3$	12-Hydroxy-cis-9-octa-decenoic				7			87		
9,12-Linoleic	$C_{18}H_{32}O_2$	cis-9,cis-12-Octadeca-dienoic	16	14	4	57	75	51	3	2	40
9,11-Linoleic	$C_{18}H_{32}O_2$	9,11-Octadeca-dienoic isomers				26					
Linolenic	$C_{18}H_{30}O_2$	cis-9,cis-12,cis-15-Octadeca-trienoic	52	64	3		1	9			
Eleostearic	$C_{18}H_{30}O_2$	cis-9,trans-11,trans-13-Octadeca-trienoic			80						

alkyd resins according to the type of oil which they contain and there are thus three groups of commercial alkyd resins, namely *drying oil resins, semi-drying oil resins and non-drying oil resins*. Alkyd resins are also conveniently classified in terms of their oil lengths, i.e. the amount of oil they contain. Thus resins which contain less than 50% oil are generally termed *short oil resins*; resins which contain 50–70% oil are considered to be *medium oil resins* whilst those with more than 70% oil are *long oil resins*.

The principal characteristics and typical applications of the various types of alkyd resins are given below.

Drying oil resins

Short oil resins: Soluble only in aromatic solvents (e.g. toluene and xylene); usually cured at elevated temperatures; give very hard, glossy finishes; used in finishes for appliances, signs and toys.

Medium oil resins: Soluble in aliphatic solvents or aliphatic-aromatic blends; may be air-dried or stoved; give durable, glossy finishes; used in finishes for farm implements, hardware and metal furniture.

Long oil resins: Soluble in aliphatic solvents (e.g. naphtha); have good brushing characteristics and dry rapidly in air; give reasonably durable, glossy films; used in household paints.

Semi-drying oil resins

Short oil, medium oil and long oil resins: Comparable to the equivalent drying oil modified resins; give films with improved resistance to yellowing on ageing; used particularly for high gloss white finishes.

Non-drying oil resins

Short oil resins: Soluble only in aromatic solvents; used mainly in conjunction with aminoresins to give improved adhesion and flexibility; used in stoving finishes for appliances.

Medium oil resins: Soluble in aromatic solvents; used mainly as plasticizers for cellulose nitrate for furniture finishes.

11.3.4 Preparation

Alkyd resins cannot be prepared by simply heating a mixture of oil, polyol and dibasic acid. Because of the preferential reaction of the polyol and acid, a heterogeneous mixture of polyester and oil is obtained which has no value as a surface coating vehicle. There are two main methods whereby useful resins are prepared commercially, namely the fatty acid process and the alcoholysis process (often called the monoglyceride process).

(a) Fatty acid process

In this process the oil is firstly hydrolysed to give free fatty acids which are then heated at 200–240°C with a mixture of polyol and dibasic acid. Simultaneous condensation of the polyol, dibasic acid and fatty acids thereby occurs and the latter become incorporated into the polymer structure. The process may be conducted in two ways. In the first procedure, which is known as the fusion or solventless method, the reactants are heated in a simple kettle under an inert atmosphere. At the end of the heating period, inert gas is blown into the resin to remove water and unreacted materials. In the second procedure, which is known as the solvent or solution method, a small amount (generally about 5%) of a solvent, usually xylene, is added to the reactants. The mixture is heated in a reactor fitted with equipment which condenses volatile vapours, separates water and returns the organic distillate to the reactor. The solvent facilitates removal of water by azeotropic distillation and, compared to the fusion method, allows much better temperature control. In addition, the solvent reduces the viscosity of the reactants; this permits more effective agitation which contributes to easier water removal and faster reaction. The solvent also continually cleans resin from the sides of the reactor and enables a more uniform product to be prepared that is free from gel particles. However, despite these advantages of the solvent method over the fusion method, the latter is widely used since it requires simpler equipment.

(b) Alcoholysis Process

In this process, the oil is firstly heated with the polyol at about 240°C in the presence of a basic catalyst (e.g. calcium hydroxide) so that alcoholysis occurs. Typically, the oil is treated with glycerol using a molar ratio of 1:2 and the principal product is monoglyceride:

$$
\begin{array}{ccccc}
CH_2\text{—}OOCR & & CH_2\text{—}OH & & CH_2\text{—}OOCR \\
| & & | & & | \\
CH\text{—}OOCR & + & 2CH\text{—}OH & \longrightarrow & 3CH\text{—}OH \\
| & & | & & | \\
CH_2\text{—}OOCR & & CH_2\text{—}OH & & CH_2\text{—}OH
\end{array}
$$

After the monoglyceride is formed, the dibasic acid is added and the mixture is treated as in the fatty acid process; again either the fusion method or the solvent method may be employed. Acidolysis is an alternative to alcoholysis, but generally requires a higher temperature and is seldom used. Compared to the fatty acid process, the alcoholysis process gives less reproducible results but is usually cheaper to run.

11.3.5 Structure

The structure of a resin obtained by the simple esterification of glycerol with phthalic anhydride may be represented as follows:

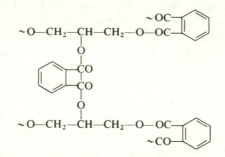

As indicated earlier, materials based on a structure of this type are brittle and of little practical use; however, the incorporation of fatty acid residues into such a structure leads to highly successful surface coating vehicles. The detailed structure of the latter materials depends on the method of preparation. In the fatty acid process (section 11.3.4(a)), the polyol, dibasic acid and fatty acids react simultaneously and it has been shown [5] that the primary hydroxyl groups of glycerol react more readily with phthalic carboxyl groups than with fatty acids whilst the reverse applies to the secondary hydroxyl groups. It is to be expected, therefore, that in an alkyd resin prepared by the fatty acid process the following type of structure would

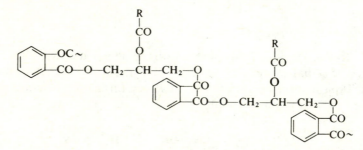

predominate. In the alcoholysis process there is no fatty acid competing in the esterification reaction. Since alcoholysis generally results in a preponderance of α-monoglycerides, it is to be expected that in an alkyd resin prepared by this process the following type of structure would predominate:

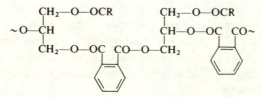

A number of reactions other than esterification may occur during the production of alkyd resin and these result in some deviation from the ideal structures shown above. For example, acid-catalysed etherification may take place; in the case of glycerol the following reaction is possible:

$$
\begin{array}{ccc}
CH_2-OH & CH_2-OH & CH_2-OH \\
| & | & | \\
2CH-OH \xrightarrow{\;H^+\;} & CH-OH & CH-OH \quad + \quad H_2O \\
| & | & | \\
CH_2-OH & CH_2-O-CH_2 &
\end{array}
$$

Esterification of this product would lead to a resin containing ether links.

11.3.6 Film formation

The mechanism by which an alkyd resin is converted from a liquid to a dry film is largely determined by the nature of the fatty acid residues present. When the fatty acid residues are derived from non-drying oils, the alkyd resin itself cannot readily form dry films. As has been noted previously, such resins are mainly used in conjunction with other film-forming materials such as aminoresins and cellulose nitrate; the alkyd resin is essentially a plasticizer and is involved in few chemical reactions. When, however, the fatty acid residues in an alkyd resin are derived from drying or semi-drying oils, the resin itself is capable of forming dry films. The drying process involves attack by oxygen in the unsaturated regions of the fatty acid residues followed by cross-linking and is essentially similar to the process which occurs when the corresponding glyceride oil is acted upon by air. (See below.) It may be noted that the molecular weight of an alkyd resin is higher than that of an oil; consequently the number of cross-links needed to give a dry film is less and the drying time is reduced. Further, since relatively little cross-linking is required to dry an alkyd film, the less unsaturated semi-drying oils can be used to prepare alkyd resins which dry readily. It may also be noted that it is possible that further polyesterification reactions might contribute to the drying of alkyd films; however, even at temperatures used for stoving (usually about 120°C) little esterification occurs.

As indicated above, the conversion of a drying or semi-drying oil alkyd resin to a solid film consists of two steps, namely aerial oxidation and cross-linking initiated by decomposition of the oxidation products. These processes are complex and the cross-linked systems are difficult to manipulate for analytical purposes. Most experimental investigations in this field therefore have been made using model systems of individual fatty acids or their esters with monohydric alcohols and the results have been extrapolated to alkyd resins. Whilst this procedure has permitted the elucidation of the principal reactions which occur during the drying of alkyd resins, some details are still in doubt. For example, oxygen and volatile decomposition products may diffuse through a liquid simple ester much more readily than through a semi-solid, partially cross-linked alkyd so that the chemistry of the two systems may differ considerably.

It has been found that there are marked differences in the behaviour of non-conjugated and conjugated unsaturated fatty acids and their esters toward oxidation. The reactions of the two types of compounds are therefore considered separately.

(i) *Reactions of non-conjugated fatty acids and esters.* Non-conjugated fatty acids and esters are susceptible to the general type of autoxidation which occurs with polyethylene and other hydrocarbon polymers (section 2.3.3). A methylene group adjacent to a double bond is particularly liable to lose a hydrogen atom since the resulting radical is resonance stabilized. This radical can react with oxygen to give a peroxidic radical which then abstracts hydrogen from a further α-methylene group to form a hydroperoxide and to propagate the reaction. In the case of, for example, linoleic acid the scheme shown in Fig. 11.1 may be envisaged. It will be noted that the shifting of the double bond enables the new bond to take up the more stable *trans*-configuration.

The hydroperoxides of the types shown (ROOH) may decompose in the following ways:

$$ROOH \longrightarrow RO\cdot \ + \ HO\cdot$$
$$2ROOH \longrightarrow RO\cdot \ + \ ROO\cdot \ + \ H_2O$$

These reactions account for the first- and second-order kinetics which have been observed for the initial decomposition of some simple hydroperoxides. Subsequent to the primary decomposition, induced decomposition may occur as follows:

$$RO\cdot \ + \ ROOH \longrightarrow ROO\cdot \ + \ ROH$$
$$ROO\cdot \ + \ ROOH \longrightarrow RO\cdot \ + \ ROH \ + \ O_2$$
$$RO\cdot \ + \ RH \longrightarrow ROH \ + \ R\cdot$$
$$HO\cdot \ + \ RH \longrightarrow H_2O \ + \ R\cdot$$
$$R\cdot \ + \ ROOH \longrightarrow ROH \ + \ RO\cdot$$

Termination of the above reaction chain may occur by combinations of the following kind:

$$RO\cdot \ + \ RO\cdot \longrightarrow ROOR$$
$$ROO\cdot \ + \ ROO\cdot \longrightarrow ROOR \ + \ O_2$$
$$R\cdot \ + \ R\cdot \longrightarrow R-R$$
$$RO\cdot \ + \ R\cdot \longrightarrow ROR$$
$$ROO\cdot \ + \ R\cdot \longrightarrow ROOR$$

Although the products shown above represent simple dimers, it will be appreciated that extension of the reaction sequence to non-conjugated un-

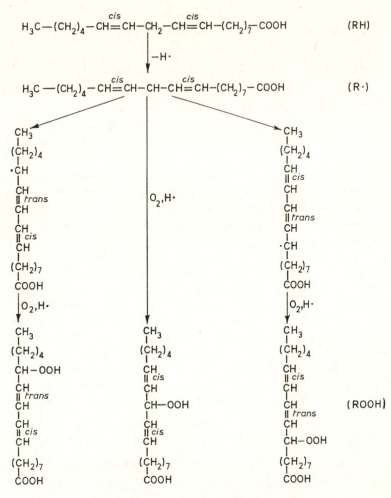

Fig. 11.1 Autoxidation of linoleic acid.

saturated oils and corresponding alkyd resins would result in the formation of a network polymer.

(ii) *Reactions of conjugated fatty acids and esters.* Conjugated fatty acids and esters do not respond to aerial oxidation in the same manner as non-conjugated fatty acids and esters. In particular, cross-linked films derived from the conjugated materials contain much more carbon-carbon bonding in the cross-links. It has been suggested [6] that in these cases the initial product of oxidation is a 1,4-peroxide which initiates a diene-type polymerization:

$$\sim CH=CH-CH=CH\sim \xrightarrow{O_2} \begin{matrix} O\text{------}O \\ | \quad\quad\quad | \\ \sim CH-CH=CH-CH\sim \end{matrix} \longrightarrow$$

$$\begin{matrix} O-O\cdot \\ | \\ \sim CH-CH=CH-CH\sim \end{matrix} \xrightarrow{\sim CH=CH-CH=CH\sim} \begin{matrix} O-O\cdot \\ | \\ \sim CH-CH=CH-CH\sim \\ | \\ \sim CH-CH=CH-CH\sim \end{matrix}$$

$$\xrightarrow{\sim CH=CH-CH=CH\sim} \begin{matrix} O-O\cdot \\ | \\ \sim CH-CH=CH-CH\sim \\ | \\ \sim CH-CH=CH-CH\sim \\ | \\ \sim CH-CH=CH-CH\sim \end{matrix} \quad \text{etc.}$$

The formation of cross-linked films by the various reactions described above is not rapid enough for practical purposes. Commercial surface coatings therefore contain catalysts or *driers* which speed up the drying process. Most commonly, driers are hydrocarbon-soluble salts (usually octoates, naphthenates and linoleates) of metals; these salts are often termed *soaps*. The metallic driers may be divided into two groups, namely primary (or participating) driers and promoter (or auxiliary) driers. Primary driers include salts of cobalt, lead and manganese and can by themselves catalyse the drying process. Promoter driers, which include calcium, barium and zinc salts, do not by themselves accelerate drying but have an activating effect on primary driers. The detailed mechanisms by which driers operate have not been elucidated although it is generally agreed that an important function of primary driers is to catalyse the decomposition of the hydroperoxides formed by the aerial oxidation of non-conjugated fatty acids and esters. In the case of cobalt salts, the following redox system may be envisaged (cf. section 11.2.4):

$$ROOH + Co^{2+} \longrightarrow RO\cdot + OH^- + Co^{3+}$$
$$ROOH + Co^{3+} \longrightarrow ROO\cdot + H^+ + Co^{2+}$$

It has also been suggested that primary driers initiate the oxidation sequence by direct interaction of the metal ion with the fatty acid or ester, e.g.

$$RH + Co^{3+} \longrightarrow R\cdot + Co^{2+} + H^+$$

The means by which promoter driers operate are less clear. One suggestion is that they form salts with the acid groups present in an oil to give ionic cross-links which contribute to the initial drying of the film.

It may be noted that oxidative processes may continue long after a paint film has dried, causing a deterioration in properties. Such decline may result from an excessive degree of cross-linking, chain scission (possibly by hydroperoxide decomposition of the type shown in section 2.3.3) or by loss of volatile products. A further type of deterioration which is sometimes encountered in alkyd films is yellowing. Yellowing appears to be associated with the unsaturated centres of the fatty acid residues and is particularly prevalent when triene acid residues are present (as in linseed oil alkyds). It is probable that discoloration arises from the formation of highly conjugated systems.

When a fatty acid residue can form a conjugated hydroperoxide, the following reaction may be envisaged:

$$\sim CH{=}CH{-}CH{=}CH{-}\underset{\underset{OOH}{|}}{CH}\sim \longrightarrow \sim CH{=}CH{-}CH{=}CH{-}\underset{\underset{O}{\|}}{C}\sim$$

$$\begin{array}{c} \sim CH{=}CH{-}CH{=}CH{-}\underset{\underset{O}{\|}}{C}\sim \\[4pt] \sim CH{=}CH{-}CH{=}CH{-}CH_2\sim \end{array} \longrightarrow \begin{array}{c} \sim CH{=}CH{-}CH{=}CH{-}\underset{\underset{}{\|}}{C}\sim \\[-2pt] \sim CH{=}CH{-}CH{=}CH{-}\underset{}{C}\sim \end{array}$$

11.3.7 Modified alkyd resins

It is common practice to modify alkyd resins with various other materials in order to bring about specific improvements. These other materials, which are frequently resinous themselves, may be present in physical or chemical combination. Examples of materials which are blended with alkyd resins include cellulose nitrate (which confers fast-drying characteristics and toughness), chlorinated rubber (which imparts fire-resistance and toughness) and silicone resins (which give improved heat and water resistance). Examples of modifying substances which are involved in chemical reactions and which are usually incorporated during the preparation of the alkyd resin are phenolic resins (which impart fast-drying characteristics and improved corrosion resistance) and aminoresins (which confer improved hardness, durability and alkali resistance). The chemistry of the reactions which occur is considered in sections 14.4.3, 15.2.4 and 15.3.4 respectively. Chemically-modified alkyd resins may also be prepared by the use of vinyl-type monomers such as styrene, vinyltoluene, methyl methacrylate and acrylonitrile, of which styrene is the most commonly used. Styrenation may be carried out in several ways but the preferred method is by reaction with the pre-formed alkyd resin. In this case, the monomer, peroxide initiator and alkyd (which usually contains conjugated unsaturation) are heated under reflux until the required viscosity is reached. The styrenation of conjugated alkyds probably involves 1,4 addition as follows:

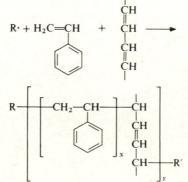

In a study of the styrenation of eleostearates, values of x and y of 4–5 and 9 respectively were obtained [7]. Styrenated alkyd resins give finishes with good colour retention and alkali resistance but with reduced solvent resistance.

11.4 POLY(ALLYL ESTER)S

The allyl radical has the structure $H_2C=CH-CH_2-$ and allyl esters may be regarded as derivatives of allyl alcohol. For the preparation of polymers, most attention has been directed towards the allyl esters of dibasic acids; such monomers give rise to cross-linked polymers. The first (1941) allyl ester to be utilized was diethylene glycol bis(allyl carbonate), prepared as follows:

$$\begin{array}{cc}
\underset{\begin{array}{c}\text{CH}_2-\text{CH}_2-\text{OH}\\|\\\text{O}\\|\\\text{CH}_2-\text{CH}_2-\text{OH}\end{array}}{}\ \xrightarrow{\text{COCl}_2}\ \underset{\begin{array}{c}\text{CH}_2-\text{CH}_2-\text{O}-\text{CO}-\text{Cl}\\|\\\text{O}\\|\\\text{CH}_2-\text{CH}_2-\text{O}-\text{CO}-\text{Cl}\end{array}}{}
\end{array}$$

diethylene glycol   diethylene glycol chloroformate

$$\xrightarrow{\text{H}_2\text{C}=\text{CH}-\text{CH}_2\text{OH}}\ \begin{array}{c}\text{CH}_2-\text{CH}_2-\text{O}-\text{CO}-\text{O}-\text{CH}_2-\text{CH}=\text{CH}_2\\|\\\text{O}\\|\\\text{CH}_2-\text{CH}_2-\text{O}-\text{CO}-\text{O}-\text{CH}_2-\text{CH}=\text{CH}_2\end{array}$$

diethylene glycol bis(allyl carbonate)

The ester was used for the production of low pressure laminates but in this application it has been displaced by polyesters of the type described in section 11.2. It is, however, now widely used for the manufacture of spectacle lenses, where its light weight, dimensional stability, abrasion resistance and dyeability are relevant.

At the present time the most important allyl esters used for the preparation of polymers are diallyl phthalate and diallyl isophthalate.

11.4.1 Diallyl phthalate polymers

Diallyl phthalate (DAP) is prepared by reaction of phthalic anhydride (section 11.2.2(c)) and allyl alcohol (section 11.3.2(a)):

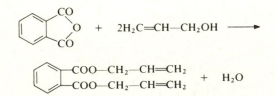

Diallyl phthalate is a colourless liquid b.p. 160°C/0.5 kPa (4 mm Hg).

In most applications, diallyl phthalate is polymerized in two stages. In the first stage the monomer is carefully heated at about 100°C with a free radical initiator (e.g. *tert*-butyl perbenzoate) to yield an essentially linear polymer (XXIV) composed of monomer units linked through one allyl group per unit. The reaction mixture is cooled when a molecular weight of about 10 000–25 000 is reached and before gelation occurs. The solid polymer is usually compounded with fillers to give a thermosetting moulding powder, but it may also be dissolved in monomer to give a casting or laminating resin. Further heating leads to a highly cross-linked structure:

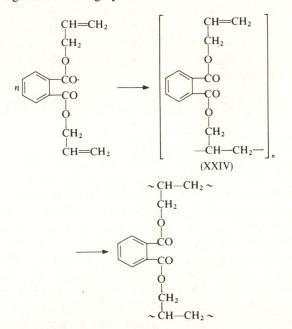

(XXIV)

Mouldings based on diallyl phthalate exhibit good electrical insulation characteristics and dimensional stability under conditions of dry and wet heat. The polymer has good thermal stability, being capable of withstanding up to about 180°C for long periods. The ester groups in the polymer provide sites for hydrolytic attack and strong alkalis cause appreciable degradation. The polymer is, however, resistant to acids other than strong oxidizing acids. Mouldings based on diallyl phthalate are expensive and are used mainly for electrical components for use under adverse operating conditions.

11.4.2 Diallyl isophthalate polymers

Diallyl isophthalate (DAIP) is prepared by esterification of isophthalic acid with allyl alcohol:

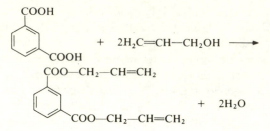

Diallyl isophthalate is a colourless liquid, b.p. 181°C/0.5 kPa (4 mm Hg).

The polymers of diallyl isophthalate are very similar in preparation, properties and application to the polymers of diallyl phthalate described above. Mouldings based on diallyl isophthalate are more expensive but have enhanced thermal stability (withstanding up to about 220°C for long periods) and resistance to organic solvents.

11.5 POLY(ETHYLENE TEREPHTHALATE)

11.5.1 Development

As mentioned in section 10.2.1., several polyesters were investigated by Carothers of E. I. du Pont de Nemours and Co. (USA) during a programme of fundamental researches into polymerization which was begun in 1929. These polyesters were mostly linear aliphatic polymers and they generally had low melting points, considerable solubility in organic liquids and poor resistance to hydrolysis. The materials therefore lacked promise as textile fibres and were soon overshadowed by the outstandingly successful polyamide, nylon 6,6. In 1941 Whinfield and Dickson of The Calico Printers Association Ltd. (UK) prepared the aromatic polyester, poly(ethylene terephthalate) and found it to have great promise as a fibre- and film-forming material. Rights for the use of the polymer were assigned to E. I. du Pont de Nemours and Co. (USA) and Imperial Chemical Industries Ltd. (UK) who both undertook large scale manufacture in 1953. Poly(ethylene terephthalate) has since become a major textile fibre. To a lesser extent, the polymer finds application in film form and as a moulding material.

11.5.2 Raw materials

Poly(ethylene terephthalate) may be obtained from ethylene glycol and either terephthalic acid or its ester, dimethyl terephthalate. Until the mid-1960s, all poly(ethylene terephthalate) was produced from the ester, mainly because the acid then available was difficult to obtain with sufficiently high purity whereas the ester was readily purified. This situation changed with the advent of fibre-grade terephthalic acid and at the present time approximately equal amounts of the polymer are made from acid and ester.

(a) Ethylene glycol

Ethylene glycol is prepared by the hydration of ethylene oxide (section 9.4.1(a)):

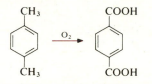

$$CH_2\text{—}CH_2 \quad + \quad H_2O \quad \longrightarrow \quad HO\text{—}CH_2\text{—}CH_2\text{—}OH$$

This reaction is carried out in a manner comparable to that described for the preparation of propylene glycol from propylene oxide (section 11.2.2(a)). Ethylene glycol is a colourless liquid, b.p. 197°C.

(b) Terephthalic acid

The major commercial route to terephthalic acid which is suitable for the direct preparation of poly(ethylene terephthalate) is from p-xylene:

p-Xylene is obtained largely from petroleum sources, being a product of the fractionation of reformed naphthas (see section 2.2.). The oxidation is carried out in the liquid phase. Typically, air is passed into a solution of p-xylene in acetic acid at about 200°C and 2 MPa (20 atmospheres) in the presence of a catalyst system containing cobalt and manganese salts and a source of bromide ions. The terephthalic acid produced contains only small amounts of impurities (mainly p-carboxybenzaldehyde), which are readily removed. The acid is dissolved in water at about 250°C and 5 MPa (50 atmospheres) and treated with hydrogen (which converts the aldehyde to p-toluic acid). The solution is then cooled to 100°C and pure terephthalic acid crystallizes.

Terephthalic acid is a white solid which sublimes at 300°C.

(c) Dimethyl terephthalate

Several processes have been developed for the preparation of dimethyl terephthalate from p-xylene, but the most important proceeds as follows:

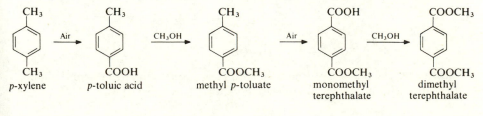

The oxidation steps are carried out in the liquid phase at about 170°C and 1.5 MPa (15 atmospheres) in the presence of a cobalt acetate or naphthenate catalyst whilst the esterifications are conducted at about 150°C.

Dimethyl terephthalate may also be produced by esterification of terephthalic acid.

Dimethyl terephthalate is a white solid, m.p. 142°C.

11.5.3 Preparation

As indicated above, poly(ethylene terephthalate) may be prepared from terephthalic acid or dimethyl terephthalate. With both starting materials, the polymerization is carried out in two steps.

In the acid-based process, the initial step is esterification to give mainly bis(2-hydroxyethyl) terephthalate:

$$\begin{array}{c}\text{COOH} \\ \text{benzene ring} \\ \text{COOH}\end{array} + 2\,\text{HO—CH}_2\text{—CH}_2\text{—OH} \longrightarrow \begin{array}{c}\text{CO—O—CH}_2\text{—CH}_2\text{—OH} \\ \text{benzene ring} \\ \text{CO—O—CH}_2\text{—CH}_2\text{—OH}\end{array} + 2\,\text{H}_2\text{O}$$

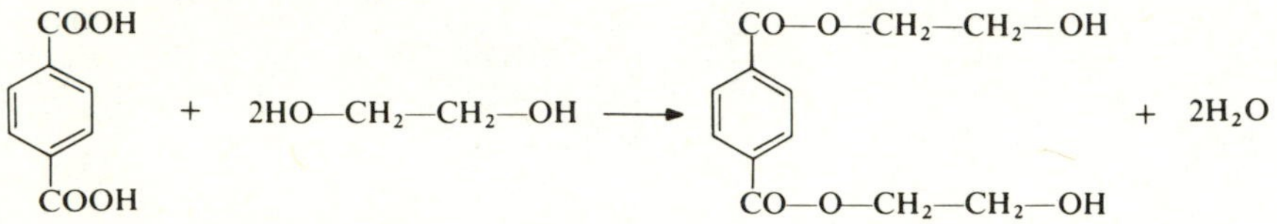

In addition to bis(2-hydroxyethyl) terephthalate, oligomers up to about the hexamer are formed; these have the general formula shown below:

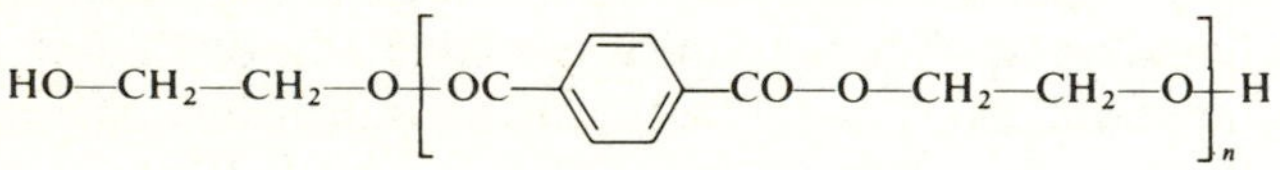

$$\text{HO—CH}_2\text{—CH}_2\text{—O}\left[\text{OC—}\langle\text{benzene}\rangle\text{—CO—O—CH}_2\text{—CH}_2\text{—O}\right]_n\text{H}$$

Typically, terephthalic acid is treated with an excess of ethylene glycol (1:1.5 molar) at about 250°C and 0.4 MPa (4 atmospheres) and water is allowed to escape as vapour as reaction proceeds.

In the ester-based process, the initial step is ester interchange and again the principal product is bis(2-hydroxyethyl) terephthalate:

$$\begin{array}{c}\text{COOCH}_3 \\ \text{benzene ring} \\ \text{COOCH}_3\end{array} + 2\,\text{HO—CH}_2\text{—CH}_2\text{—OH} \longrightarrow$$

$$\begin{array}{c}\text{CO—O—CH}_2\text{—CH}_2\text{—OH} \\ \text{benzene ring} \\ \text{CO—O—CH}_2\text{—CH}_2\text{—OH}\end{array} + 2\,\text{CH}_3\text{OH}$$

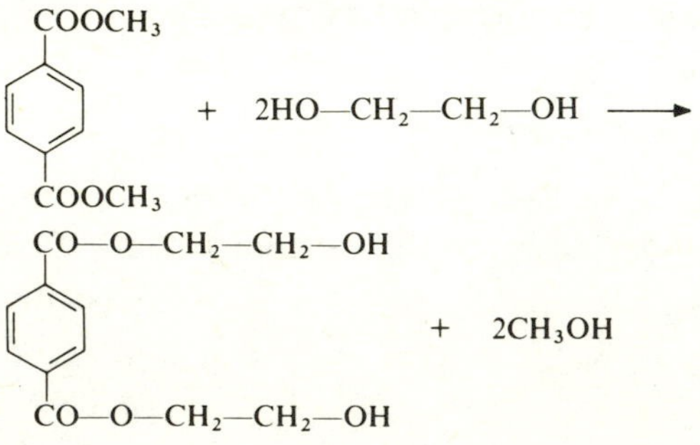

Also, some oligomers are formed. In a typical process, dimethyl terephthalate is heated with an excess of ethylene glycol (1:2.2 molar) at 140–220°C and

atmospheric pressure in the presence of a catalyst (usually manganese acetate). Methanol is removed as reaction proceeds.

Whether the starting material is terephthalic acid or dimethyl terephthalate, the second step in the polymerization sequence is the same. An ester
interchange reaction occurs, in which the bis(2-hydroxyethyl) terephthalate
serves as both ester and alcohol. Successive interchanges result in the
formation of a polyester, as represented in the following scheme:

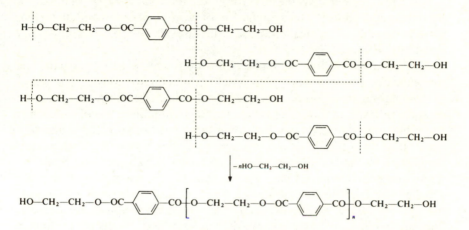

Under the reaction conditions employed, the ethylene glycol which is eliminated is removed continuously from the system; thus there is a progressive
increase in molecular weight. Of course, ester interchange reactions also
occur at non-terminal ester linkages, but such reactions do not lead to the
formation of ethylene glycol and so do not change the average molecular
weight of the mixture.

The bis(2-hydroxyethyl) terephthalate is heated at about 290°C in the
presence of a catalyst such as antimony trioxide while the pressure is lowered
to about 0.1 kPa (1 mm Hg). Ethylene glycol is distilled off as reaction
proceeds. Polymerization is continued until the molecular weight reaches the
desired level (20 000 for fibre and film; 30 000 for moulding material). The
molten polymer is either spun directly into fibre or extruded and disintegrated.

Compared to the ester-based process, the acid-based process avoids costs
involved in using and recycling methanol and gives higher yields. On the
other hand, the ester-based process is somewhat easier to carry out.

11.5.4 Properties

Poly(ethylene terephthalate) is a colourless rigid substance. Because of its
structural regularity, the polymer readily crystallizes and the physical pro-

perties of the bulk material are greatly affected by the extent of crystallinity, which, in turn, is largely determined by the previous history of the material. In the production of fibre, the molten polymer is extruded through spinnerets into air at normal temperature. The filaments are thus rapidly cooled and are largely amorphous and are weak. The filaments are then drawn at a temperature above the glass-transition temperature (80°C), whereby molecular orientation and crystallinity are induced. Finally, the filaments are heated at about 200°C under tension to give a dimensionally-stable material of maximum crystallinity. The manufacture of poly(ethylene terephthalate) film closely resembles that of fibre and also involves extrusion, orientation and stabilization.

Because of its rather high glass-transition temperature, poly(ethylene terephthalate is not ordinarily a satisfactory material for conventional injection moulding. Amorphous mouldings are obtained since little crystallization can occur during cooling after moulding. Such mouldings are of little value since they are relatively weak and when heated above 80°C, crystallization occurs leading to shrinkage, distortion and clouding. Some improvement can be effected by the addition of nucleating agents (which promote crystallization) and plasticizers (which lower Tg) and by the use of heated moulds. Thus poly(ethylene terephthalate) remained of little interest as a moulding material until the technique of biaxial stretching used in film production was applied to bottle manufacture. Bottles are produced by firstly forming a substantially amorphous parison by injection into a cold mould. The parison is then removed from the mould, heated and blown under high pressure into the bottle mould. In this way the polymer is biaxially stretched to give a thin-walled container of high strength, toughness and clarity combined with low permeability to carbon dioxide, oxygen and water vapour. Poly(ethylene terephthalate) bottles are very extensively used for carbonated beverages and other food products.

The values of some properties of various forms of poly(ethylene terephthalate) which are given in Table 11.3 illustrate the influence of crystallinity. The crystalline melting point of poly(ethylene terephthalate) is 265°C. Although a polar polymer, poly(ethylene terephthalate) has good electrical insulating properties at room temperature (even at high frequencies) since dipole orientation is restricted at temperatures below the glass-transition temperature (80°C). The principal application of poly(ethylene terephthalate) film is for electrical insulation.

Poly(ethylene terephthalate) is most usually encountered in the crystalline form and, as such, it is soluble at normal temperatures only in proton donors which are capable of interaction with the ester group. Effective solvents of this kind are chlorinated and fluorinated acetic acids, phenols and anhydrous hydrofluoric acid. The polymer is soluble at elevated temperatures in various organic liquids, which include anisole, aromatic ketones, dibutyl phthalate

Table 11.3 Comparative properties of various forms of poly(ethylene terephthalate)

	Fibre	Film	Amorphous moulding	Crystalline moulding	Stretch-blown moulding	
Specific gravity	1.38	1.38	1.30–1.34	1.32–1.38	1.36	
Tensile strength (MPa)	690	170	55	76	120 (axial)	190 (hoop)
(1bf/in^2)	100 000	25 000	8 000	11 000	17 000	28 000
Elongation at break (%)	15–50	70	250	250	125 (axial)	30 (hoop)
Impact strength, Izod (J/m)	–	–	53	43	–	–
(ft 1bf/in)	–	–	1.0	0.8	–	–

and dimethyl sulphone. Chloroform has the peculiar property of dissolving amorphous poly(ethylene terephthalate) at temperatures below 0°C, but on warming such solutions the polymer separates in crystalline form. Chloroform is without effect on polymer which has already been crystallized.

Poly(ethylene terephthalate) has good resistance to water and dilute mineral acids but is degraded by concentrated nitric and sulphuric acids. The polymer is also sensitive to basic reagents. Ionic bases such as aqueous sodium hydroxide attack the surface of the polymer, leaving the interior unaffected, whilst organic bases such as methylamine diffuse into the material and attack it in depth.

The resistance of poly(ethylene terephthalate) to photochemical degradation is very good. Some thermal degradation occurs when the polymer is heated above the melting point, the principal products being carbon dioxide, acetaldehyde and terephthalic acid. The following molecular reaction is involved:

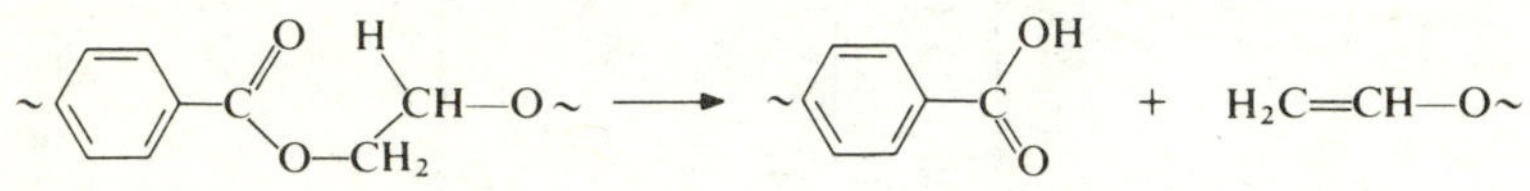

Carbon dioxide may arise by decarboxylation of the carboxylic acid whilst acetaldehyde may be formed by cleavage of the vinyl ether group through ester interchange.

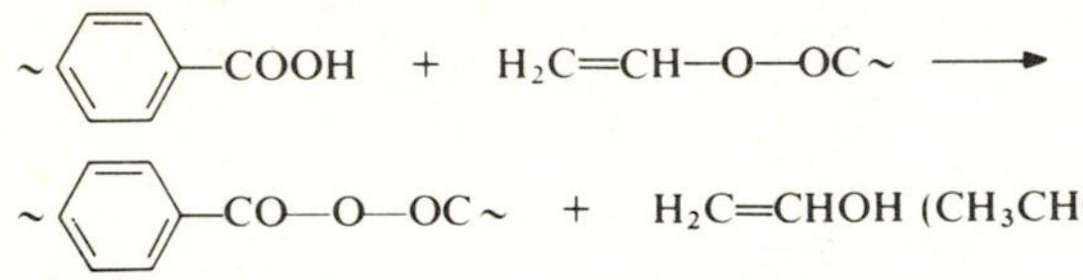

11.6 POLY(BUTYLENE TEREPHTHALATE)

Poly(butylene terephthalate) (PBT) (also known as poly(tetramethylene terephthalate) has the following structure:

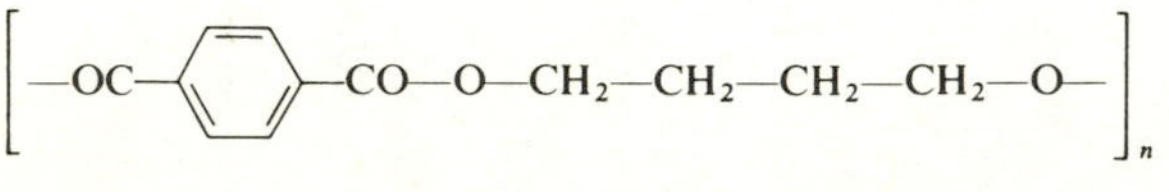

It is prepared from 1,4-butanediol and terephthalic acid or dimethylterephthalate by processes which are essentially the same as those used for poly(ethylene terephthalate) and described in the preceding section.

Due to the longer sequence of methylene groups, the polymer chains are more flexible and less polar than in poly(ethylene terephthalate). This results in a lower crystalline melting point (225°C) and glass-transition point (22–43°C). The low glass-transition temperature facilitates rapid crystallization in the mould and most poly(butylene terephthalate) is processed by injection moulding.

The chemical resistance of poly(butylene terephthalate) is good, being similar to that of poly(ethylene terephthalate). The less polar nature of the polymer results in lower water absorption, leading to improved dimensional stability and electrical insulating properties.

Poly(butylene terephthalate) is used as an engineering material in a variety of applications which include automotive parts, electrical components and domestic appliance assemblies.

11.7 CYCLOHEXYLENEDIMETHYLENE TEREPHTHALATE POLYMERS

A third class of terephthalate polyester consists of polymers based on 1,4-dimethylolcyclohexane (1,4-cyclohexylene glycol). Three polymers of this type are commercially available and are described in this section.

1,4-Dimethylolcyclohexane is prepared from dimethyl terephthalate by the following route:

COOCH3
COOCH3
H2
Pd
COOCH3
CH
H2C
CH2
H2C
CH2
CH
COOCH3
H2
Cu chromite
CH2OH
CH
H2C
CH2
H2C
CH2
CH
CH2OH

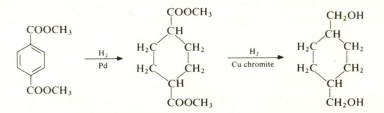

11.7.1 Poly(1,4-cyclohexylenedimethylene terephthalate)

Poly(1,4-cyclohexylenedimethylene terephthalate) is prepared from dimethyl terephthalate and 1,4-dimethylolcyclohexane and has the following structure:

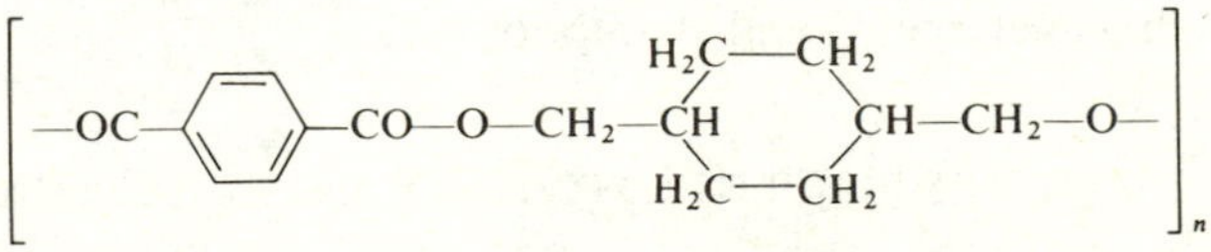

—OC—
—CO—O—CH2—CH
H2C—CH2
H2C—CH2
CH—CH2—O—
n

This polymer has a slightly stiffer chain than poly(ethylene terephthalate) and hence the glass-transition temperature (about 130°C) and melting point (about 290°C) are somewhat higher. The polymer is available in both fibre and film form and is of interest mainly because of its superior resistance to water and weathering.

11.7.2   Isophthalic acid copolymers

In the isophthalic acid copolymers, a portion of the terephthalate units in the chain is replaced by isophthalate units. Copolymers of this type are commonly known as copolyesters. A segment of such a polymer might be as follows:

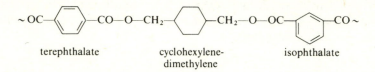

terephthalate   cyclohexylene-   isophthalate
                       dimethylene

The presence of two isomeric units in the chain results in an amorphous polymer which is easily processed. The main use of the material is for extrusion into film and sheet for packaging. The material has brilliant clarity, good toughness, high tear strength and good chemical resistance.

11.7.3   Ethylene glycol copolymers

In the ethylene glycol copolymers, ethylene units are introduced into the chain by reaction of a mixture of ethylene glycol and 1,4-dimethylolcyclohexane with terephthalic acid. A segment of this type of copolyester might be as follows:

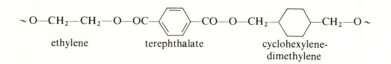

ethylene   terephthalate   cyclohexylene-
                                         dimethylene

The irregular chain results in a material which is usually amorphous. Thus conventional processes such as injection moulding, blow moulding and extrusion are readily performed. The material is clear and virtually colourless, even in thick sections. It has good toughness, high stiffness and good chemical resistance. Applications include packaging film and bottles for products such as detergents and shampoos.

11.8   POLYESTER THERMOPLASTIC ELASTOMERS

As discussed previously, thermoplastic elastomers are materials which have the functional properties of conventional vulcanized rubbers but which may be processed as normal thermoplastics (see section 2.9).

Polyester thermoplastic elastomers are block copolymers made up of alternating hard and soft segments. The hard segments are alkylene terephthalate blocks and the soft segments are poly(alkylene ether) terephthalate blocks. In current commercial products, the alkylene group is most commonly 1,4-butylene. Such copolymers have the following structure:

$$-\left[-OC-\left\langle\bigcirc\right\rangle-CO-O-(CH_2)_4-O-\right]_x\left[-OC-\left\langle\bigcirc\right\rangle-CO-O-[(CH_2)_4-O-]_n\right]_y-$$

The copolymers are prepared by methods which are entirely analogous to those described previously for poly(ethylene terephthalate) (section 11.5), except that two diols are used rather than one. For the preparation of the block copolymers the preferred diols are 1,4-butanediol and a poly(oxytetramethylene) glycol (section 9.4.6) of molecular weight in the range 600–3000. Thus dimethyl terephthalate is transesterified in two steps using a mixture of the two diols. The stoichiometry is such that relatively long sequences of tetramethylene terephthalate occur to form the hard segments. The copolymers have a molecular weight (M_n) of 25 000–30 000.

Polyester thermoplastic elastomers are tough materials which exhibit useful mechanical properties over the range -40–$130°C$. As polar polymers they have good oil and fuel resistance; they are attacked by chlorinated solvents and by concentrated acids and bases. Despite their rather high cost relative to most other rubbers, polyester thermoplastic elastomers have found various uses, including automotive parts, gear wheels and boot soles.

11.9 POLYARYLATES

Polyarylates are highly aromatic polyesters which are prepared from diphenols and aromatic dicarboxylic acids. Most commonly, polyarylates are produced from bisphenol A and mixtures of terephthalic acid and isophthalic acid. A segment of such polymer might be as follows:

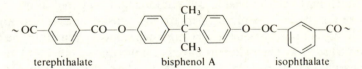

terephthalate   bisphenol A   isophthalate

Details of the manufacture of polyarylates have not been disclosed but the polymers may be made by interfacial polymerization of the diacid chlorides (dissolved in a chlorinated hydrocarbon) and bisphenol A (in aqueous alkali).

The use of two isomeric acids in the synthesis results in an irregular chain which precludes crystallization. Thus the material can be processed at much lower temperatures than would be possible with a crystalline polymer derived from only one of the acids. Polyarylates can, in fact, be processed by standard methods such as injection moulding, extrusion and blow moulding.

Polyarylates are engineering thermoplastics which show high impact strength, good flexural recovery, good surface hardness and high dimensional stability. Mechanical properties are retained up to about 140°C. The polymer has good electrical insulation properties and is self-extinguishing; it also has very good resistance to ultraviolet light. Polyarylates thus find use in a number of outdoor applications such as traffic light components and solar collectors.

11.10   POLY(p-HYDROXYBENZOATE)S

The homopolymer of p-hydroxybenzoic acid is a polyester with the following structure:

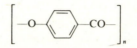

This polymer has a very high level of thermal stability and chemical resistance. It is, however, extremely difficult to process and this factor has severely limited use of the homopolymer.

Much wider interest has been aroused by the development of copolymers of p-hydroxybenzoic acid which are commonly known as liquid crystal polymers. Polyesters of this type were introduced commercially in 1985 and details of their nature have not been fully disclosed. Two commercial products are currently available: one is based on p-hydroxybenzoic acid and a hydroxynaphthoic acid whilst the other is based on p-hydroxybenzoic acid, terephthalic acid and p,p'-dihydroxybiphenyl.

The characteristic feature of liquid crystal polymers is that the stiff, rod-like molecules become highly ordered in the melt when subjected to a process such as injection moulding. This morphology is retained on cooling in the mould. The highly ordered aggregates impart a wood-like structure to the solid, which may be described as self-reinforcing. Compared to other moulded materials, liquid crystal polymers exhibit considerable anisotropy.

Liquid crystal polymers have very high mechanical strength in the direction parallel to flow. The materials have good thermal stability and mechanical properties are retained up to about 200°C. The materials are unaffected by most organic solvents but are susceptible to attack by alkalis. Electrical insulation properties and weatherability are also very good. Applications for liquid crystal polymers include automotive parts, electrical components and pump parts.

11.11   POLYESTER PLASTICIZERS

Polyester plasticizers are linear saturated polyesters of low molecular weight, i.e. less than 10 000. It may be noted that low molecular weight saturated polyesters of a different kind are also of commercial importance: these are hydroxy-terminated polyesters which are used for the preparation of poly-urethanes. These intermediate polymers are described in section 16.3.2(b).

Polyester plasticizers are normally prepared by a diol-dicarboxylic acid condensation. Often a monohydric alcohol or a monocarboxylic acid is included in the reactants; in this way, reactive end-groups are eliminated and also the molecular weight of the product may be controlled. Commonly used reactants are given in Table 11.4. In a typical process, a mixture of the reactants is heated at 200–250°C in an inert atmosphere for several hours with continuous removal of water. By such processes, polyesters with molecular weights up to about 1000 may be prepared. If higher molecular weight material is required, an excess of diol is used in the reaction and the initial product is subsequently heated at 200–250°C under reduced pressure (about 0.1 kPa). Ester interchange involving diol-terminated chains occurs and free diol is formed:

$$\sim CO\!-\!O\!-\!R\!-\!OH \; + \; HO\!-\!R\!-\!O\!-\!CO \sim \; \rightleftharpoons$$
$$\sim CO\!-\!O\!-\!R\!-\!O\!-\!CO \sim \; + \; HO\!-\!R\!-\!OH$$

Under the conditions used, the diol distils from the reaction mixture; thus the average molecular weight of the latter increases. Linear polyesters intended for use as plasticizers generally have average molecular weights (M_n) within the range 500–8000. These polymers are mostly viscous liquids.

Polyester plasticizers are employed mainly in poly(vinyl chloride) compositions for use in applications where minimal plasticizer loss is required. The polymeric plasticizers are less volatile and more resistant to solvent extraction than conventional monomeric plasticizers.

Table 11.4 Reactants used for the preparation of polyester plasticizers

Diols	Dicarboxylic acids	Monohydric alcohols	Monocarboxylic acids
Ethylene glycol	Adipic acid	n-Butanol	Capric acid
Diethylene glycol	Azelaic acid	n-Octanol	Lauric acid
Triethylene glycol	Sebacic acid	2-Ethylhexanol	Pelargonic acid
Poly(ethylene glycol)	Phthalic acid	n-Decanol	Benzoic acid
($M_n \simeq 200$)	(anhydride)	Isooctyl alcohol*	
Propylene glycol		Nonanol*	
1,3-Butanediol		Isodecyl alcohol*	
2,2-Dimethyl-1,			
3-propanediol			

*Mixture of isomeric alcohols made by the oxo-process.

REFERENCES

1. Hayes, B. T. *et al.* (1957) *Chem. Ind.*, 1162.
2. Karnojitzki, V. (1958) *Les Peroxydes Organiques*, Herman, Paris.
3. Criegee, R. *et al.* (1949) *Ann. Chem.* **565**, 7.
4. Horner, L. and Schlenk, E. (1949) *Angew, Chem.*, **61**, 411.
5. Goldsmith, H. A. (1948) *Ind. Eng. Chem.*, **40**, 1205.
6. Faulkner, R. N. (1958) *J. Appl. Chem.*, **8**, 448.
7. Redknap, E. F. (1960) *J. Oil Colour Chemists' Assoc.*, **43**, 260.

BIBLIOGRRAPHY

Martens, C. R. (1961) *Alkyd Resins*, Reinhold Publishing Corporation, New York.

Lawrence, J. R. (1962) *Polyester Resins*, Reinhold Publishing Corporation, New York.

Boenig, H. V. (1964) *Unsaturated Polyesters: Structure and Properties*, Elsevier Publishing Co., Amsterdam.

Goodman, I. and Rhys, J. A. (1965) *Polyesters, Vol. 1: Saturated Polymers*, Iliffe Books Ltd, London.

Raech, H. (1965) *Allylic Resins and Monomers*, Reinhold Publishing Corporation, New York.

Parkyn, B., Lamb, F. and Clinton, B. V. (1967) *Polyesters, Vol. 2: Unsaturated Polyesters and Polyester Plasticizers,* Iliffe Books Ltd, London.

Bruins, P. F. (ed.) (1976) *Unsaturated Polyester Technology*, Gordon and Breach Science Publishers, New York.

Burns, R. (1982) *Polyester Molding Compounds*, Marcel Dekker, Inc., New York.

OTHER AROMATIC POLYMERS CONTAINING p-PHENYLENE GROUPS

12.1 SCOPE

Several of the polymers previously considered, e.g. poly(p-phenylene tere-phthalamide), poly(ethylene terephthalate), poly(butylene terephthalate) and polyarylates have in common, p-phenylene groups as part of the main chain. Several other linear aromatic polymers possessing this structural feature are commercially available and it is these polymers that are dealt with in this chapter. The p-phenylene group has a stiffening effect so that the polymers have high softening temperatures. As a consequence, the polymers are generally rigid at room temperature and have high heat deformation temperatures. Such polymers are counted as engineering thermoplastics, i.e. materials which lend themselves to use in engineering design, being capable of substituting for traditional materials, particularly metals.

12.2 POLY-p-PHENYLENE

Poly-p-phenylene is the simplest polymer containing p-phenylene groups. Several methods have been described for the synthesis of the polymer. One of the more successful routes is by dehydrogenation of poly-1,3-cyclohexadiene obtained by polymerization of 1,3-cyclohexadiene with a Ziegler-Natta catalyst [1]:

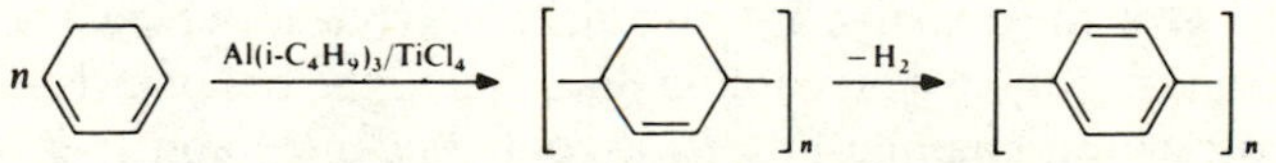

A second method is the direct polymerization of benzene in the presence of combinations of Lewis acids with inorganic oxidizing agents [2]:

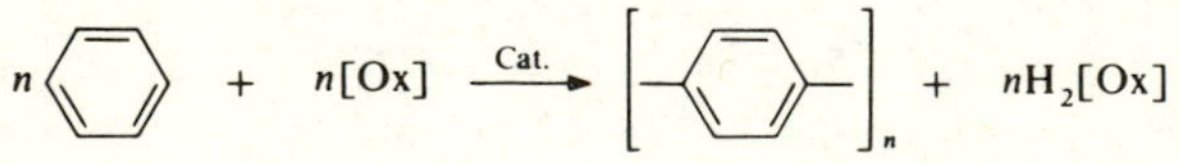

For example, benzene polymerizes in the presence of aluminium chloride-cupric chloride under remarkably mild conditions, i.e. 35–50°C. Polymerization is thought to be initiated by a proton (formed by reaction of traces of water and the Lewis acid) and to proceed through successive electrophilic substitution and oxidative dehydrogenation reactions:

Initiation

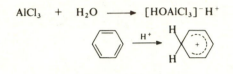

Propagation

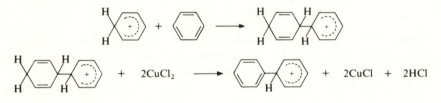

Termination

Poly-*p*-phenylene is of interest because of its outstanding thermal stability. The thermal properties depend to some extent on the method of preparation, but generally there is little change on heating in air at 300–400°C. Since the material is highly insoluble and infusible (m.p. 530°C with decomposition), it is difficult to fabricate and there has been little commercial development.

12.3   POLY-*p*-XYLYLENE

Poly-*p*-xylylene is synthesized from *p*-xylene as shown on page 267. When di-*p*-xylylene is pyrolysed, both methylene bridges are homolytically cleaved to give the reactive intermediate *p*-xylylene which spontaneously forms the linear polymer on cooling. Generally, the polymer is deposited (in the polymerization chamber) on to a cold condenser and then removed as film; alternatively, the polymer is deposited directly on to an object which is required to be coated. Typically, an average molecular weight of about 500 000 is attained.

The most significant property of poly-*p*-xylylene is its thermal stability. The crystalline melting point is 375–425°C and in an inert atmosphere the

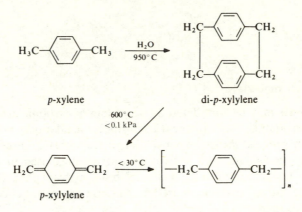

p-xylene   di-p-xylylene

600°C
<0.1 kPa

H_2C=⟨⟩=CH_2 < 30°C $[$—H_2C—⟨⟩—CH_2—$]_n$

p-xylylene

material is claimed to have a useful life of 10 years at 220°C. The thermal stability in air is not exceptional. Poly-p-xylylene also maintains useful properties at temperatures as low as -200°C. The polymer has outstandingly good electrical insulation characteristics and is used as a dielectric in special capacitors.

It may be noted that poly-p-xylylene is considerably less stable than poly-p-phenylene owing to the presence of the relatively weak dibenzyl linkages. It may also be noted that the isomeric polyxylylenes have been prepared but poly-o-xylylene has a polymer melt temperature of 110°C whilst poly-m-xylylene has a softening temperature of only 60°C; these polymers are therefore of no immediate interest.

12.4 POLYCARBONATES

Polycarbonates may be defined as polymers containing recurring carbonate groups (–O–CO–O–) in the main chain. Such polymers may also be considered to be polyesters but since only polycarbonates containing p-phenylene groups are produced commercially at present, these polymers are dealt with in this chapter.

Interest in aromatic polycarbonates dates from 1956 when Farbenfabriken Bayer A. G. (Germany) and the General Electric Co. (USA) independently described polycarbonates based on 4,4'-dihydroxydiphenyl alkanes (I). In general these polymers have high melting points together with thermal and hydrolytic stability. The polycarbonates of many 4,4'-dihydroxydiphenyl alkanes have been investigated but only that based on 2,2-bis(4'-hydroxyphenyl)propane has achieved commercial importance and only this product is described in this chapter. Full-scale production of the material began in 1959 in Germany and in 1960 in the USA.

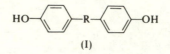

(I)

12.4.1 Raw materials

Polycarbonates may be considered as polyesters of carbonic acid and poly-hydroxy compounds. Carbonic acid itself does not take part in the normal esterification process; instead reactive derivatives must be used. Phosgene and diphenyl carbonate are the two derivatives most used in commercial operations. As mentioned previously, 2,2-bis(4'-hydroxyphenyl)propane is the only polyhydroxy compound used for the manufacture of polycarbonates at the present time.

(a) *Phosgene*

Phosgene (or carbonyl chloride) is obtained by the action of chlorine on carbon monoxide at about 200°C in the presence of charcoal as catalyst:

$$CO \ + \ Cl_2 \longrightarrow COCl_2$$

Phosgene is a gas, b.p. 8°C; it has a pungent, unpleasant odour.

(b) *Diphenyl carbonate*

Diphenyl carbonate is prepared by passing phosgene into a solution of phenol in aqueous sodium hydroxide in the presence of an inert solvent such as methylene chloride:

$$2 \text{C}_6\text{H}_5\text{—ONa} \ + \ COCl_2 \longrightarrow \text{C}_6\text{H}_5\text{—O—CO—O—C}_6\text{H}_5 \ + \ 2NaCl$$

The organic phase, which contains the diphenyl carbonate, is separated; the solvent is stripped off and the diphenyl carbonate is purified by distillation. The reaction is accelerated by tertiary amines and is analogous to the interfacial polycondensation of 2,2-bis(4'hydroxyphenyl)propane and phosgene described in section 12.4.2(a) Diphenyl carbonate is a white crystalline solid, m.p. 78°C.

(c) *2,2-Bis(4'-hydroxyphenyl)propane*

The production of 2,2-bis(4'-hydroxyphenyl)propane (commonly referred to by the trivial name bisphenol A) is described in section 18.3.1.(a). It may be noted that high-purity bisphenol A (m.p. 154–157°C) is used for the preparation of polycarbonates. Less pure material (such as is commonly used in the manufacture of epoxy resins) results in a polycarbonate with poor colour and physical properties. Bisphenol A is the preferred dihydroxy compound for polycarbonate production on the grounds of availability and all-round balance of properties of the resulting polymer.

12.4.2 Preparation

There are two main methods for the manufacture of poly(2,2-bis(4'-phenylene)propane carbonate) (commonly called bisphenol A polycarbonate), namely direct phosgenation and ester interchange.

(a) Direct phosgenation

In this method, the polymer is obtained by a Schotten-Baumann reaction by treating bisphenol A directly with phosgene in the presence of base:

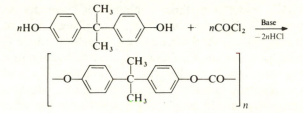

The simple method of passing phosgene into a solution of bisphenol A in aqueous sodium hydroxide is unsatisfactory because the growing polymer becomes insoluble in alkali and only low molecular weight material is obtained. Two techniques have been devised to circumvent this difficulty, namely the solution method and the interfacial method.

Solution method. In this method the reaction is carried out in pyridine, which is a solvent for the reactants and polymer; the pyridine also acts as a catalyst and combines with the hydrogen chloride formed. Because of the high cost of pyridine, the process is generally carried out using a mixture of pyridine and a cheaper solvent such as chloroform, methylene chloride or 1,1,2,2-tetrachloroethane. Typically, phosgene is passed into the bisphenol A solution at 25–35°C. Pyridine hydrochloride is precipitated and a viscous solution of polymer is rapidly formed. The solution is then washed with dilute hydrochloric acid (which converts any free pyridine into the corresponding water-soluble salt) and then with water until the washings are free from ionic contaminants. Practically, effective washing of the viscous polymer solution is difficult. The polymer is then isolated either by precipitation with a non-solvent such as methanol or by evaporation. The product is finally extruded and pelletized. The pyridine and other solvents used are recovered and recycled.

In the above procedure, the reactive species is an ionic adduct of pyridine and phosgene and reaction occurs as follows:

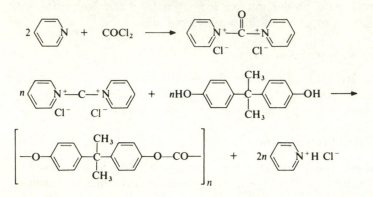

Interfacial method.   In this method a solution of bisphenol A in aqueous sodium hydroxide is dispersed in an organic solvent such as methylene chloride by rapid stirring. A small quantity of tertiary amine (e.g. triethylamine) or quaternary ammonium base (e.g. tetramethylammonium hydroxide) is added to the system as catalyst and then phosgene is passed in at about 25°C. When reaction is complete the organic phase, which contains the polymer, is separated and the polymer is isolated as in the solution method described above.

In this process, it is thought that a two-step polymerization reaction is involved [3]. In the first step an adduct of the catalyst and a growing polycarbonate chain is formed in the organic phase, e.g.

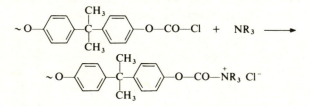

In the second step the adduct reacts at the interface (adducts of this kind are not stable in water) with the sodium salt of bisphenol A or a growing chain to form a new carbonate group and regenerate the catalyst, e.g.

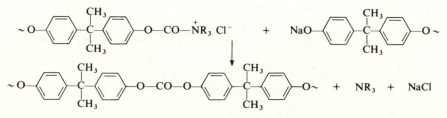

(b)   Ester interchange

In this method, the polymer is obtained by ester interchange between bisphenol A and diphenyl carbonate. In order to obtain high yields of

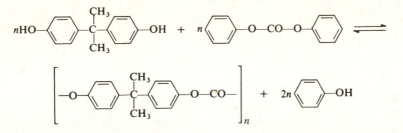

polymer and high molecular weights, almost complete removal of phenol from the reaction mixture is required (cf. section 11.5.3).

In a typical process, a mixture of bisphenol A and diphenyl carbonate together with a basic catalyst (e.g. lithium hydride, zinc oxide or antimony oxide) is melted and agitated at about 150°C under nitrogen. The temperature is then raised to about 210°C over 1 hour and the pressure is reduced to about 3 kPa (20 mm Hg). By the end of this time most of the phenol has been distilled off. The reaction mixture is then heated for a further period of 5–6 hours during which time the temperature is raised to about 300°C and the pressures is lowered to about 0.1 kPa (1 mm Hg). During this period the melt becomes increasingly viscous and the reaction is eventually stopped while the material can still be forced from the reactor under inert gas pressure. The extruded material is then pelletized.

Theoretically, the formation of polycarbonate by ester interchange between bisphenol A and diphenyl carbonate, as shown above, requires equimolar quantities of reagents. However, if equimolar quantities are used the polymer is liable to be discoloured. The reason for this lies in the thermal instability of bisphenol A. In the presence of alkali at temperatures above 150°C, bisphenol A decomposes into p-isopropenyl phenol and phenol:

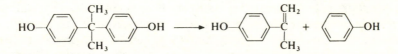

The unsaturated phenol is highly reactive and gives rise to coloured products. This decomposition may be repressed by the use of a molar excess of diphenyl carbonate in the ester interchange reaction. In this way, initial esterification and consequent stabilization of the bisphenol A occurs more rapidly. When an excess of diphenyl carbonate is used, the initial products are low molecular weight polycarbonate chains terminated with phenyl carbonate groups. When these low molecular weight polymers are heated above 250°C under reduced pressure they are converted to high molecular weight polycarbonate with the elimination of diphenyl carbonate:

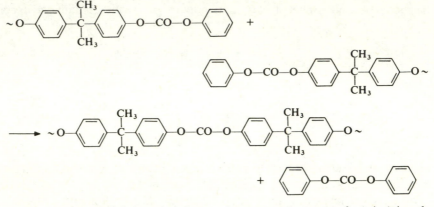

The use of excess of diphenyl carbonate is a convenient way of minimizing the discoloration of the product but this technique does have the disadvantage that the less volatile diphenyl carbonate is more difficult to remove than phenol.

Compared to direct phosgenation, the ester interchange method of preparing the polycarbonate has the following advantages: solvents (and attendant recovery operations) are not required; purification of the polymer is unnecessary; and the product does not require densification. On the other hand, the ester interchange method does require more complex equipment (since high temperatures and low pressures are involved) and the high viscosity of the polymer melt makes it impracticable to handle products with molecular weights greater than about 50 000. Nevertheless, the ester interchange method is usually preferred to direct phosgenation in commercial operations.

12.4.3   Properties

Bisphenol A polycarbonate is a transparent rigid substance. Typical values for some properties of the polycarbonate are given in Table 12.1. The material has outstanding rigidity and toughness, which properties are retained at both low and elevated temperatures; the maximum permissible service temperature is about 135°C. The resistance of the polycarbonate to deformation under load and the dimensional stability in humid atmospheres are also outstanding. Because of the aforementioned characteristics, it was initially thought that the polycarbonate would become an important engineering material. This hope has not been fully realized because of the liability of the polycarbonate to craze or crack under strain or on ageing. Bisphenol A polycarbonate has good electrical insulating properties; although it is a polar

Table 12.1 Properties of a typical commercial grade of polycarbonate

Specific gravity	1.20
Tensile strength (MPa)	55–69
(lbf/in^2)	8000–10 000
Impact strength, Izod	
($\frac{1}{8}$ in $\times \frac{1}{2}$ in bar) (J/m)	640–850
(ft lbf/in)	12–16
Elongation at break (%)	60–100
Water absorption (max. at 23°C) (%)	0.35
Glass-transition temperature (°C)	149
Crystalline melting point (°C)	220–230

polymer, dipole orientation is restricted at temperatures below the glass-transition temperature (149°C).

The bisphenol A polycarbonate chain is very stiff because of the presence of the aromatic rings. This stiffness leads to the high glass-transition temperature noted above and also restricts crystallization in normal mouldings. Highly crystalline material can be obtained only by special techniques such as heating the polymer at 180°C for several days or slow evaporation of solutions. Although the polycarbonate normally encountered is substantially amorphous, the material exhibits solubility characteristics resembling those of crystalline polymers. Thus although there are several solvents with approximately the same solubility parameter as the polymer which dissolve the amorphous polymer, the resulting solution may then precipitate the polymer in a crystalline form (which is thermodynamically more favourable). If, however, there is some interaction between the solvent and polymer then solution may be maintained. Effective solvents for bisphenol A polycarbonate are proton donors such as chloroform, *cis*-1,2-dichloroethylene, methylene chloride, 1,1,1-trichloroethane and 1,1,1,2-tetrachloroethane. Solvents which have limited dissolving power include acetophenone, cyclohexanone, dimethylformamide, dioxan and tetrahydrofuran. Swelling effects are observed with many organic liquids, including acetone, benzene, carbon tetrachloride and ethyl acetate. Aliphatic hydrocarbons, ethers, and alcohols (with the exception of methanol) do not dissolve or swell the polycarbonate. It will be noted that bisphenol A polycarbonate is affected by a wide range of solvents and this feature is a limitation of the material. The polycarbonate also exhibits environmental stress cracking in various media such as hydrocarbon vapours, moisture at elevated temperatures and soap solutions.

The stability of bisphenol A polycarbonate towards aqueous solutions of inorganic and organic acids, salts and oxidizing agents is very good. The ester group is, however, readily hydrolysed by bases. Amines and aqueous ammonia lead to rapid and complete breakdown to bisphenol A; aqueous solutions of strong alkalis attack the surfaces of specimens causing progressive saponification.

Bisphenol A polycarbonate has good oxidative stability, largely because of the absence of secondary and tertiary carbon atoms. The polymer is stable in air up to 150°C over long periods; at higher temperatures some oxidation and cross-linking occur. Ultraviolet light is strongly absorbed by the polycarbonate and causes crazing and degradation. However, such effects are restricted to the surface and in-depth deterioration does not occur. Thus, whilst film may become brittle on weathering, moulded parts are not seriously affected.

Bisphenol A polycarbonate may be melt-processed by all of the standard techniques although its melt-viscosity is rather high. Despite its fairly high cost, the polymer has found a wide variety of uses. The largest application is in the electronics/business machine field for such parts as connectors, terminals and covers. An important new application is for compact discs. Other uses include glazing, safety glasses, medical devices and domestic appliance housing.

12.5   PHENOXIES

Phenoxies are polyethers obtained by reaction of bisphenol A and epichlorhydrin:

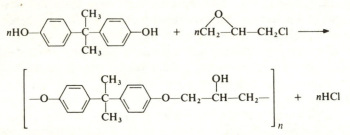

Although phenoxies are based on the same reactants and have structure similar to the better-known epoxy resins (see Chapter 18), the two materials are quite distinct by virtue of differing molecular weights. Phenoxies have a molecular weight (M_n) of about 25 000 whilst epoxy resins generally have molecular weights of a few hundred.

In the preparation of high molecular weight polymers from bisphenol A and epichlorhydrin there is an increasing probability of branching through reaction of secondary hydroxyl groups along the chain and epoxy groups. This reaction is catalysed by the presence of the relatively large amounts of sodium hydroxide which are necessary to effect dehydrochlorination. Branching may be minimized by using a two-step process. In this method, pure pre-formed diglycidyl ether of bisphenol A is treated with the stoichiometric amount of bisphenol A in the presence of a small quantity of basic catalyst such as benzyltrimethylammonium hydroxide:

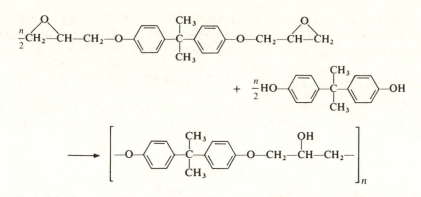

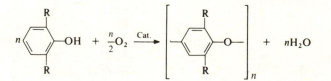

Phenoxies have a structure resembling that of bisphenol A polycarbonate (section 12.4) and the properties of the materials also show similarities. Phenoxies are tough and rigid and have good creep resistance. The mechanical properties show little change over the temperature range $-60°C$ to $80°C$; the glass transition temperature is $100°C$. Since phenoxies are slightly branched, they are amorphous; they are therefore soluble at room temperature in solvents of similar solubility parameter, e.g. esters, ethers and ketones. In contrast to polycarbonates, phenoxies have very good resistance to alkalis. Phenoxies were available commercially for some years but are no longer manufactured.

12.6   POLY(2,6-DIMETHYL-1,4-PHENYLENE OXIDE)

Phenols may be polymerized to high molecular weight products by the technique of oxidative coupling. The overall reaction is as follows:

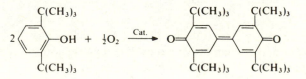

Under the reaction conditions used the o- and p- positions are reactive. If two of these positions are blocked (as shown) linear polymers are obtained; if fewer blocking groups are used cross-linked polymers result. The blocking groups must not be too bulky or oxidative coupling leads to diphenoquinones rather than polyethers, e.g.

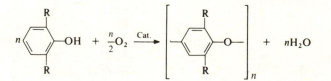

The methyl group is a very satisfactory blocking substituent and poly(2,6-dimethyl-1,4-phenylene oxide) (commonly called poly(phenylene oxide) or poly(phenylene ether)) is a well-established material.

12.6.1 Raw materials

Although 2,6-xylenol can be obtained from coal tar and petroleum refining streams, it is mainly obtained by the alkylation of phenol with methanol:

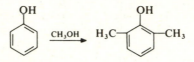

In one process, the reaction is conducted in the gas phase at 350°C using magnesium oxide as catalyst.

12.6.2 Preparation

Oxidative coupling is readily accomplished by passing oxygen into a reaction mixture containing 2,6-xylenol, pyridine and cuprous chloride. (The molar ratio of pyridine to cuprous ion is generally in the range 10:1 to 100:1.) External heating is unnecessary; during the course of the reaction the temperature rises to about 70°C. The polymer is precipitated with dilute hydrochloric acid and collected by filtration.

It is generally accepted that aryloxy radicals are intermediates in the polymerization reaction since ESR studies have shown the presence of both monomeric and polymeric aryloxy radicals in polymerizing solutions of 2,6-xylenol [4]. The simplest mechanism that can be suggested for the reaction is aromatic substitution:

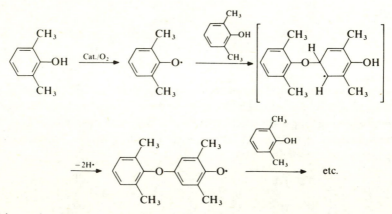

If this mechanism operated, the presence of compounds such as 2,6-dimethyl-anisole would lead to chain termination:

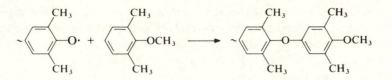

However, it has been found that the addition of 2,6-dimethylanisole to a polymerization mixture does not result in a lower molecular weight product. It is therefore generally supposed that polymerization proceeds by free radical coupling. The simplest mechanism of this type involves the successive coupling of monomer units to polymeric aryloxy radicals. Although reactions

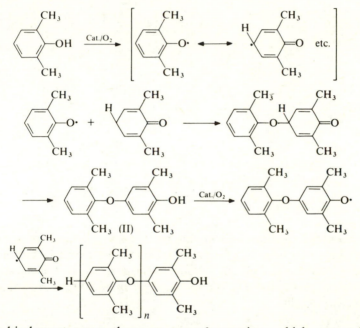

of this kind must occur, there are two observations which cannot be explained on this basis. Firstly, the dimer (II) forms a polymer identical with that obtained from 2,6-xylenol; since no monomer is present in this case, the above scheme cannot explain polymerization completely. Secondly, when 2,6-xylenol is polymerized, there is a sharp increase in average molecular weight near the end of the reaction; this type of behaviour is typical of stepwise reactions in which polymer molecules couple with each other (section 1.4.5). However, there is no immediately apparent way by which two polyarylene ether molecules can be coupled. There is experimental evidence (see Reference 5 for an account of this work) that coupling takes place through two processes, namely redistribution and rearrangement with the relative contribution of each depending on reaction conditions.

In the redistribution process it is suggested that two aryloxy radicals couple to give an unstable quinone ketal (III). This rapidly decomposes to yield either the radicals from which it was formed or two different aryloxy radicals, as illustrated below for two dimer radicals:

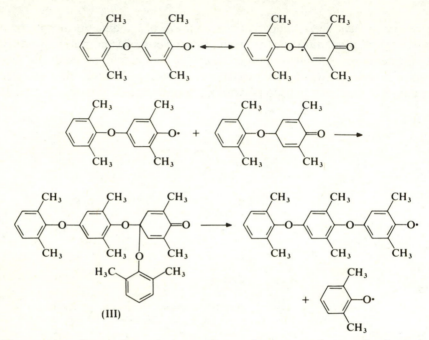

(III)

The reaction shown above does not change the average degree of polymerization of the system since one fragment has one more unit than before whilst the other fragment has one unit less. However, if one of the radicals is a monomer radical (as above) it is possible for coupling to result in an increase in molecular weight:

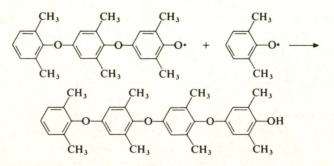

Thus the redistribution process can account for the production of polymer from low molecular weight radicals but is unlikely to account for the production of polymer from high molecular weight radicals because of the large number of steps required to produce a monomer radical.

The rearrangement process also involves quinone ketals formed by the coupling of two aryloxy radicals. It is suggested that the carbonyl oxygen of a ketal is within bonding distance of the *p*-position of the next succeeding ring; thus rearrangement can occur to give a new ketal in which the second ring carries the carbonyl oxygen:

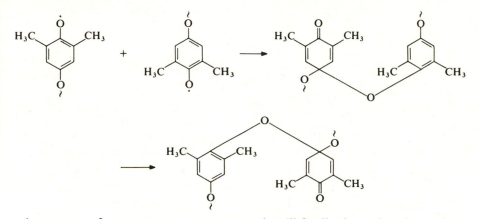

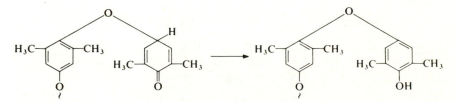

A sequence of rearrangements can proceed until finally the carbonyl oxygen is located on a terminal ring; enolization then gives a polymeric phenol which is identical to that which would be formed by the direct head-to-tail coupling of two aryloxy radicals:

It will be appreciated that redistribution and rearrangement reactions may occur within the same polymer molecule. Dissociation into aryloxy radicals can occur at any point during the rearrangement. Thus redistribution need not take place solely by transfer of a single unit; rearrangement followed by dissociation provides a means by which any number of monomer units may be transferred in what is essentially a single step.

Commercial poly(2,6-dimethyl-1,4-phenylene oxide) generally has a molecular weight in the range 25 000–60 000. The polymer is essentially linear but may contain a few branches or cross-links arising from thermal oxidation. Some commercial grades of the polymer contain a small amount of copolymerized 2,3,6-trimethylphenol.

12.6.3   Properties

Poly(2,6-dimethyl-1,4-phenylene oxide) is characterized by high tensile strength, stiffness, impact strength and creep resistance; it also has good dielectric properties. These properties are maintained over a broad temperature range (about -45–$120°C$). The polymer is self-extinguishing. The polymer has a high glass-transition temperature ($Tg = 208°C$) which restricts crystallization in normal moulding techniques. Crystalline material may be obtained by heating the amorphous material or from solutions. One notable feature of the polymer is its good dimensional stability; it has a very low coefficient of thermal expansion and low water absorption.

Poly(2,6-dimethyl-1,4-phenylene oxide) is soluble in aromatic hydrocarbons and chlorinated solvents and several aliphatic hydrocarbons cause environmental stress cracking. The polymer is outstandingly resistant to most aqueous reagents, being unaffected by acids, alkalis and detergents.

At the present time, most poly(2,6-dimethyl-1,4-phenylene oxide) intended for commercial use is blended with polystyrene (principally in the form of high-impact polystyrene). The two materials are miscible over the complete composition range. The effect of blending is to facilitate melt-processing (by lowering melt viscosity) and to lower cost whilst the desirable properties of the straight polymer are largely retained. In one recent product, the polystyrene is grafted on to the polyether. Polystyrene-modified poly(2,6-dimethyl-1,4-phenylene oxide) (commonly referred to as PPO) has found a variety of uses, which include telecommunication and business equipment housings and components and automotive parts.

12.7   POLY(*p*-PHENYLENE SULPHIDE)

Poly(*p*-phenylene sulphide) (commonly called poly(phenylene sulphide) or PPS) is the sulphur analogue of poly(*p*-phenylene oxide) and is prepared from *p*-dichlorobenzene and sodium sulphide:

The reaction is carried out in *N*-methylpyrrolidone at 270°C.

Poly(phenylene sulphide) is a strong, stiff although somewhat brittle material. Mouldings are generally processed so as to be crystalline; since the glass-transition temperature is high ($Tg = 90°C$), heated moulds are used. The outstanding characteristic of the polymer is its thermal stability, useful mechanical properties being retained up to about 250°C. Electrical insulation characteristics are also good. The polymer has excellent solvent resistance, being unaffected by all common organic solvents. The polymer is resistant to most acids and alkalis, although it is attacked by concentrated sulphuric acid. Poly(phenylene sulphide) is mostly processed by injection moulding and is used in a variety of applications such as electrical components, automotive parts and processing equipment exposed to corrosive environments.

12.8   AROMATIC POLYSULPHONES

Polysulphones are polymers which contain recurring sulphone groups ($-SO_2-$) in the main chain.

The sulphone group is highly polar and polymers containing both sulphone and p-phenylene groups might be expected to have high softening points. In fact, the simplest polymer of this kind, poly(p-phenylene sulphone) (Fig. 12.1A) melts at 520°C with decomposition and so is not a useful thermoplastic material. In order to obtain a material which can be processed on conventional equipment it is necessary to make the polymer chain less stiff and in all currently available commercial polysulphones this is achieved by the incorporation of ether links.

Three basic types of aromatic polysulphone are produced commercially (Fig. 12.1B, C and D). These products differ only slightly in molecular structure and have similar properties but, nevertheless, it has become common practice to differentiate between them by use of the designations shown. This differentiation is not justified on the basis of chemical structure since all of the products may be described as polysulphones, polyarylsulphones or polyethersulphones.

Structural unit	Tg	Designation

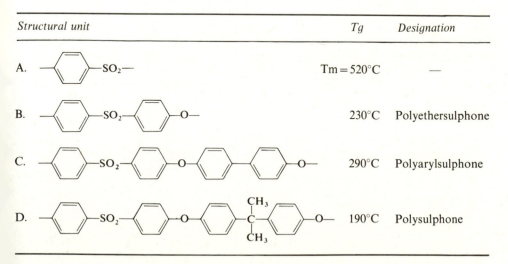

A.	Tm = 520°C	—
B.	230°C	Polyethersulphone
C.	290°C	Polyarylsulphone
D.	190°C	Polysulphone

Fig. 12.1 Types of aromatic polysulphones.

12.8.1   Preparation

There are two principal methods for the large scale preparation of aromatic polysulphones, i.e. polyetherification and polysulphonylation.

Polyetherification involves nucleophilic substitution of aromatic halogen by phenoxy ions. The functional groups may be present in the same monomer or in different monomers, as illustrated by the following schemes:

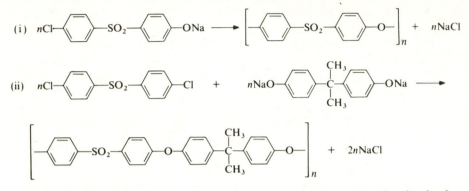

The reactions are carried out at 130–250°C in solvents which dissolve both reactants and polymer, e.g. chlorobenzene/dimethylsulphoxide.

Polysulphonylation involves electrophilic substitution of aromatic hydrogen by sulphonylium ions. Again, two approaches are possible, as illustrated by the following schemes:

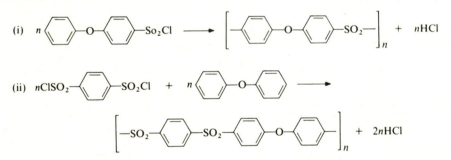

The reactions are carried out at 100–250°C in solvents such as nitrobenzene or tetrachloroethylene in the presence of a Friedel-Crafts catalyst such as ferric chloride.

12.8.2   Properties

The commercial aromatic polysulphones are always amorphous materials. Despite their regular structure, the polymers cannot be crystallized. The differences between the various commercial products are quite small. They are all rigid materials with outstanding creep resistance which are characterized by the ability to withstand high temperatures (up to about 180°C) in long-term use. Their dielectric properties are also very good. The commercial polymers are insoluble in aliphatic hydrocarbons but do dissolve in solvents such as aniline, dimethyl acetamide and pyridine. Many solvents cause environmental stress-cracking. The polymers are generally resistant to aqueous acids and alkalis but are attacked by concentrated sulphuric acid. The polysulphones may be melt-processed by standard techniques although processing temperatures are high. The polymers are used in a variety of applications, including electrical components such as coil formers, connectors and printed circuit boards, sterilizable medical equipment and components of domestic appliances such as coffee makers, heaters and microwave ovens.

12.9 POLYARYLETHERKETONES

The polyar29letherketones of commercial interest are polymers which are made up of *p*-phenylene groups joined by carbonyl and ether links. As with the aromatic polysulphones described in the preceding section, various arrangements are possible, each with a common designation (see Fig. 12.2).

At the present time, the only polyar29letherketone which is commercially available is polyetheretherketone (PEEK). Polyetherketone was available for some years but has now been withdrawn. Several other polyar29letherketones are said to be in various stages of market introduction.

12.9.1 Preparation

Polyar29letherketones may be prepared by methods which are analogous to those used for polysulphones, namely polyetherification and polyacylation.

Polyetherification occurs when fluorinated aromatic ketones react with phenates, as illustrated by the following reactions:

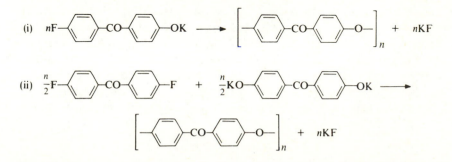

A difficulty in the preparation of polyar29letherketones is that the polymers readily crystallize (unlike the polysulphones) and thus precipitate from most organic solvents as they are being formed giving a low molecular weight

Structural unit	Tg	Tm	Designation
A. ![structure A]	154°C	367°C	Polyetherketone
B. ![structure B]	144°C	335°C	Polyetheretherketone

Fig. 12.2   Types of polyaryletherketones.

product. The problem is overcome by carrying out the reaction in arylsulphones at 280–340°C: the polymer remains in solution and high molecular weight material is produced.

Polyacylation is carried out by a Friedel-Crafts acylation of an aromatic ether by an acid chloride, as shown in the following example:

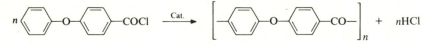

The reaction is performed in hydrofluoric acid at 20°C with boron trifluoride as catalyst. Under these conditions, the polymer remains dissolved and high molecular weights are attained.

12.9.2   Properties

Polyetheretherketone is a strong rigid material with outstanding heat resistance. It withstands long periods at 250°C with little change in mechanical properties and can be used up to 315°C. This level of heat resistance is the highest of current commercial melt-processable thermoplastics. The polymer also has excellent electrical insulation characteristics. Polyetheretherketone is resistant to all organic solvents although a few can cause environmental stress-cracking. The material is soluble in concentrated sulphuric acid. Polyetheretherketone can be injection moulded and extruded on conventional equipment and can be used to prepare glass-fibre and carbon-fibre composites. Applications include coatings for high performance wiring used in the aeronautics and nuclear fields and advanced structural composites.

REFERENCES

1. Marvel, C. S. and Hartzell, G. E. (1959) *J. Am. Chem. Soc.*, **81**, 448.
2. Kovacic, P. and Lange, R. M. (1963) *J. Org. Chem.*, **28**, 968.
3. Schnell, H. (1964) *Chemistry and Physics of Polycarbonates*, Interscience Publishers, New York, p. 38.

4. Huysmans, W. G. B. and Waters, W. A. (1967) *J. Chem. Soc.*, (B) 1163.
5. Cooper, G. D. and Katchman, A. (1969) in *Addition and Condensation Polymerization Processes*, American Chemical Society, Washington, ch. 43.

BIBLIOGRAPHY

Christopher, W. F. and Fox, D. W. (1962) *Polycarbonates*, Reinhold Publishing Corporation, New York.

Schnell, H. (1964) *Chemistry and Physics of Polycarbonates*, Interscience Publishers, New York.

Lee, H., Stoffey, D. and Neville, K. (1967) *New Linear Polymers*, McGraw-Hill Book Company, New York, ch. 3, (Poly(phenylene oxide)).

Lee, H., Stoffey, D. and Neville, K. (1967) *New Linear Polymers*, McGraw-Hill Book Company, New York, ch. 4, (Poly-*p*-xylylene).

Frazer, A. H. (1968) *High Temperature Resistant Polymers*, Interscience Publishers, New York, ch. 2, (Poly-*p*-xylylene).

Elias, H.-G. and Vohwinkel, F. (1986) *New Commercial Polymers* 2, Gordon and Breach Science Publishers, New York, ch. 8, (Aromatic polysulphones).

CELLULOSE AND RELATED POLYMERS

13.1 SCOPE

Cellulose is the most abundant of naturally occurring polymers and is of immense technological importance. In addition, cellulose may be subjected to numerous chemical modifications and several cellulose derivatives are also of industrial importance. Cellulose itself and the derivatives which find commercial use form the principal contents of this chapter. The carbohydrate, starch, is also briefly considered.

13.2 CELLULOSE

13.2.1 Sources

Cellulose is a very widely distributed polymer since it is the main constituent of the cell-wall of all plants. Within the context of this book, the cellulosic materials which are of technological importance are cotton fibre and wood together with a few other substances of less significance. These various materials are briefly considered below.

(a) Cotton fibre

The cotton plant has two types of seed hair, namely long fibres (cotton) and short fibres (linters). The long fibres (which range from about 1 to 5 cm in length) are separated from the seed by 'ginning', in which closely set high-speed circular saws pull the fibres through apertures small enough to retain the seeds. Practically the whole cotton output is used by the textile and allied industries. The short fibres remaining on the seeds are then removed by 'delinting', which is performed in much the same way as ginning. Cotton linters are used principally for the preparation of chemical cellulose, i.e. purified cellulose of reagent quality.

Oven-dried cotton contains about 90% cellulose whilst cotton linters contain 80–85%. In the preparation of chemical cellulose, cotton linters are processed in the following manner. Firstly, the linters are heated at 130–180°C under pressure with 2–5% aqueous sodium hydroxide for 2–6 hours; this treatment solubilizes particles of seed-hull and other contaminants present in the linters. The liquor is drained off. The residual linters are washed with water, bleached with gaseous chlorine or calcium hypochlorite, re-washed and dried at about 70°C. The final product has a cellulose content of about 99%.

(b)   Wood

Wood is, of course, much used as such as a material of construction. A great deal of wood is also subjected to further processing. The chief raw material for such processing is wood pulp, i.e. finely divided wood.

Besides cellulose, wood has several other constituents, the most important of which are hemi-celluloses and lignin. The chemical composition of wood varies widely according to source but a typical hardwood analysis might be as follows:

Cellulose	50.0% (dry weight)
Hemi-celluloses	24.0
Lignin	22.5
Extractable and inorganic materials	3.5

Hemi-celluloses are lower molecular weight polysaccharides, which on hydrolysis yield various pentoses, hexoses and uronic acids. They are not usually appreciably soluble in water but do dissolve in aqueous sodium hydroxide. Lignin is not a polysaccharide. Its nature has not been completely elucidated but degradation studies indicate that the material possesses a highly branched polymeric structure based on the propylbenzene unit. Since appreciable quantities of guaiacol (o-methoxyphenol) and its derivatives are frequently found in the degradation products of lignin, it has been suggested that biogenesis is through unsaturated alcohols such as coniferyl alcohol (I). An idealized lignin structure derived from this compound may be envisaged as shown overleaf.

Wood pulp is manufactured by purely mechanical means or by chemical treatment or by a combination of these two basic methods. Mechanical pulp (or groundwood) is produced by forcing barked bolts of wood lengthwise against a rapidly rotating abrasive surface in the presence of water; alternatively, chips of wood may be ground. The chemical composition of mechanical pulp is little different from that of the original wood. Mechanical pulp is used mainly for the production of cheap paper such as newsprint. The

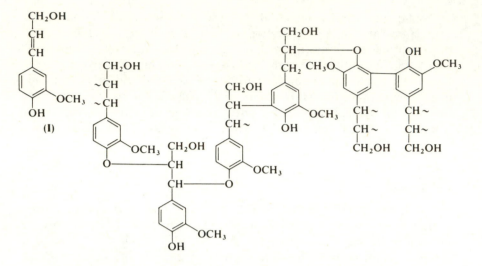

(I)

presence of lignin accounts for the yellowing and general deterioration of newsprint with age. For the manufacture of both better grades of paper and of chemical cellulose, it is necessary to remove the non-cellulosic constituents by chemical treatment. There are two principal methods used for producing chemical pulp, namely sulphite pulping and sulphate pulping.

In sulphite pulping, wood chips are treated under pressure in a digester with an aqueous solution of sulphur dioxide and calcium, magnesium or sodium bisulphite at about 130–150°C and 0.7 MPa (7 atmospheres) for 6–12 hours. The lignin is converted into water-soluble sulphonates, probably by displacement of phenolic groups:

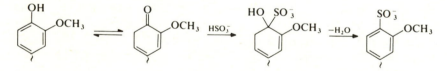

The resulting liquor is run off and the remaining pulp is then washed, bleached, sheeted and dried.

In sulphate (or kraft) pulping, wood chips are treated in a digester with an aqueous solution of sodium hydroxide and sodium sulphide at about 160–180°C and 1 MPa (10 atmospheres) for 3–5 hours. (The process is called *sulphate* pulping because sodium sulphate is used as the source of sodium sulphide.) The lignin is degraded and passes into solution as the corresponding sodium phenates. The resulting liquor is run off and the remaining pulp is washed, bleached (if required), sheeted and dried.

Until about 1940, approximately equal amounts of chemical pulp were made by these two processes. Sulphate pulp had the advantage of giving paper of higher strength but was limited by its dark brown colour which could not be satisfactorily bleached. Thus sulphate pulp was used mainly for containers and wrappers whilst sulphite pulp went mostly into printing papers. Sulphite pulp also represented a source of chemical cellulose. After 1940, sulphate pulp became dominant, mainly because of its ease of production from several species of wood not suitable for sulphite pulping and because effective bleaches (especially chlorine dioxide) were found. Furthermore, the introduction of an additional step in the sulphate process resulted in pulps with a very high cellulose content. In this modification the wood chips are 'pre-hydrolysed', i.e. given an acid-steam treatment before the alkali digestion; this treatment facilitates the removal of hemi-celluloses. The level of hemi-celluloses may be further reduced by submitting the pulp to extraction with aqueous sodium hydroxide at room temperature. Chemical pulps commonly have a cellulose content of about 92–98%, the remainder being mainly hemi-celluloses.

Up until about 1940, the use of chemical cellulose derived from wood was restricted mainly to the manufacture of regenerated cellulose whilst cotton linters were preferred as the source of cellulose for making cellulose derivatives. At this time, production of chemical cellulose from the two sources was approximately equal but since then the production of chemical cellulose from wood has grown and it now accounts for over 95% of the total. The reasons for this growth are based on price stability, uniformity of product and continued improvement in quality.

(c)   Other sources

Cellulosic fibres from vegetable sources other than cotton and wood are used in a variety of textile and industrial products. These fibres are mostly obtained either from the leaves of tropical plants or from the stems of reed-like plants. Leaf fibres are generally stiff ('hard fibres') and are used mostly for cordage. Stem fibres (also known as bast fibres) are usually finer ('soft fibres') and find use in textile applications. The cellulose content of these materials is usually in the range 70–90% (on dry weight). The more important commercial products and their principal usage are shown below.

Leaf fibres
> Sisal: cordage, paper
> Abaca: cordage, paper

Bast fibres
> Jute: sacking, webbing, twine
> Flax: linen

13.2.2 Structure

Cellulose is a carbohydrate, the structure of which may be deduced as follows. The molecular formula of cellulose is $(C_6H_{10}O_5)_n$, where n is a few thousand (see later). Hydrolysis of cellulose by boiling with concentrated hydrochloric acid yields D-glucose (II) in 95–96% yield. Thus cellulose is a polyanhydro-glucose. When cellulose is subjected to acetolysis (i.e. simultaneous acety-lation and hydrolysis) by treatment with a mixture of acetic anhydride and concentrated sulphuric acid, cellobiose octa-acetate is formed. Thus the structure of cellulose is based on the cellobiose unit. Cellobiose is known to be the disaccharide, 4-O-β-D-glucopyranosyl-D-glucopyranose (III). Finally, very careful acetolysis of cellulose produces a cellotriose, a cellotetraose and a cellopentaose and in each of these all the 1,4-links have been shown to be β-links (from calculations of the optical rotations). Thus it follows that all the 1,4-links in cellulose are β-links; cellulose may therefore be represented by structure (IV).

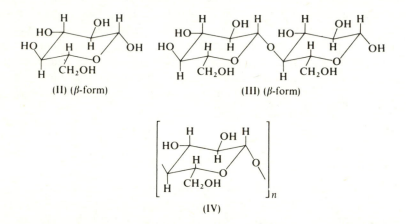

The molecular weight of cellulose has been determined by several investi-gators; their results have not always been in agreement, possibly due to the occurrence of degradation during the course of their work and the use of cellulose from different sources. Probably the most reliable results are those of Schulz and Marx [1] who converted various carefully obtained cellulose specimens to corresponding nitrates and determined their molecular weight by sedimentation-diffusion and viscometry. These workers showed that no degradation occurred during this procedure; some of their results are shown in Table 13.1 together with the ranges of degree of polymerization which are likely in cellulose from industrial sources.

The regularity of the cellulose chain and extensive hydrogen bonding between hydroxyl groups in adjacent chains cause cellulose to be a tightly-packed crystalline material. As a result, it is insoluble, even in hydrogen

Table 13.1 Average degree of polymerization of cellulose from various sources

Source	Average degree of polymerization (n)
Cotton, American	6700
Cotton, Egyptian	6200
Cotton linters	6500
Pine wood	3300
Flax fibre	8000
Chemical cellulose (cotton)	500–3000
Chemical cellulose (wood pulp)	500–2000

bonding solvents (but it does swell in water). Further, although cellulose is a linear polymer, it is infusible; decomposition occurs before the melting point is reached. Thus cellulose cannot be processed in the melt or in solution. However, several derivatives in which there is less hydrogen bonding are processable. Numerous derivatives of cellulose may be obtained by reaction of the hydroxyl groups but esterification and etherification are the commonest means of preparing processable polymers. The more important commercial cellulosic polymers are described below.

13.3 REGENERATED CELLULOSE

As mentioned above, cellulose itself is insoluble whereas many of its derivatives are soluble. It is common practice to process (usually by extrusion) a solution of a cellulose derivative and then, having manipulated the polymer into the desired shape (commonly fibre or film), to remove the modifying groups to reform unmodified cellulose. Such material is known as regenerated cellulose.

13.3.1 Development

The first form of regenerated cellulose was developed in 1884 by de Chardonnet. Cellulose was nitrated and dissolved in a mixture of ether and ethanol. The solution was extruded through a small orifice and the solvent evaporated to leave a fibre. The fibre (which was too inflammable for direct use) was then passed through aqueous ammonium hydrogen sulphide at about 40°C; this treatment resulted in denitration. The product (often called *Chardonnet silk*) was popular for many years but was gradually displaced by other synthetic fibres. Commercial production of the material ceased in 1949.

A second process for producing regenerated cellulose fibre was introduced in 1897 in Germany. In this method, cellulose is treated with an ammoniacal solution of cupric hydroxide ($Cu(NH_3)_4(OH)_2$) to form a soluble complex. The solution is then spun into dilute sulphuric acid to regenerate the cellulose. This process is relatively expensive because of the need to recover copper and the product, called *cuprammonium rayon*, is no longer of commercial importance.

The origins of present-day methods of producing regenerated cellulose may be traced to the discovery in 1892 by Cross, Bevan and Beadle that cellulose can be rendered soluble by treatment with sodium hydroxide and carbon disulphide and regenerated by acidification of the solution. In 1900 Topham devised means of producing fibres and full-scale production was begun by Samuel Courtauld and Co. Ltd. (UK) in 1907. This product, known as *viscose rayon*, is now a major commercial textile fibre. In 1908 Brandenberger developed a technique for the production of continuous film (and coined the familar name *cellophane* (from *cellu*lose and dia*phane* (French: transparent)). This product is widely used in packaging applications but is facing serious competition from the polyolefins. Viscose rayon and cellophane are considered below.

A further type of regenerated cellulose is prepared by the hydrolysis of stretched cellulose acetate fibre. The resulting fibre, known as *saponified acetate rayon*, has high strength and low elongation (see Table 13.2).

Mention may be made here of a product which has been manufactured for many years under the name of *vulcanized fibre*. In the preparation of this material, sheets of absorbent paper are soaked in aqueous zinc chloride. The zinc chloride causes the cellulose fibres to swell and be covered with a gelatinous layer. The sheets are pressed together and then the zinc chloride is removed by prolonged washing. The laminate is then dried and consolidated under pressure to give a rigid product. The material may be shaped by first softening in hot water or steam and then pressing in moulds. Vulcanized fibre is used for the manufacture of such articles as suitcases and protective guards.

13.3.2   Preparation

As indicated previously, the most important current method of preparing regenerated cellulose is by the viscose (or xanthate) process. The reactions involved in this process may be represented as follows:

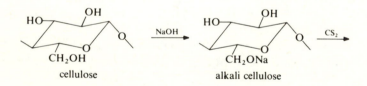

cellulose   alkali cellulose

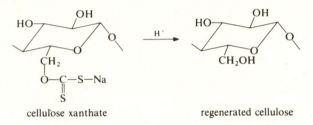

cellulose xanthate   regenerated cellulose

In the first step, cellulose is steeped in approximately 20% aqueous sodium hydroxide at room temperature for 20–60 minutes. This treatment results in the formation of 'alkali cellulose', which is a sodium alcoholate; reaction probably takes place at each hydroxymethyl group. (This reaction is also the basis of the important textile process of *mercerization* in which cotton is treated with aqueous sodium hydroxide to give a fibre of enhanced lustre and dyeability.) The alkali cellulose is pressed to remove liquid and is then shredded and allowed to age at room temperature for 2–3 days. During this time, oxygen is absorbed from the air and there is a reduction of molecular weight (which facilitates the subsequent spinning operation). The alkali cellulose is then treated with carbon disulphide in a rotary drum at 25–30°C for 2–4 hours to give cellulose xanthate (in which there is approximately 1 xanthate group per 2 glucose units). This product is dissolved in aqueous sodium hydroxide and the resulting solution (which is known as *viscose*) is allowed to age at room temperature for 2–5 days. During this period, some hydrolysis occurs and about half of the xanthate groups are removed; as a result the solution can be coagulated more readily in subsequent operations. The solution is then fed either from spinnerets with many small holes (in the production of fibre) or from hoppers with slits (in the production of film) through a bath containing 10–15% sulphuric acid and 10–20% sodium sulphate at 35–40°C. In this step the viscose is firstly coagulated and then hydrolysed.

Subsequent operations depend on the type of product involved. Film is washed, bleached, plasticized (with ethylene glycol or glycerol) and dried; sometimes a coating of cellulose nitrate is applied to give heat-sealability and lower moisture permeability. Fibre is subjected to such treatments as washing, bleaching, twisting and crimping. It is also possible to carry out a drawing operation on fibre as it passes through the coagulating bath. In this process the filaments are stretched by 50–150% of their original length and crystalline orientation occurs. The product, known as *high-tenacity rayon*, has high strength and low elongation (see Table 13.2).

13.3.3 Properties

Comparative values for some properties of typical commercial fibres prepared from regenerated cellulose are given in Table 13.2; cotton is also shown for comparison. Some properties of cellophane are shown in Table 13.3. It will be noted that although regenerated cellulose is chemically identical to

Table 13.2 Typical values for various properties of cellulosic fibres

	Cotton	Cuprammonium rayon	Regular viscose rayon	High-tenacity viscose rayon	Saponified acetate rayon	Cellulose acetate (secondary)	Cellulose tri-acetate
Specific gravity	1.54	1.52	1.51	1.52	1.52	1.30	1.32
Tensile strength (g/denier):							
Sample at 21°C, 65% RH	3.0–4.9	1.7–2.3	1.5–2.4	3.0–4.6	5.0–8.0	1.0–1.5	1.2–2.1
Sample wet	3.3–6.4	0.9–1.3	0.7–1.4	1.9–3.0	4.5–7.0	0.7–1.1	0.8–1.4
Elongation at break (%):							
Sample at 21°C, 65% RH	3–7	10–17	15–30	9–26	6–6.5	25–40	20–30
Sample wet	8–10	17–35	20–40	14–32	6–6.5	30–45	30–40

Table 13.3 Typical values for various properties of cellulosic films

	Regenerated cellulose (cellophane)	Cellulose acetate (secondary)	Cellulose triacetate	Cellulose acetate-butyrate	Ethyl cellulose
Tensile strength (MPa) (1bf/in²)	30–130 4400–18 600	37–96 5400–13 900	62–110 9000–16 000	28–67 4100–9700	47–73 6800–10 600
Elongation at break (%)	15–45	25–45	10–40	40–100	30–40
Tear Strength (kgf/m) (1bf/in)	2000–9200 110–515	980–7400 55–415	980–7100 55–395	1400–1900 80–105	3800–7100 215–395
Water absorption, 24 hr. immersion (%)	45–115	3.6–6.8	3.5–4.5	0.1–3.4	2.5–7.5

'native' cellulose, its mechanical properties are generally inferior. This is a consequence of the reduction in crystallinity and molecular weight which inevitably occurs during processing. Regenerated cellulose in the form of rayon or cellophane normally has a degree of polymerization of about 300–400.

As mentioned previously, cellulose is insoluble in organic solvents but the polymer is sensitive to acids and alkalis, which bring about chain scission and eventual solution. Cellulose is degraded when heated in air above about 140°C; heating in an inert atmosphere at 170°C has little effect. Cellulose is readily oxidized at its hydroxyl groups by a wide range of reagents and various oxygenated derivatives have been prepared. Cellulose is also readily grafted and several graft copolymers have been obtained [2].

13.4   CELLULOSE NITRATE

13.4.1   Development

Although there are earlier references to the nitration of starch and paper, Schönbein is generally credited with the first preparation of cellulose nitrate in 1845. Schönbein certainly established the conditions for controlled nitration; he also recognized the potential use of cellulose nitrate as an explosive and the material became known as *guncotton*. In 1848, a solution of cellulose nitrate in a mixture of ether and ethanol was introduced as a medical aid. The solution, called *collodion*, was applied to wounds; evaporation of the solvent left a protective film. Collodion also found application in the manufacture of photographic film base. At the Great International Exhibition held in London in 1862, the English inventor Alexander Parkes showed solid articles moulded from a mixture of cellulose nitrate and various oils and solvents. This event is generally regarded as representing the birth of the plastics industry. In 1866, Parkes began commercial production of his material but in 1868 his company was liquidated. The factory-produced material was not up to the standard of the exhibition material, probably due to attempts to cut costs by using cheaper materials and by omitting to remove completely volatile solvents. Spill (an associate of Parkes) continued operations and produced modest amounts of acceptable material.

In 1865, J. W. Hyatt in the United States began to investigate the production of billiard balls from materials other than ivory and in 1869 he patented the use of collodion for coating billiard balls. In 1870, the brothers J. W. and I. S. Hyatt took out a patent describing the preparation and moulding of a material based on cellulose nitrate and camphor. Although Parkes and Spill mentioned camphor in their work, it was the Hyatts who appreciated the unique value of camphor as a plasticizer for cellulose nitrate. Thus, using techniques which form the basis of current practice, the Hyatt brothers overcame the difficulties which beset the earlier workers; in particular they eliminated shrinking and warping of articles through evaporation of volatile solvents. In 1872, the now familiar name *celluloid* was coined to describe the material which quickly became a commercial success. The manufacture of celluloid was soon established in several countries and the material became very widely used in applications such as collars, combs, dentures, knife handles, spectacle frames, toothbrushes, toys and, later, cine-film. Celluloid is now manufactured only on a limited scale, having been largely displaced by other plastics; nevertheless, it certainly may be counted as the first successful commercial plastics material. Present day uses include fashion accessories, guitar facings, table-tennis balls and dice.

Cellulose nitrate also finds use in surface coatings, which have their origins in the previously mentioned collodion. A tremendous growth of cellulose nitrate lacquers, especially for finishing automobiles, occurred immediately after the First World War. Cellulose nitrate lacquers continue to be of importance as furniture finishes but automobile finishes are now largely based on acrylic polymers (section 6.7.2).

13.4.2 Preparation

The nitration of cellulose is usually carried out with a mixture of nitric and sulphuric acids. The reaction conditions used depend somewhat on the type of product required (see later). Typically, dried cellulose is stirred with the acid mixture at about 30–40°C for 20–60 minutes. In the prepration of, for example, plastics-grade material, the acid mixture is made up of nitric acid (25%), sulphuric acid (55%) and water (20%). During the acid treatment, the fibrous structure of the cellulose is retained and there is little change in appearance. Acid is then removed by centrifuging and the residual cellulose nitrate is dropped into a 'drowning' tank where retained acid is diluted with a large volume of water. The resulting slurry is washed with water and then stabilized (see section 13.4.3) by boiling for several hours. The product is bleached with sodium hypochlorite and washed. At this point, material intended for use in lacquers is often subjected to a controlled degradation reaction in which the molecular weight is reduced so that solutions are not too viscous for convenient application. Degradation is carried out by heating a slurry of cellulose nitrate in water under pressure at about 130–160°C for up to 30 minutes. The product (degraded or otherwise) is then drained to give a material containing about 40% water. This material is then dehydrated by forcing ethanol through it under pressure. The final product contains about 30% ethanol and 5% water; cellulose nitrate is commonly handled in this form rather than in the less stable dry state.

The reaction between cellulose and nitric acid is one of esterification. By careful control of the reaction conditions (especially the water content of the

Table 13.4   Properties of typical commercial grades of cellulose nitrate

Nitrogen content (%)	Degree of esterification (no. of nitrate groups/ glucose residue)	Average degree of polymerization	Field of application
10.7–11.1	1.9–2.0	200–500	Plastics
10.7–11.1	1.9–2.0	70–850	Lacquers
11.2–12.3	2.0–2.4	70–850	Lacquers
12.4–13.5	2.4–2.8	3000–5000	Explosives, propellants

acid mixture) it is possible to regulate the extent of reaction, i.e. various degrees of esterification may be achieved. The various products may be characterized by nitrogen content; typical values of this and other quantities for commercial grades of cellulose nitrate are given in Table 13.4. The precise location of the nitrate groups at various degrees of esterification is not known (all hydroxyl groups may not be equally available) and the values given for the degree of esterification should be regarded as statistical averages over the whole material. The esterification of cellulose to give a material corresponding to cellulose dinitrate might be represented as follows:

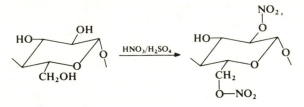

It may be noted that although cellulose nitrate is very commonly referred to as nitrocellulose, this terminology is incorrect since nitro-compounds properly contain a C–N bond.

(a)   Celluloid

As mentioned previously, celluloid is cellulose nitrate plasticized with camphor. At the present time, the material is produced mainly in sheet form. In a typical process alcohol-wet cellulose nitrate is kneaded at about 40°C with camphor (about 30%); the ethanol-camphor mixture constitutes a powerful solvent for the polymer and a gelatinous mass is obtained. Colourants, if any, are added at this stage. The dough is then heated at about 80°C on rollers until the alcohol content is reduced to about 15%. The calendered sheets are laid up in a press and consolidated into a block. The block is then sliced into sheets of thickness 0.01–2 cm; various multi-colour effects may be obtained by plying sliced sheets of different colours and then re-slicing on the bias. The sheets are then allowed to season for several days at about 50°C so that the volatile content is reduced to about 2%.

Although the plasticizing effect of camphor (V) was first appreciated over a hundred years ago, a better plasticizer for cellulose nitrate (for use in bulk form) has not been found. Many alternatives have been investigated but the products invariably exhibit inferior mechanical properties. It may be noted that more conventional plasticizers perform satisfactorily in cellulose nitrate surface coatings. The solubility parameters of cellulose nitrate and camphor differ considerably. This indicates there is specific interaction between the materials, probably hydrogen bonding between unsubstituted hydroxyl groups in the polymer and the camphor carbonyl group.

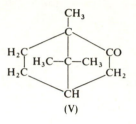

(V)

13.4.3 Properties

Dry cellulose nitrate is a colourless solid, much like cellulose in appearance but with a harsher feel. The straight material has a glass transition temperature of about 53°C but only plasticized material finds commercial use. In the case of surface coatings, plasticizers (such as dibutyl phthalate and tritolyl phosphate) are necessary to give films of acceptable flexibility and adhesion. For use in bulk form, the material is plasticized (with camphor) to permit compounding and shaping operations. Typical values of some physical properties of celluloid are shown in Table 13.5.

The solubility characteristics of cellulose nitrate depend to some extent on nitrogen content and molecular weight. All types of cellulose nitrate are soluble in a wide range of solvents, including esters such as ethyl, butyl and amyl acetates; ethers such as dioxan; and ketones such as acetone and methyl ethyl ketone. Cellulose nitrate of low nitrogen content and low molecular weight is soluble in ethanol. Cellulose nitrate with nitrogen content in the range 11.2–12.3% has the important characteristic that solutions show maximum tolerance for non-solvents such as aliphatic and aromatic hydrocarbons. Thus grades of cellulose nitrate with nitrogen contents within this range are generally preferred for the preparation of lacquers since they allow wider scope in formulating cheap solvent systems with desired properties.

Table 13.5 Typical values for various properties of cellulosic plastics [3]

	Cellulose nitrate (celluloid)	Cellulose acetate (secondary)	Cellulose acetate-butyrate	Cellulose acetate-propionate	Cellulose tripro-pionate	Ethyl cellulose
Specific gravity	1.35–1.40	1.27–1.32	1.15–1.22	1.19–1.23	1.18–1.24	1.12–1.15
Tensile strength (MPa)	34–69	24–76	17–52	24–50	14–41	41–62
(lbf/in^2)	5000–10 000	3500–11 000	2500–7500	3500–7300	2000–6000	6000–9000
Elongation at break (%)	10–40	5–55	8–80	30–100	45–65	10–40
Impact strength, Izod (J/m)	110–430	27–270	21–270	48–270	64–590	53–350
(ft lbf/in)	2.0–8.0	0.5–5.0	0.4–5.0	0.9–5.0	1.2–11.0	1.0–6.5
Flow temperature (°C)	145–152	115–165	115–165	150–180	145–180	100–150
Water absorption, 24 hr. immersion (%)	0.6–2.0	1.0–3.0	0.9–2.4	1.5–2.8	1.6–2.0	0.5–1.5

Ranges of figures are given since the value of a particular property is very dependent on formulation.

Cellulose nitrate is relatively water resistant. It is little affected by dilute acids but concentrated acids and alkalis cause chain scission. Dilute alkalis slowly hydrolyse the material and cause surface crazing. Alkali hydrosulphides bring about rapid hydrolysis.

Cellulose nitrate is highly inflammable. The most highly nitrated grades burn with great rapidity if ignited in the open and explode upon detonation; these grades find use in explosives and propellants. The less nitrated grades burn readily but cannot be detonated. Cellulose nitrate with nitrogen content in the range 10.7–11.1% is the least inflammable and is preferred for the production of celluloid. The thermal stability of cellulose nitrate depends to a large extent on the sulphate content of the polymer. The use of sulphuric acid in the nitration of cellulose results in the formation of hydrogen sulphate ester groups and it is thought these are hydrolysed on heating in the presence of moisture to give sulphuric acid which initiates a rapid degradative process. In order to remove as many of these ester groups as possible before the polymer is put into service, cellulose nitrate is boiled with water for several hours after nitration has been effected (see section 13.4.2). Despite this treatment, celluloid is not sufficiently stable to be processed by conventional injection moulding techniques and this factor has contributed to the decline of the material. Cellulose nitrate is discoloured on exposure to ultraviolet light and there is a decrease in tensile and impact strengths.

13.5   CELLULOSE ACETATE

13.5.1   Development

The acetylation of cellulose was first carried out in 1865 by Schutzenberger. The product remained of mainly academic interest until 1894 when Cross and Bevan described an acetylation process suitable for industrial use. However, commercial development was very limited because only relatively toxic and costly solvents, such as chloroform and carbon tetrachloride, could be used to prepare solutions for such operations as fibre spinning or film casting. A breakthrough occurred in 1905 when Miles discovered that mild hydrolysis of the primary acetylation product yielded a cellulose acetate which was soluble in the cheap solvent acetone. Commercial development was still slow, although some film was being manufactured by 1910. During the First World War cellulose acetate 'dope' became of great importance for weather-proofing and stiffening the fabric of aircraft wings. After the war there was a large surplus production capacity and renewed efforts were made to find civilian outlets. During the 1920s, the brothers C. and H. Dreyfus formed the British Celanese Co. (UK) and developed a process for the production of cellulose acetate fibre (known as *acetate rayon* or *acetate*). Initially, the fibre was difficult to dye by existing techniques but after this difficulty had been overcome acetate rayon became an important textile fibre. It is also now used extensively for cigarette filters.

In 1921 Eichengruen designed the modern injection moulding machine and cellulose acetate became the principal thermoplastic moulding material. This position was maintained until after the Second World War but since this time cellulose acetate has been largely displaced by polyethylene and polystyrene. The injection moulding of cellulose acetate is now mainly restricted to such items as toothbrush handles and combs. Cellulose acetate film is still employed for packaging purposes and sheet is used for display boxes and spectacle frames. Cellulose acetate lacquers find limited use for coating materials such as fabrics, glass, metals and paper. It may also be noted that after the war, methylene chloride became available on a large scale and represented a low-cost solvent for cellulose triacetate (primary cellulose acetate). As a result, it became feasible to use this polymer for both fibre and film, commercial production of which began in 1954. Cellulose triacetate film is used for photographic film base and motion picture sound tape.

13.5.2 Preparation

The acetylation of cellulose is usually carried out with acetic anhydride in the presence of sulphuric acid as catalyst. Although the reaction bears some resemblance to nitration there is an important difference. As mentioned previously (section 13.4.2), the degree of nitration may be regulated by choice of reaction conditions but it is not practicable to stop acetylation short of the essentially completely esterified triacetate. If acetylation is interupted before esterification is complete a heterogeneous product is obtained, which in effect consists of a mixture of cellulose triacetate and cellulose and is of little use. However, it is possible to partially hydrolyse cellulose triacetate so that products with various degrees of esterification are, in fact, obtainable, although only by a two-step process. It may be noted that cellulose triacetate is often known as 'primary cellulose acetate' (since it is the first product in the two-step process) whilst partially hydrolysed material is called 'secondary cellulose acetae' (or, commonly, just 'cellulose acetate'). Commercial materials are usually characterized by 'acetic acid yield' or 'combined acetic acid' (i.e. the weight of acetic acid produced by complete hydrolysis of the ester) or, alternatively, by 'acetyl content' (i.e. the weight of acetyl radical (CH_3CO-) in the material). Typical values of these quantities for various commercial grades of cellulose acetate are shown in Table 13.6, together with the corresponding degrees of esterification.

(a) Secondary cellulose acetate

In a typical process, dried cellulose is given a pre-treatment with glacial acetic acid at about 50°C in order to open up the fibre structure so that subsequent

Table 13.6 Properties of typical commercial grades of cellulose acetate

Acetic acid yield (%)	Acetyl content (%)	Degree of esterification (no. of acetate groups/ glucose residue)	Average degree of polymerization	Field of application
52–54	37.3–38.7	2.2–2.3	200–250	Injection moulding
54–56	38.7–40.1	2.3–2.4	250–300	Fibre, film and lacquers
56–58	40.1–41.6	2.4–2.6	—	Injection moulding
61–62.5	43.7–44.8	2.9–3.0	—	Triacetate fibre and film

acetylation proceeds more rapidly and uniformly. The pre-treated cellulose (100 parts cellulose, 40 parts acetic acid) is then added to the acetylating mixture, which consists of acetic anhydride (300 parts) and sulphuric acid (1 part) together with methylene chloride (400 parts). The acetic acid and methylene chloride are solvents for cellulose triacetate and make for more rapid and homogeneous acetylation. In addition these diluents permit better temperature control; an advantage of the low-boiling methylene chloride (b.p. 40°C) is that any large heat build-up is effectively alleviated by evaporation of the solvent. The reaction mixture is agitated, with cooling, at about 25–35°C for 5–8 hours. At this point the product is a viscous solution of cellulose triacetate (primary cellulose acetate). In the second stage of the process the triacetate is partially hydrolysed, without isolation. To the acetylation product is added water, usually as 50% aqueous acetic acid (to prevent precipitation of the polymer), and then the solution is allowed to stand (age) until the required degree of hydrolysis is reached; typically this requires 72 hours at room temperature. Sodium acetate is added to neutralize the sulphuric acid and then the methylene chloride is distilled off and recovered. The partially hydrolysed material (secondary cellulose acetate) is isolated by precipitation with water and thoroughly washed with water. The product is then stabilized by boiling with very dilute sulphuric acid (about 0.02%) for about 2 hours; this treatment removes from the polymer hydrogen sulphate ester groups which, if left, cause instability (cf. cellulose nitrate, section 13.4.3). Finally the product is dried to a moisture content of about 2%.

The preparation of secondary cellulose acetate might be represented as follows:

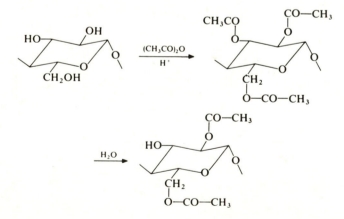

(b)   Cellulose triacetate

As indicated previously, cellulose triacetate is necessarily a precursor of secondary cellulose acetate. It is possible, therefore, to obtain the triacetate by acetylating cellulose as described above and then immediately precipitating the product rather than carrying out the hydrolysis reaction. In one process, acetylation is completed and then the sulphuric acid is neutralized with magnesium acetate. A small amount of water is then added to the mixture at 50–100°C whereby hydrogen sulphate ester groups are removed and replaced by acetate groups. The triacetate is then precipitated with water, washed and dried. Cellulose triacetate is also conveniently prepared by *heterogeneous acetylation*, in which the product does not pass into solution and there is no need for isolation by precipitation. (This is in contrast to the process described above which involves *homogeneous acetylation*.) In one process, pre-treated cellulose is acetylated at 25–40°C with acetic anhydride in the presence of the non-solvent, benzene. During this reaction, the fibrous nature of the cellulose is retained and the product is easy to handle. Also, perchloric acid is used as catalyst; this acid does not esterify cellulose and the need for subsequent stabilization is avoided. When acetylation is complete, the product is filtered off, washed and dried.

13.5.3 Properties

Many physical and chemical properties of the products of acetylation of cellulose are strongly dependent on the degree of esterification and it is important to distinguish between the commercial products cellulose acetate (52–56% acetic acid yield), 'high-acetyl' cellulose acetate (56–59% acetic acid yield) and cellulose triacetate (61–62.5% acetic acid yield).

Cellulose acetate with a low degree of acetylation (52–54% acetic acid yield) is usually preferred for use in general purpose injection moulding compounds since increasing acetylation is accompanied by a decrease in hardness and impact strength and by an increase in the temperature required for satisfactory moulding. Even at this degree of acetylation, cellulose acetate decomposes below its softening point and it is necessary to add a plasticizer (e.g. dimethyl phthalate or triphenyl phosphate) (usually 25–35%) in order to obtain a mouldable material. The type and amount of plasticizer used greatly affect the physical properties of the resultant moulding. Typical ranges for some properties are indicated in Table 13.5.

Cellulose acetate with a slightly higher degree of esterification (54–56% acetic acid yield) is generally preferred for the preparation of fibres, films and lacquers on account of the enhanced water resistance which results from a lower hydroxyl content. Plasticizers are usually present in films and lacquers. Some physical properties of a typical fibre and film are shown in Tables 13.2 and 13.3 respectively. Compared to viscose rayon, the fibre has better handle, drape and crease-resistance but has lower strength and is more difficult to dye.

   High-acetyl cellulose acetate (56–58% acetic acid yield) has found occasional use in injection moulding compounds where improved dimensional stability is required but processing is more difficult.

   Cellulose triacetate (61–62.5% acetic acid yield) is difficult to plasticize adequately enough to permit melt processing; hence the material finds little use in moulding compositions. The principal applications have been for fibres and films, some typical properties of which are shown in Tables 13.2 and 13.3 respectively. Compared to cellulose acetate of lower degree of esterification, cellulose triacetate gives fibres which are more crease and heat resistant and which can be more durably pleated.

   As indicated previously the degree of esterification of the various types of cellulose acetate has a significant influence on solubility. This effect is illustrated in Table 13.7 wherein solubilities of various types of cellulose acetate are compared. The most important solvent for materials with lower degrees of esterification is acetone; about 56% acetic acid yield generally represents the upper limit of acetone solubility for commercial products. In the manufacture of fibre, a solution of cellulose acetate in acetone is extruded through a spinneret into a heated chamber, wherein the acetone is evaporated to leave the fibre. In the production of packaging film, a solution of cellulose acetate in acetone is cast on to a band and the solvent is evaporated, leaving the film. A methylene chloride/methanol mixture is the most common solvent used for the preparation of cellulose triacetate fibre and film.

Table 13.7   Solubilities of various types of cellulose acetate

Solvent	Cellulose acetate (52–56% AcOH)	High-acetyl cellulose acetate (58–59% AcOH)	Cellulose triacetate (61–62.5% AcOH)
n-Hexane	I	I	I
Benzene	I	I	I
Acetic acid	S	S	Sw
Ethanol	I	I	I
Methylene chloride	pS	S	S
Ethylene chloride	Sw	Sw	I
Chloroform	Sw	pS	S
Methyl acetate	S	Sw	Sw
Ethyl acetate	Sw	Sw	I
Acetone	S	Sw	I
Methyl ethyl ketone	S	I	I
Methylene chloride/ methanol (4:1)	S	S	S

I = insoluble   Sw = swollen
S = soluble   pS = partially soluble

As mentioned earlier, cellulose acetates of increasing degree of acetylation have lower water absorption. Thus cellulose triacetate is preferred for photographic film where dimensional stability is required. All types of cellulose acetate are readily hydrolysed by aqueous acids and alkalis; cellulose triacetate is initially more resistant to hydrolysis since it is less easily wetted.

13.6 OTHER CELLULOSE ESTERS

Various other cellulose esters have been investigated and cellulose tripropionate and the mixed esters, cellulose acetate-propionate and cellulose acetate-butyrate are commercially available. Of these materials, which all have similar properties and applications (see Table 13.5 for some comparative properties), cellulose acetate-butyrate is probably the best known and is described below. These polymers have larger side-chains than cellulose acetate and with equal degrees of esterification, molecular weight and plasticizer content they have lower density and are softer and easier to mould. The larger hydrocarbon side-chain also results in slightly lower water absorption.

13.6.1 Cellulose acetate-butyrate

Cellulose acetate-butyrate (CAB) is prepared in a manner similar to that described previously for cellulose acetate. Esterification is carried out using a mixture of acetic anhydride and butyric anhydride with sulphuric acid as catalyst and then the product is slightly hydrolysed. Depending on the reaction conditions, various products may be obtained; commercial materials generally have about 1–2 butyryl groups per glucose residue. (See Table 13.8 for typical analyses.)

Generally, an increase in butyryl content increases flexibility, moisture resistance, solubility and compatibility with resins; as the butyryl content increases, hardness, tensile strength and heat resistance decrease. Cellulose

Table 13.8 Properties of typical commercial grades of cellulose acetate-butyrate

Acetyl content (%)	Butyryl content (%)	Hydroxyl content (%)	Degree of esterification (no. of groups/glucose residue)			Field of application
			Acetate	Butyrate	Hydroxyl	
29.5	17	1	2.1	0.7	0.2	Lacquers
20.5	26	2.5	1.4	1.1	0.5	Lacquers
13	37	2	0.95	1.65	0.4	Plastics, lacquers
6	48	1	0.5	2.3	0.2	Melt coatings

acetate-butyrate injection mouldings (see Table 13.5 for typical properties) are particularly tough and are used for such applications as tabulator keys and tool handles; film (see Table 13.3 for typical properties) is employed as a puncture-resistant packaging material. Cellulose acetate-butyrate is also used in lacquers for such items as cables, fabrics and furniture and in melt and strip coatings.

13.7   CELLULOSE ETHERS

The commercial history of cellulose ethers dates from 1912 when Dreyfus, Leuchs and Lilienfeld independently and almost simultaneously patented processes for the production of cellulose ethers soluble in organic solvents. Since this time, several cellulose ethers have become commercially available and have found a variety of uses. In general, however, cellulose ethers have not assumed such commercial importance as cellulose esters.

In general, cellulose ethers have found application mostly in solution and thus their solubility characteristics are of some interest. The introduction of a small number of alkyl (e.g. methyl or ethyl) groups into the cellulose molecule sufficiently opens up the structure to permit solubility in aqueous sodium hydroxide. As substitution increases, the products become soluble in decreasing concentrations of alkali and at length they are soluble in water. As the number of alkoxy groups (which are less hydrophilic than hydroxyl groups) increases, the products become less soluble in water and more soluble in polar organic solvents. At higher degrees of substitution, solubility in polar organic solvents declines whilst solubility in non-polar solvents increases. These principles are illustrated by the solubility characteristics of ethylcellulose shown in Table 13.9. It may be noted that if the substituent group in a cellulose ether is too hydrophobic, the ether may be insoluble in water at all levels of substitution; such is the case with butyl- and benzylcellulose. On the

Table 13.9   Solubility of ethylcellulose of various degrees of substitution

Number of ethoxy groups per glucose residue	Effective solvents
about 0.5	Aqueous sodium hydroxide (approx. 5%)
0.8–1.3	Water
1.5–2.0	Methanol
2.1–2.4	Ethyl acetate, ethylene dichloride, benzene/methanol.
2.4–2.6	Ethyl acetate, benzene, toluene, carbon tetrachloride
2.6–2.8	Benzene, toluene

other hand, ethers with very hydrophilic groups may be water-soluble at high degrees of substitution; hydroxyethylcellulose is a material of this kind.

The more common cellulose ethers are described below.

13.7.1   Methylcellulose

The preparation of methylcellulose from cellulose may be represented as follows:

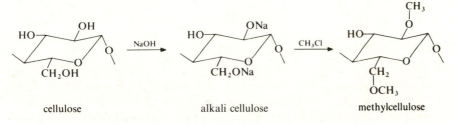

cellulose   alkali cellulose   methylcellulose

In the first step, cellulose is treated with approximately 50% aqueous sodium hydroxide at 60°C for about 20 minutes. The resulting alkali cellulose is then placed in an autoclave and heated with methyl chloride at about 70°C and 0.5 MPa (5 atmospheres) for 12 hours. The autoclave is cooled and unreacted methyl chloride is vented off. The product is washed with water at 85–90°C to remove sodium hydroxide and sodium chloride and then dried. The degree of etherification of the methylcellulose produced may be regulated by control of the reaction conditions; commercial materials generally have methoxy contents ranging from about 5 to 30% (0.3–1.8 methoxy groups per glucose residue).

The introduction of the relatively small methyl group into the cellulosic structure is not sufficient to render methylcellulose fusible without decomposition and the polymer cannot be melt processed. However, the disruption of the cellulosic structure is sufficient to make methylcellulose soluble. The solubility characteristics of methylcellulose are similar to those of ethylcellulose. The following solvents are effective at the degrees of substitution indicated: aqueous alkali (0.1–0.6), cold water (1.3–2.4), oxygenated solvents (2.4–2.7) and hydrocarbons (2.6–2.8). As indicated previously only products with degrees of substitution in the range 0.3–1.8 are of commercial interest. Aqueous solutions of methylcellulose are unusual in that they reversibly gel on heating. In the cold, the polymer chains are hydrated by layers of water molecules which allow the chains to move over one another so that a smooth-pouring viscous solution is obtained. On heating to about 50–55°C the water molecules break away, their lubricating action is lost and the chains interlock; the solution therefore gels. Methylcellulose is also soluble in some organic solvents, including dimethyl sulphoxide, ethyl lactate and propylene glycol. Methylcellulose finds application mainly on account of its water-solubility; the material is used as a thickener and emulsifier in adhesives, cosmetics and latex paints and for coating pharmaceutical tablets.

A mixed ether, hydroxypropyl-methylcellulose, is also available. This is prepared by treating alkali cellulose with a mixture of methyl chloride and propylene oxide. Commercial grades of the polymer generally have a methoxy content of about 30% and a hydroxypropoxy content of about 5–10%. The mixed ether has increased organo-solubility and thermoplasticity over the methylcellulose counterparts.

13.7.2 Ethylcellulose

Ethylcellulose is prepared from alkali cellulose and ethyl chloride by procedures entirely analogous to those described above for methylcellulose. Similarly, the degree of etherification can be regulated by control of reaction conditions; commercial materials generally have ethoxy contents in the range 44–50% (2.2–2.6 ethoxy groups per glucose residue).

Unlike methylcellulose, ethylcellulose can be melt-processed; the larger ethyl group spaces the cellulosic chains further apart and thus lowers the softening point. (It may be noted, however, that the fully etherified product is not thermoplastic.) The ethyl group also renders ethylcellulose soluble in various solvents, according to the degree of substitution (see Table 13.9).

Ethylcellulose finds limited use as film and in injection moulding. In these applications an ethoxy content of about 45.5–47% (2.3–2.4 ethoxy groups per glucose residue) is generally preferred. Some typical properties of the material are shown in Tables 13.3 and 13.5. Ethylcellulose is tough and flexible and, when suitably plasticized (e.g. 10–20% tritolyl phosphate), retains these characteristics at temperatures as low as $-40°C$. Ethylcellulose mouldings have never become well known in Europe. Although the material has good water resistance and electrical insulating properties for a cellulosic polymer, this is of little significance since, when these properties are important, there are several superior non-cellulosic alternatives available. The principal use of ethylcellulose injection mouldings is in applications where good impact strength at low temperatures is required, e.g. refrigerator bases and ice-crusher parts.

Ethylcellulose finds rather greater use in various coating compositions. For this application ethylcellulose with an ethoxy content of about 47.5–49% (2.4–2.5 ethoxy groups per glucose residue) is generally preferred since this material has the widest solubility range. Ethylcellulose is also compatible with a wide range of plasticizers and other polymers and diverse types of coatings can be formulated. Compared to cellulose nitrate, ethylcellulose shows better alkali resistance, is more heat resistant and is not so readily degraded by sunlight (provided an antioxidant has been added). Ethylcellulose compositions find use in such applications as paper and textile coatings, gel lacquers and strippable coatings.

A mixed ether, ethyl-hydroxyethylcellulose is also available. This is similar to ethylcellulose but has broader solubility characteristics.

13.7.3   Benzylcellulose

Benzylcellulose may be prepared from alkali cellulose and benzyl chloride in a manner analogous to that described previously for methylcellulose. Benzyl-cellulose is not manufactured commercially, although there was some output in the 1930s. This material contained about two benzyl groups per glucose residue and found use mainly in injection moulding. Benzylcellulose is similar to ethylcellulose; it has better water resistance but significantly lower soften-ing point and poorer stability to heat and light. The latter properties and high production costs contributed to the demise of the material.

13.7.4   Hydroxyethylcellulose

Hydroxyethylcellulose is prepared from alkali cellulose and ethylene oxide. A segment of the product might be represented as follows:

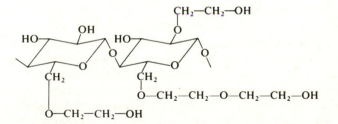

It may be noted that the hydroxyethyl group itself can react with ethylene oxide so that side-chains of varying length may be present in the product. Commercial materials generally contain between 1.4 and 2.0 ethylene oxide residues per glucose residue and have a degree of substitution of about 0.8–1.0.

Commercial hydroxyethylcellulose is readily soluble in water. It is in-soluble in most organic solvents but will dissolve in a few such as dimethyl-formamide and dimethyl sulphoxide. Hydroxyethylcellulose finds application mainly on account of its water-solubility; the material is used as a thickener and emulsifier in adhesives, cosmetics and latex paints and as a paper and textile size.

13.7.5   Sodium carboxymethylcellulose

Sodium carboxymethylcellulose is prepared from alkali cellulose and sodium monochloroacetate. The reaction may be represented as follows:

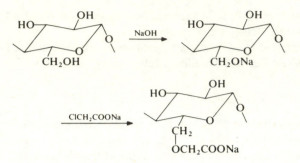

The extent of carboxymethylation may be regulated; commercial materials generally have a degree of substitution within the range 0.5–0.85.

Sodium carboxymethylcellulose is readily soluble in water and is insoluble in organic solvents. The material is used as a soil-suspending agent in detergents, suspending agent in latex paints and as an adhesive; it is also used as a stabilizer in food products such as ice cream.

13.8 STARCH

Starch is a widely distributed material, occurring in the roots, seeds and fruits of plants. Seeds usually contain 40–70% starch whilst roots and fruits contain about 5–25%. The principal commercial source of starch is corn; sources of lesser importance include tapioca, wheat and potato. In the extraction of starch, the plant material is ground with water; the resulting slurry is filtered to remove coarse tissue fragments, and a suspension of starch granules is obtained. The granules are collected by centrifuging and dried in warm air.

Starch consists of two compounds – amylose and amylopectin. The ratio of these components varies according to the source of the starch; generally, amylose accounts for 15–30% of starch and amylopectin for 70–85%. The fractionation of starch has been carried out in several ways. In one method, n-butanol is added to a colloidal solution of starch in hot water and the mixture is allowed to cool slowly. Amylose is precipitated and amylopectin is obtained from the liquor by the addition of methanol.

Amylose is converted by the enzyme, β-amylase (diastase) into maltose (VI) in about 70% yield. From this and other evidence it is supposed that amylose is a linear polymer consisting mainly of glucose residues joined by α-1,4-links

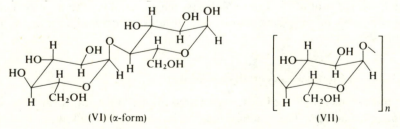

(VI) (α-form)   (VII)

(VII). The average degree of polymerization of amylose depends upon the source and method of isolation; values between about 250 and 8000 have been recorded.

Amylopectin gives only about 50% of maltose when hydrolysed by β-amylase whilst other enzymes yield small amounts of isomaltose (VIII). From this and other evidence it is supposed that amylopectin is a highly branched polymer consisting mainly of glucose residues joined by α-1,4-links with 1,6-branch points (IX). Amylopectin has a very high molecular weight and a very

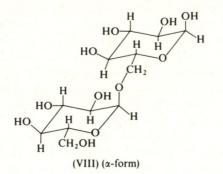

(VIII) (α-form)

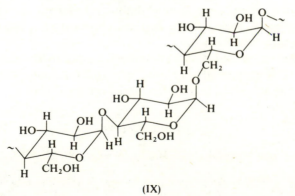

(IX)

broad distribution of molecular weight. For example, potato amylopectin gives fractions ranging in average molecular weight (M_w) from 7 million to 73 million [4]. The branch points are believed to occur at intervals of about 20–25 glucose residues.

When an aqueous suspension of starch is heated to about 60–80°C the starch granules swell and rupture, yielding a viscous colloidal dispersion (paste) with concurrent solution of some polymer molecules. On cooling, the hot colloidal dispersion increases markedly in viscosity and eventually forms a gel. Gelation is due to the aggregation of linear amylose chains; this process is essentially a crystallization phenomenon and is known as 'retrogradation'. (It may be noted that the highly branched amylopectin chains cannot align and associate so readily and amylopectin solutions do not gel under normal conditions.) Most technological applications of starch involve pastes and retrogradation is an important factor in their use. Starch may also be modified in various ways and a wide range of products is available commercially [5]. For example, *thin-boiling starches* are prepared by heating a slurry of starch at 50°C for 6–24 hours with a little sulphuric acid; hydrolysis occurs to give products which form pastes of low viscosity. *Dextrins* are prepared by roasting starch at 100–200°C for 3–20 hours either alone or in the presence of hydrochloric acid; a complex combination of hydrolysis, rearrangement and repolymerization occurs to give products which form low-viscosity pastes with good film-forming characteristics. *Oxidized starches* are prepared by heating a slurry of starch at 20–40°C for 5–24 hours in the presence of sodium hypochlorite/hydroxide; chain scission and oxidation of hydroxyl groups to carbonyl and carboxyl groups occur to give products which form pastes of low viscosity and good stability. Several starch derivatives are also available. Since starch is inherently water-dispersible, it is not necessary to prepare highly substituted derivatives to impart solubility (in contrast to cellulose). Thus commercial derivatives normally have very low degrees of substitution, typically in the range 0.02–0.05. Starch derivatives have various specific properties such as improved paste-stability and film properties. The more common starch derivatives include acetates, phosphates, hydroxyalkyl ethers and tertiary aminoalkyl ethers.

Unmodified starch and modified starches find approximately equal tonnage usage. These materials are used principally in paper-making, paper coatings, paper adhesives, textile sizes and food thickeners.

REFERENCES

1. Schultz, G. V. and Marx, M. (1954) *Makromol. Chem.*, **14**, 52; (1958) *J. Polymer Sci.*, **30**, 119.
2. Arthur, J. C. (1985) in *Encyclopedia of Polymer Science and Engineering*, Vol. 3, John Wiley & Sons, New York, p. 68.
3. Brydson, J. A. (1982) *Plastics Materials*, 4th edn, Butterworth Scientific, London, p. 554.
4. Witnauer, L. P. *et al.* (1955) *J. Polymer Sci.*, **16**, 1.
5. Ruttenberg, M. W. (1980) in *Handbook of Water-Soluble Gums and Resins*, (ed. R. L. Davidson), McGraw-Hill Book Company, New York, 1980, ch. 22.

BIBLIOGRAPHY

Ott, G., Spurlin, H. M. and Grafflin. M. W. (1954) *Cellulose and its derivatives*, 2nd edn, Interscience, New York,
Paist, W. D. (1958) *Cellulosics*, Reinhold Publishing Corp., New York.
Yarsley, V. E., Flavell, W., Adamson, P. S. and Perkins, N. G. (1964) *Cellulosic Plastics*, Iliffe Books Ltd, London.

Davidson, R. L. (ed.) (1980) *Handbook of Water-Soluble Gums and Resins*, McGraw-Hill Book Company, New York.

Kennedy, J. F. *et al.* (ed.) (1985) *Cellulose and Its Derivatives*, Ellis Horwood Ltd, Chichester.

Wurzburg, M. S. (ed.) (1987) *Modified Starches: Properties and Uses*, CRC Press, Inc., Boca Raton.

PHENOL-FORMALDEHYDE POLYMERS

14.1 SCOPE

Phenol-formaldehyde polymers are polymers formed by the interaction of a phenol, or a mixture of phenols, and formaldehyde. Commercial materials are most commonly based on phenol itself; other phenols such as cresols, xylenols and resorcinol are used to a limited extent. It may be noted that several aldehydes other than formaldehyde have been used to prepare phenolic polymers but none has attained appreciable commercial significance.

14.2 DEVELOPMENT

That phenols and aldehydes react to give resinous substances was first recorded in 1872 by A. Bayer and round the turn of the century there was some evaluation of such materials as electrical insulators. The first successful commercial exploitation of phenol-formaldehyde products was by Baekeland, who took out his first patent in 1907 and formed the General Bakelite Co. in the USA in 1910. Baekeland discovered that the base-catalysed reaction between phenol and formaldehyde can be carried out in two parts. If the reaction is carefully controlled, an intermediate product (which is either liquid or solid depending on the extent of reaction) may be isolated; this product is soluble in various solvents and is fusible. Baekeland found that on heating at high temperatures (namely about 150°C) under pressure, the intermediate is rapidly converted into a hard infusible, insoluble solid. (If the system is not subjected to pressure, gases are evolved and form bubbles in the mass, making it porous and weak.) Baekeland produced a variety of articles, ranging from bonded laminates prepared from paper impregnated with the liquid intermediate to compression mouldings made from a mixture of pulverized solid intermediate and filler (e.g. wood flour, asbestos or cotton). Phenol-formaldehyde materials became established in several fields within a few years; they remain important commercial materials although they have been somewhat eclipsed by newer polymers. The principal current uses include thermosetting moulding powders (which are widely used in such applications as general-purpose electrical mouldings, heated appliance components and automotive parts), laminates (which are extensively used for printed circuit boards and for the core of decorative laminates), adhesives, binders and surface coatings.

Phenol-formaldehyde polymers are also of significance in that they were the first wholly synthetic polymers to be utilized. They were truly man-made, being prepared from low molecular weight reagents. Prior to this time, the only polymers available were those based on naturally occurring polymers.

14.3   RAW MATERIALS

14.3.1   Phenol

Phenol occurs in coal tar and at one time this source satisfied commercial demands. Nowadays, however, natural phenol accounts for very little of the total world production of phenol; the majority is synthesized from benzene. Since benzene (which also occurs in coal tar) is now extensively obtained from petroleum sources, phenol may be counted as a major petroleum chemical. A number of processes for the manufacture of phenol have been operated over the years but now the cumene process accounts for at least 90% of the current output of phenol and only this process is described in this section. The method proceeds from benzene as follows:

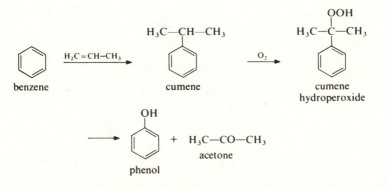

In the first step, benzene is alkylated with propylene to give isopropyl benzene (cumene). The reaction may be carried out in either the vapour or liquid phase. In the vapour phase process (which is more common), propylene and an excess of benzene are passed over a catalyst of phosphoric acid supported on kieselguhr at about 250°C and 2.5 MPa (25 atmospheres). The exit gases are fractionated; cumene is taken off and unreacted benzene is recycled. The conversion of benzene is restricted to about 15% per pass to limit the formation of diisopropylbenzenes. In the liquid phase alkylation, the catalyst is aluminium chloride (promoted with hydrogen chloride or isopropyl chloride) and the reaction is carried out at 50–100°C at a pressure marginally above atmospheric. Cumene is recovered by distillation and unreacted benzene is recycled; up to 50% of the benzene is converted per pass. In the second step in the synthesis of phenol, cumene is oxidized to cumene hydroperoxide with air. The reaction is carried out either in an anhydrous system or in the presence of a small amount of water; the reaction mixture is maintained at pH 7 by the addition of small quantities of alkali. The oxidation is usually conducted in a series of towers at about 95–130°C. The temperature is highest in the first tower and falls progressively along the series as the hydroperoxide concentration increases. The conversion of cumene is restricted to 35–50% per pass to reduce by-product formation. Unreacted cumene is removed by distillation under reduced pressure and recycled. The oxidation of cumene proceeds by a free radical mechanism, of which the following are the more important reactions (cf. section 2.3.3):

Initiation

$$I\cdot \; + \; RH \longrightarrow R\cdot \; + \; IH$$
$$ROOH \longrightarrow RO\cdot \; + \; HO\cdot$$
$$RO\cdot \; + \; RH \longrightarrow ROH \; + \; R\cdot$$
$$HO\cdot \; + \; RH \longrightarrow H_2O \; + \; R\cdot$$

Propagation

$$R\cdot \; + \; O_2 \longrightarrow ROO\cdot$$
$$ROO\cdot \; + \; RH \longrightarrow ROOH \; + \; R\cdot$$

Termination

$$2\,ROO\cdot \longrightarrow ROOR \; + \; O_2$$

where I· is an initiating radical and R =

In the third step in the synthesis of phenol, the cumene hydroperoxide undergoes cleavage to phenol and acetone. This is usually accomplished by feeding the hydroperoxide, with sulphuric acid as catalyst, continuously into previously decomposed material maintained at about 50–90°C. The product is then neutralized and fractionated by distillation. The cleavage reaction proceeds as follows:

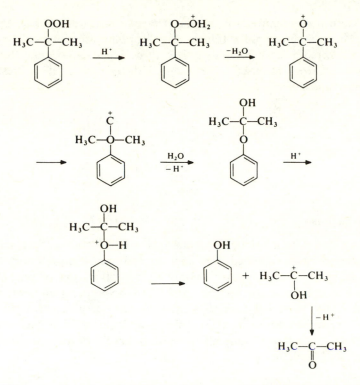

In addition to phenol and acetone, there are obtained minor amounts of other products such as acetophenone, α-methylstyrene and phenyl dimethyl carbinol. The yield of phenol on benzene is about 85%. The economics of the process are bound up with the value of the co-product, acetone and a decline in the use of acetone for the production of methyl methacrylate (see section 6.2.2) could have a major effect on the profitability of the cumene process.

Phenol is a colourless crystalline solid, m.p. 41°C.

14.3.2   Other phenols

At least 95% of the output of commercial phenolic resins is based on phenol itself. For the remainder, other phenols are utilized in order to confer specific properties. Some of these other phenols are briefly mentioned below.

(a)   Cresols

There are three isomeric cresols (methylphenols) (I). Originally, varying mixtures of the three isomers (known as cresylic acid) were obtained from coal tar and as by-products of petroleum refining operations but these sources are now unimportant. At the present time, most cresols are synthesized from either phenol or toluene. For the preparation of resins, *m*-cresol is the preferred isomer because it has three reactive positions (see section 14.4.1). Cresols are used for making alkali-resistant grades of resins for moulding powders.

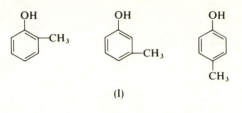

(I)

(b)   Xylenols

There are six isomeric xylenols (dimethylphenols) (II). At one time, mixtures of the isomers were obtained from coal tar and petroleum refining streams but now xylenols are obtained mainly by the alkylation of phenol with methanol (see, for example, section 12.6.1). For the preparation of resins, 3,5-xylenol is the preferred isomer because it has three reactive positions. Xylenols are used for making oil-soluble phenolic resins for surface coatings. They are also used for making alkali-resistant grades of resins for moulding powders and laminates.

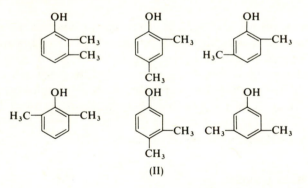

(II)

(c)   Resorcinol

Resorcinol (III) is produced mostly by the sulphonation of benzene. The 2-, 4- and 6- positions in resorcinol are all extremely reactive and resorcinol finds some use in the preparation of resins for cold-setting adhesives.

(III)

(d)   Higher homologues of phenol

A few synthetic higher homologues of phenol are used for making laminating resins which give products with enhanced flexibility and oil-soluble resins for surface coatings. They include *p-tert*-butylphenol (IV), *p-tert*-amylphenol (V), *p-tert*-octylphenol (VI) and *p*-phenylphenol (VII).

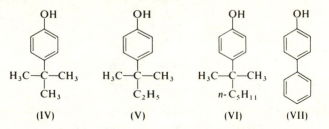

(IV)	(V)	(VI)	(VII)

14.3.3   Formaldehyde

The manufacture of formaldehyde is described in section 9.3.2. For the preparation of phenolic resins, formaldehyde is normally used as the aqueous solution, formalin. Paraformaldehyde is occasionally used in the production of resins and as a hardener for adhesives.

14.3.4   Furfural

As indicated previously, commercial phenolic resins are almost exclusively based on formaldehyde. Furfural is occasionally used to produce resins with good flow properties (see section 9.4.1(e)).

14.3.5   Hexamethylenetetramine

Hexamethylenetetramine (also known as hexamine, hexa or HMT) is prepared from ammonia and formaldehyde:

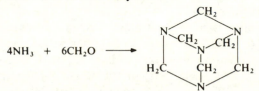

$$4NH_3 + 6CH_2O \longrightarrow$$

Ammonia is passed into formalin at 20–30°C, with agitation. The resulting solution is evaporated under reduced pressure until most of the water is removed and the hexamethylenetetramine crystallizes. Yields of about 95% on both ammonia and formaldehyde can be achieved. Hexamethylenetetramine is a colourless crystalline solid which, on heating, sublimes with decomposition.

The formation of hexamethylenetetramine probably involves the inter-mediates shown below. The molecule contains four equivalent rings, each of the four nitrogen atoms being at a bridgehead joining two rings.

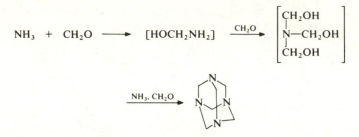

Hexamethylenetetramine is much used as a cross-linking agent for novolak resins in moulding powders (see section 14.5.2).

14.4 RESIN PREPARATION

As indicated previously, phenol-formaldehyde polymers find practical utiliz-ation mainly in the form of network polymers. The polymerization is nor-mally carried out in two separate operations. The first operation involves the formation of a low molecular weight fusible, soluble resin and the second operation involves curing reactions which lead to the cross-linked product. Various types of initial low molecular weight resins are produced commer-cially and are considered in this section.

14.4.1 Resol resins

Resol resins (resols or resoles) are prepared by the interaction of a phenol with a molar excess of formaldehyde (commonly about 1:1.5−2) under alkaline conditions. This procedure corresponds to Baekeland's original technique. Resols are now used mainly in the preparation of adhesives, binders and laminates.

Typically, reaction is carried out batch-wise in a stirred reactor, jacketed for heating and cooling. The reactor is also fitted with a condenser such that either reflux or distillation may take place, as required. A mixture of phenol, formalin and ammonia (about 1–3% on the weight of phenol) is heated under reflux at about 100°C for 0.25–1 hour and then water is removed by distillation under reduced pressure (to limit further reaction). Distillation is continued until a cooled sample of the residual resin has a melting point in the range 45–50°C. The resin is then quickly discharged and cooled to give a hard, brittle solid. In an alternative procedure, the removal of water is not taken to completion but is halted when the resol content of the residual aqueous solution reaches about 70%. The solution is then used directly, mainly in the preparation of paper laminates. In the preparation of such aqueous solutions, the preferred catalyst is usually sodium hydroxide (about 1% on the weight of phenol) since it leads to resols with a greater water-solubility than those given by ammonia. Sometimes the sodium hydroxide catalysed resol solution is spray-dried to give a powder which is readily redispersible in water; in this way, transportation costs are reduced.

In undissociated phenol, delocalization of an unshared electron pair on the oxygen atom results in increased electron densities at the *o*- and *p*-positions. This effect is larger than the decrease in electron density which results from the inductive effect of the hydroxyl group (VIII). Thus substitution of phenol by electrophilic reagents takes place at the *o*- and *p*-positions. In the phenoxide ion, similar delocalization occurs but because of the negative charge on the oxygen atom the inductive effect is greatly increased and reversed in direction so that electron density in the benzene ring is notably increased (IX). Thus phenol is more reactive and more *o*-/*p*-directing in alkaline solution than in neutral or acid solution.

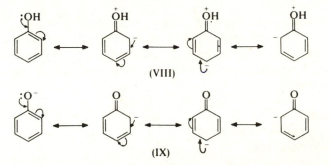

The reaction of phenol and formaldehyde in alkaline conditions therefore results in the formation of *o*- and *p*-methylol groups, e.g.

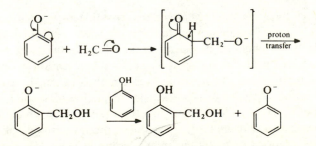

(It should be noted, however, that the identity of the actual hydroxyalkylating species has not been established and it is not clear how formaldehyde, as methylene glycol, reacts with the phenoxide ion.) The resulting *o*- and *p*-methylolphenols are more reactive towards formaldehyde than the original phenol and rapidly undergo further substitution with the formation of di- and trimethylol derivatives. Since virtually no substitution occurs at the *m*-position, the possible products are as shown below:

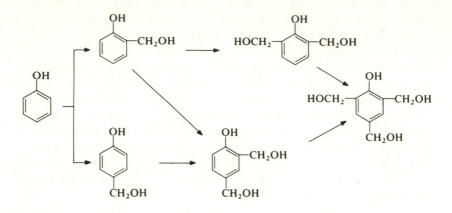

All of these compounds have been obtained from aqueous alkaline solutions of phenol and formaldehyde. For example, when a mixture of phenol, formalin and sodium hydroxide (1 : 3 : 1 molar respectively) is heated at 30°C for 5 hours, the molar composition of the resulting mixture is approximately as follows [1]:

Phenol	3%
o-Methylolphenol (saligenin)	12
p-Methylolphenol	17
2,4-Dimethylolphenol	24
2,6-Dimethylolphenol	7
2,4,6-Trimethylolphenol	37

The methylolphenols obtained are relatively stable in the presence of alkali but can undergo self-condensation to form dinuclear and polynuclear phenols in which the phenolic nuclei are linked by methylene groups. These products can arise from two different types of self-condensation; one involves two methylol groups and the other involves a methylol group and a hydrogen atom at an unsubstituted *o*- or *p*-position in the methylolphenol. Examples of such reactions are provided by an investigation [2] in which *o*- and *p*-methylolphenols were heated with aqueous sodium hydroxide at 80°C and the products analysed by paper chromatography. *o*-Methylolphenol gave 3-methylol-2′,4-dihydroxydiphenylmethane (X) by reaction between a methylol group and a *p*-hydrogen atom together with traces of 3-methylol-2,2′-dihydroxydiphenylmethane (XI) by reaction between a methylol group and an *o*-hydrogen atom:

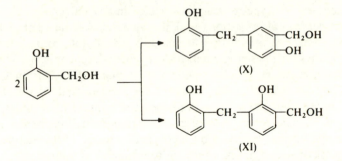

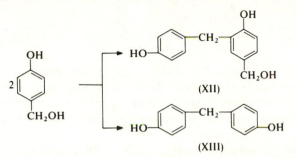

p-Methylolphenol gave 5-methylol-2,4'-dihydroxydiphenylmethane (XII) by reaction between a methylol group and an *o*-hydrogen atom and 4,4'-dihydroxydiphenylmethane (XIII) by reaction between two methylol groups:

Possible mechanisms of these two types of self-condensation are shown below. Both cases involve attack on a methylol group with displacement of a hydroxyl group. In the first case a proton is subsequently eliminated and in the other a methylol group is subsequently eliminated (as formaldehyde):

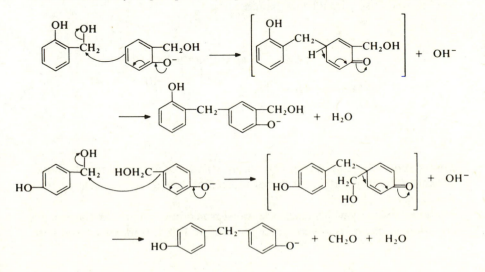

Which of these two types of reactions occurs in any particular case depends on the experimental conditions and the structure of the methylolphenols involved. Generally, *p*-hydrogen atoms and *p*-methylol groups are the predominant sites for the self-condensation of methylolphenols.

The above reactions, which are shown as leading to dinuclear phenols, may be repeated so that trinuclear phenols are formed, and so on. Thus the product obtained by the reaction of phenol and formaldehyde under alkaline conditions is a complex mixture of mono- and polynuclear phenols in which the phenolic nuclei are linked by methylene groups. The structures of the components of such a mixture may be represented as follows:

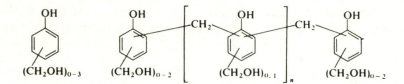

$(CH_2OH)_{0-3}$ $(CH_2OH)_{0-2}$ $(CH_2OH)_{0.1}$ $(CH_2OH)_{0-2}$

In these general formulae, the methylol groups and methylene bridges are restricted to *o*- and *p*-positions and there is a tendency for *o*-methylol groups and *p*-methylene bridges to predominate. Typical commercial liquid resols have an average of less than 2 aromatic rings per molecule whilst solid resols have 3–4.

As indicated previously, resols prepared with ammonia as catalyst are different in some respects from those produced with metal hydroxides. The ammonia probably enters into the resin structure since various nitrogen-containing compounds have been obtained by treating methylolphenols with ammonia. Included in the compounds which have been obtained in this way are bis- and tris(hydroxybenzyl) amines, e.g.

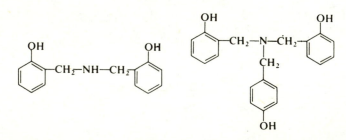

The presence of such nitrogen-containing structures in a resol results in loss of water-solubility and allows higher molecular weights to be reached before the resin gels.

The reactions by which resols are transformed into high molecular weight, cross-linked materials are considered in section 14.5.1. It may be noted here, however, that the conversion may be accomplished by simply heating a resol, it being unnecessary to add any further reagent. The preparation of resols is often referred to as a 'one-stage' process, since a quantity of formaldehyde sufficient to permit the formation of highly cross-linked material is present from the outset. Thus during the preparation of a resol, polymerization is steadily advancing and there is a danger of gelation in the reactor if reaction is taken too far. This disadvantage may be overcome by the use of a 'two-stage' process as described in section 14.4.2.

14.4.2 Novolak resins

Novolak resins (novolaks or novolacs) are normally prepared by the interaction of a molar excess of phenol with formaldehyde (commonly about 1.25:1) under acidic conditions.

The reaction is commonly carried out batch-wise in a reactor of the type described in the previous section. A mixture of phenol, formalin and acid (usually oxalic acid (about 0.5–2% on the weight of phenol)) is heated under reflux at about 100°C. Heating is continued for 2–4 hours until the resin becomes hydrophobic and the mixture appears turbid. Water is then distilled off, usually at atmospheric pressure, until a cooled sample of the residual resin has a melting point in the range 65–75°C. The resin is then discharged and cooled to give a hard, brittle solid.

The reaction between phenol and formaldehyde under acidic conditions proceeds through a mechanism different from that described previously for the base-catalysed reaction. The initial step involves the protonation of formaldehyde (methylene glycol) to give a carbenium ion:

$$HO-CH_2-OH \ + \ H^+ \rightleftharpoons \overset{+}{C}H_2-OH \ + \ H_2O$$

The phenol then undergoes electrophilic substitution with the formation of *o*- and *p*-methylol groups, e.g.

In the presence of acid the initial products, *o*- and *p*- methylolphenols, are present only transiently in very small concentrations. They are converted to benzylic carbenium ions which rapidly react with free phenol to form dihydroxydiphenylmethanes, e.g.

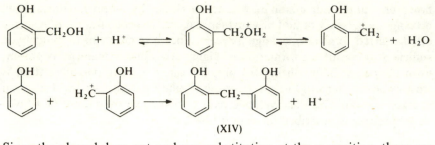

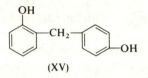

(XIV)

Since the phenol does not undergo substitution at the *m*-position, there are possible three dihydroxydiphenylmethanes, namely 2,2'-(XIV), 2,4'-(XV) and 4,4'-dihydroxydiphenylmethane (XIII). The proportions in which the three

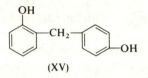

(XV)

isomers are formed depend on the conditions used, particularly on pH. For example, in one study [3], aqueous mixtures of phenol and formaldehyde (5.5:1 molar) of varying pH were heated and the dihydroxydiphenylmethanes isolated. Under strongly acidic conditions (below pH 2), such as are normally used in making novolaks, only the 2,4'- and 4,4'-isomers were isolated in any appreciable yield whereas at a higher pH (3–6) the 2,2'-isomer predominated. In the acid-catalysed reaction between phenol and formaldehyde the first substitution in the phenolic nucleus substantially deactivates the ring against further substitution (in contrast to the base-catalysed reaction). Also, although the dihydroxydiphenylmethanes formed are activated in the same manner as the single phenolic nucleus, the activation is not so strong. Because of the aforementioned factors, there is an initial build-up in the concentration of dihydroxydiphenylmethanes.

The concentration of dihydroxydiphenylmethanes subsequently falls as polynuclear phenols are produced by further methylolation and methylene link formation, e.g.

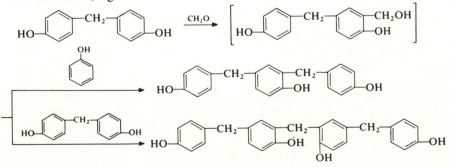

Reactions of the above type continue until all the formaldehyde has been consumed. The final product therefore consists of a complex mixture of polynuclear phenols linked by o- and p-methylene groups. The average molecular weight which is eventually attained is governed by the initial molar ratio of phenol and formaldehyde. The average molecular weight (M_n) of a typical commercial novolak is about 600, which corresponds to about six phenolic nuclei per chain; the number of nuclei in individual chains is typically in the range 2–13. The potentially reactive third positions in the nuclei of a novolak are deactivated and thus the chains are essentially linear, although a small amount of branching occurs. The complexity of novolaks is illustrated by the fact that for a chain of eight phenolic nuclei there are 1485 possible unbranched isomers and around 11 000 branched isomers [4]. A typical novolak chain might be represented as follows:

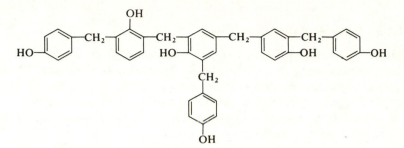

As evidence for the above conclusions regarding the structure of novolaks, the following may be cited:

(i) Crystalline polynuclear methylene-bridged phenols, which have been identified by unequivocal synthesis, have been isolated from novolaks. All three possible dinuclear and all seven possible trinuclear compounds have been separated in reasonable purity [5], showing that they are constituents of the resins. It has also been established that these compounds react with formaldehyde to give resins, indicating that they also function as intermediates in the formation of the higher molecular weight compounds.

(ii) The products of decomposition of novolaks are in keeping with the postulated structure. For example, phenol and o- and p-cresol have been obtained by heating a resin at 450°C [6]. These products are to be expected from the following type of chain segment:

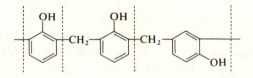

The essential feature of novolaks is that they represent a completed reaction and themselves have no ability to continue increasing in average molecular weight. The resins are therefore permanently fusible and there is no danger of gelation during production. This is in contrast to resols which contain reactive methylol groups and are capable of cross-linking on heating. In order to convert novolaks into network polymers the addition of an auxiliary chemical cross-linking agent is necessary. Because of this require-ment, the preparation of novolaks is often referred to as a 'two-stage' process.

In conclusion, it may be noted that novolaks can be prepared using mildly acidic catalysts such as zinc acetate. The resulting products have a high proportion of o–o'-links and are commonly called 'high-*ortho* novolaks'. Such novolaks are more reactive than the normal resins and are sometimes preferred.

14.4.3　Surface coating resins

The use of straight resol and novolak resins derived from phenol in surface coatings is very limited, mainly because of the brittleness of the films. Further, vegetable oils cannot be used as plasticizers since they are incompatible with the resins. However, there are a number of ways in which oil-solubility can be achieved and large quantities of phenolic resins suitable for surface coatings are produced by such methods as the following:

(i) *Rosin modification.* In this method, a simple resol (i.e. based on phenol or cresols) is heated at 150–300°C with an excess (about 10:1 by weight) of rosin (the main constituent of which is abietic acid). It is probable that chroman derivatives are produced by interaction of abietic acid and phenolic methylol groups via quinone methides:

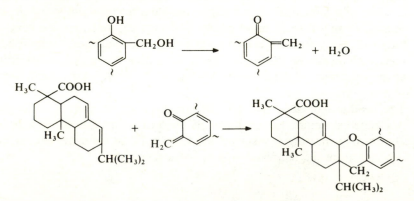

The resulting complex acid is usually esterified with glycerol or penta-erythritol and then heated with a drying oil, e.g. linseed oil. The product finds application in low-cost paints used for the protection of metal chemical plant.

(ii) *Use of substituted phenols.* Both resol and novolak resins prepared from the phenols given in section 14.3.2(d) are soluble in drying oils. Besides conferring solubility of the resin by providing a large oleophilic group, the *p*-substituent also reduces the functionality of the phenol and lowers the cross-link intensity of the final product. Typically, the resin is heated with the oil (about equal weight) at 150–200°C for 0.5 hours. In the case of a resol, chroman derivatives are probably formed by the interaction of *o*-methylol-phenol structures (via, quinone methides) with the unsaturated centres in the drying oil in a manner similar to that described above for rosin. The cooking of novolaks with drying oils is less well understood, although some reaction appears to occur. Resins based on substituted phenols are also modified with alkyd resins (section 11.3). In this case, methylol groups may be involved both in chroman formation, as described previously, and in etherification by hydroxyl groups in the alkyd resin:

$$
\begin{array}{ccc}
\text{CH}_2\text{OH} & \longrightarrow & \text{CH}_2 \\
| & & | \\
\text{OH} & & \text{O}
\end{array}
\quad + \ \text{H}_2\text{O}
$$

Oil- and alkyd-modified phenolic resins are used mainly in corrosion resistant coatings.

14.4.4 Casting resins

Phenolic casting resins were at one time important materials, being used for umbrella handles, artificial jewellery, knobs and such objects where decorative effects were required. They are now little used in such applications, having been displaced by more readily processed plastics.

Casting resins are prepared by the interaction of phenol with a large molar excess of formaldehyde (commonly about 1:2.3). In a typical process, a mixture of phenol, formalin and sodium hydroxide is heated at 70°C for 3 hours and then water is removed by distillation under reduced pressure. The resulting resin is mixed with a strong acid, e.g. benzenesulphonic acid, immediately poured into a mould and allowed to harden either at room temperature or at 60–80°C.

The structure of a casting resin resembles that of a resol (section 14.4.1) but a large number of methylol groups is present owing to the greater amount of formaldehyde used in the reaction.

14.5 CROSS-LINKING

In the preceding section the nature of various low molecular weight resins is considered. This section deals with the means whereby these resins are converted into high molecular weight network polymers.

14.5.1 Resol resins

Resols, which are produced under alkaline conditions, are generally neutralized or made slightly acid before cure is carried out. Network polymers are then obtained simply by heating.

The cure of a resol is extremely complex, involving a number of competing reactions each of which may be influenced by reaction conditions and it is not easy to unravel precisely what takes place. Also, the cured product, being infusible and insoluble, is not amenable to chemical investigation. Information regarding the mechanism of cure has thus been obtained mainly by investigation of simpler systems. In particular, pure mononuclear methylolphenols have been used in place of the complex mixtures present in typical resins. This approach is fundamentally sound since functional groups generally undergo the same reactions in both monomeric and polymeric systems. From this work, two types of primary reactions of methylolphenols in acidic conditions have been established, namely those wherein a methylol group reacts with an *o*- or *p*-hydrogen atom to form a methylene bridge and those wherein two methylol groups react with one another to form an ether linkage.

Both these reactions probably proceed through benzylic carbenium ions formed by protonation of methylol groups (see also section 14.4.2), e.g.

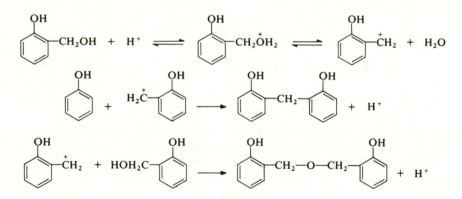

Thus methylene and ether links arise through competitive reactions. Ether formation will be favoured when there are few free *o*- and *p*-positions available for attack by carbenium ions. Thus when these active positions are blocked by non-reactive substituents or if complete methylolation has occurred (as in trimethylolphenol) then ether formation is to be expected. This is in accordance with the finding that treatment of the methylol derivative of *p*-*tert*-butyl-*o*-cresol with dilute hydrochloric acid yields the corresponding dibenzyl ether [7]. It also follows that ether formation will be favoured when the number of free methylol groups in the system is great compared to the number which have been consumed to produce benzylic carbenium ions.

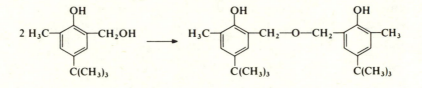

Thus ethers should form most readily when the system is most nearly neutral. This is in agreement with the observation that when pure (i.e. neutral) o-methylolphenol is heated at 110°C the main product is 2,2′-dihydroxydibenzyl ether (salireton) (XVI) with very little of the methylene compound (X) [8]:

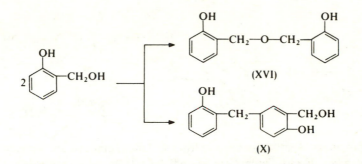

It was also found in this investigation that when o-methylolphenol is heated at 70°C in the presence of hydrochloric acid both compounds (XVI) and (X) are produced but in the presence of sodium hydroxide only the methylene compound (X) is formed.

Thus when a resol is cured, it is likely that both methylene and ether links are present in the product. However, since commercial resins are not based on blocked phenols (and hence have free o- and p-positions) and since curing conditions are seldom neutral it is probable that there is a predominance of methylene over ether links.

It has been found that whilst methylene compounds are generally rather stable, dibenzyl ethers are not so stable and at temperatures above about 150°C (i.e. at temperatures commonly used for curing resols) they undergo a number of ill-defined reactions. The mechanisms of these secondary reactions have been extensively studied but are still rather poorly understood. Various workers have put forward conflicting suggestions, which are briefly summarized below.

Zinke and his co-workers have postulated that the main reaction involved is the breakdown of ether links to give methylene links with loss of formaldehyde. This suggestion is supported by the isolation of diphenylmethane derivatives from the reaction mixtures obtained by heating dibenzyl ethers, e.g. 2,2′-dihydroxy-3,5,3′,5′-tetramethyldibenzyl ether:

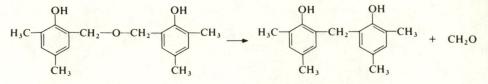

To account for the fact that less than one mole of formaldehyde is evolved for each ether link broken it is supposed that some of the formaldehyde produced reacts with the methylene compound, resulting in cross-linking.

On the other hand, Hultzsch and von Euler and co-workers have suggested that the main reaction involved in the breakdown of ether links is the formation of quinone methides e.g.

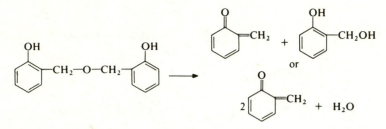

This postulate is in agreement with the small amount of formaldehyde evolved when dibenzyl ethers are heated. Also, stilbenes and diphenylethanes have been isolated from the complex mixtures obtained by heating various blocked dibenzyl ethers and it is suggested that these products are derived from quinone methides, e.g.

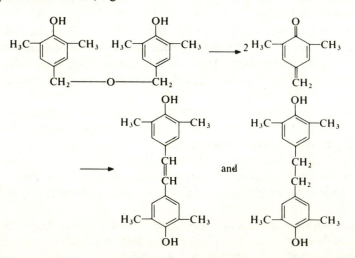

It is also thought that compounds derived from quinone methides are responsible for the dark colour of the cured material.

To summarize, it may be said that the network polymer obtained from a resol is composed principally of phenolic nuclei joined by methylene groups but there is the possibility of other types of linkages, the nature and extent of which depend on the nature of the resol and the conditions of cure. The possible structure of such a network might be represented as follows:

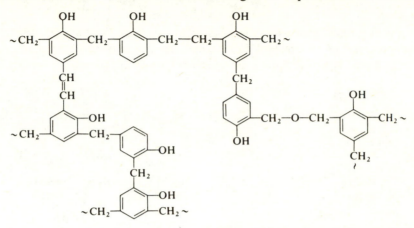

It is to be noted that the relative amounts of the various linkages shown above are not intended to have any quantitative significance. It may also be noted that phenol-methylene chains are extremely irregular and their geometry precludes a large proportion of the theoretically possible cross-links.

It may be mentioned here that it is common practice to refer to resols as *A stage resins* and to the final network polymers as *resits* or *C stage resins*; intermediate materials are called *resitols* or *B stage resins*.

14.5.2 Novolak resins

As noted previously (section 14.4.2), the conversion of novolaks into network polymers can be accomplished only after the addition of a cross-linking agent. Although novolaks can be cross-linked by reaction with additional formaldehyde or with paraform, hexamethylenetetramine is almost invariably used for this purpose. The mechanism of the curing process is not fully understood.

In an investigation [9] involving blocked phenols, it was shown that phenols having free *o*-positions generally react with hexamethylenetetramine at 130–150°C to give dibenzylamines, e.g.

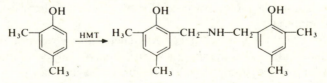

and phenols with free *p*-positions tend to give tribenzylamines, e.g.

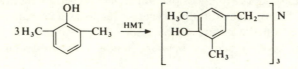

It was also found that when these benzylamines are heated at 180–190°C in the presence of the phenol from which they are derived, ammonia or methylamine is evolved and the corresponding diphenylmethane is formed:

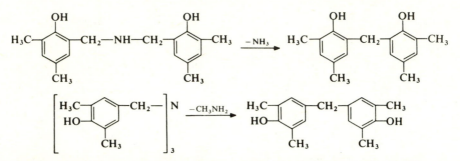

When the benzylamines are heated in the absence of free phenol, different products are obtained. For example, the above dibenzylamine gives rise to an azomethine:

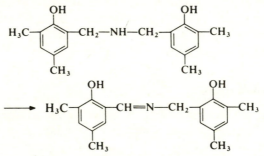

On the basis of this work, it is possible that the primary reaction between a novolak and hexamethylenetetramine (which in a typical commercial mould-ing powder is present to the extent of only about 12% on the weight of resin) leads to a complex structure containing secondary and tertiary amine links. On further heating most of these links break down to give methylene links and the resulting product has only a small nitrogen content. Some azomethine links may be formed and may account for the characteristic brown colour of the cured material.

A study [10] of the novolak-hexamethylenetetramine reaction by differen-tial thermal analysis and thermogravimetric analysis supports the view that the cross-linking reaction proceeds through dibenzylamines.

To summarize, it may be said that the network polymer obtained from a hexamethylenetetramine-cured novolak is composed of phenolic nuclei joined mainly by methylene groups with small numbers of various nitrogen-containing links. The possible structure of such a network might be represented as follows:

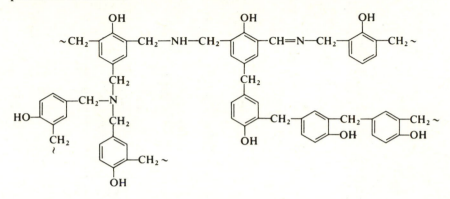

It is to be noted that the relative amounts of the various linkages shown above are not intended to have any quantitative significance. Thus the network polymer obtained from the novolak-hexamethylenetetramine reaction has a structure which is predominantly similar to that of the network polymer derived from a resol.

14.5.3 Surface coating resins

As discussed in section 14.4.3, phenolic surface coatings normally contain substantial quantities of drying oils. The reactions which occur on cross-linking are essentially those described in section 11.3.6.

14.5.4 Casting resins

As indicated in section 14.4.4, casting resins have a resol-type structure but contain a greater proportion of methylol groups. It may therefore be anticipated (cf. section 14.5.1) that an appreciable number of ether links are produced when a casting resin is cured. Further, since castings are normally cured at relatively low temperatures these ether links are likely to persist in the final product. It may be noted that it is possible to prepare castings which are colourless, whereas resols cured at about 150°C are dark coloured. This observation is in accordance with the suggestion that it is the thermal decomposition of ether links into quinone methides which leads to colour formation in cured resols.

Table 14.1 Typical values for various properties of cured phenol-formaldehyde polymers, unfilled and filled [11]

	Unfilled casting	Moulding, wood flour filled	Moulding, cotton fabric filled	Laminate, paper filled
Specific Gravity	1.3	1.3–1.4	1.3–1.4	1.3–1.4
Tensile strength (MPa) (1bf/in^2)	21–69 3000–10 000	34–55 5000–8000	34–55 5000–8000	55–170 8000–25 000
Bending strength (MPa) (1bf/in^2)	48–100 7000–15 000	55–100 8000–15 000	55–100 8000–15 000	100–210 15 000–30 000
Compression strength (MPa) (1bf/in^2)	69–210 10 000–30 000	100–280 15 000–40 000	140–240 20 000–35 000	140–280 20 000–40 000
Impact strength, Izod (J/m) (ft 1bf/in)	5.3–27 0.1–0.5	5.3–27 0.1–0.5	16–160 0.3–3.0	11–110 0.2–2.0
Water absorption (mg)	2–20	70–150	200–400	15–300

14.6 PROPERTIES OF CROSS-LINKED POLYMERS

When phenolic resins have been cross-linked they are rigid, infusible and insoluble. The mechanical properties of the network polymers are considerably influenced by the incorporation of fillers. This point is illustrated by Table 14.1, in which are given comparative values of some properties of cured phenol-formaldehyde polymers. The polymers have good heat stability, showing little weight loss up to about 200°C. Since the polymer is polar, its electrical insulating properties are not outstanding but they are adequate for many purposes. Also, phenolics have relatively poor tracking resistance under conditions of high humidity, i.e. there is a tendency to form a conductive path through carbonization along the surface of material situated between two metal electrodes.

Cured phenol-formaldehyde polymers are very resistant to most chemical reagents. They are unaffected by all ordinary organic solvents and water, although the presence of cellulosic fillers results in water absorption and consequent swelling. The polymers are dissolved slowly with decomposition by boiling phenols such as α-naphthol. Simple phenol-formaldehyde materials are readily attacked by aqueous sodium hydroxide but cresol and xylenol based polymers are more resistant. The polymers are resistant to acids except formic acid and strong oxidizing acids but the presence of cellulosic fillers increases the sensitivity of mouldings towards acids.

REFERENCES

1. Freeman, J. H. and Lewis, C. W. (1954) *J. Am. Chem. Soc.*, **76**, 2080.
2. Yeddanapi, L. M. and Francis, D. J. (1962) *Makromolekulare Chem.*, **55**, 74.
3. Finn, S. R. and Musty, J. W. G. (1950) *J. Soc. Chem. Ind.*, **69**, S49.
4. Megson, N. J. L. and Hollingdale, S. H. (1955) *J. Appl. Chem.*, **5**, 616.
5. Bender, H. L. *et al.* (1950) Preprint Bull. *Div. Paint, Varnish and Plastics Chem.*, *Am. Chem. Soc.* Meeting, **73**.
6. Megson, N. J. L. (1936) *Trans. Faraday Soc.*, **32**, 336.
7. Rodia, J. S. and Freeman, J. H. (1959) *J. Org. Chem.*, **24**, 21.
8. Reese, J. (1955) *Kunstoffe*, **45**, 137.
9. Zinke, A. and Pucher, St. (1948) *Monatsh*, **79**, 26; Zinke, A. *et al.* (1949) *ibid.*, **80**, 148.
10. Orrell, E. W. and Burns, R. (1968) *Plastics & Polymers*, **36**, 469.
11. Whitehouse, A. A. K., Pritchett, E. G. K. and Barnett, G. (1967) *Phenolic Resins*, Iliffe Books Ltd, London, pp. 56, 57.

BIBLIOGRAPHY

Carswell, T. S. (1947) *Phenoplasts*, Interscience Publishers, New York.
Martin, R. W. (1956) *The Chemistry of Phenolic Resins*, John Wiley & Sons, Inc., New York.

Megson, N. J. L. (1958) *Phenolic Resin Chemistry*, Butterworths Publications Ltd, London.

Gould, D. F. (1959) *Phenolic Resins*, Reinhold Publishing Corporation, New York.

Whitehouse, A. A. K., Pritchett, E. G. K. and Barnett, G. (1967) *Phenolic Resins*, Iliffe Books Ltd, London.

Knop, A. and Scheib, W. (1979) *Chemistry and Application of Phenolic Resins*, Springer-Verlag, New York.

Knop, A. and Pilato, L. A. (1985) *Phenolic Resins*, Springer-Verlag, Berlin.

15

AMINOPOLYMERS

15.1 SCOPE

For the purposes of this chapter, aminopolymers are defined as polymers formed by the interaction of amines or amides with aldehydes. Of the various polymers of this type which have been investigated, only two are currently of appreciable commercial importance, namely urea-formaldehyde and melamine-formaldehyde polymers. In addition, melamine-phenol-formaldehyde and benzoguanamine-formaldehyde polymers find limited application. In the past there has been some commercial interest in thiourea-formaldehyde and aniline-formaldehyde polymers but these products are now of little importance. The aforementioned polymers form the contents of this chapter.

15.2 UREA-FORMALDEHYDE POLYMERS

15.2.1 Development

That urea reacts with formaldehyde was first noted in 1884 by Hölzer but commercial interest in the polymers stems from a patent of 1918 to John. John treated urea with formalin to give viscous solutions which were suggested for use as adhesives and textile impregnants. Between 1920 and 1924 there appeared several patents which disclosed many of the factors governing polymerization and casting resins and thermosetting moulding compositions were envisaged; however, at this time mouldings based on urea-formaldehyde resins were very water-sensitive. This limitation was overcome by the British Cyanides Co. Ltd. (later British Industrial Plastics Ltd) (UK) by the use of urea-thiourea-formaldehyde resins and moulding powders were marketed in 1926. Successful moulding powders based on straight urea-formaldehyde resins were made in about 1930 by the incorporation of an acidic accelerator, which resulted in fast, adequate cures. Since this time, urea-formaldehyde moulding powders have become firmly established for the production of domestic electrical fittings, bottle caps and sundry other items. Whilst moulding powders were being developed, urea-formaldehyde resins were investig-

ated in other applications such as wood adhesives and surface coatings (I. G.
Farbenindustrie (Germany) patents, 1925) and textile finishes (Tootal, Broad-
hurst Lee Co. Ltd (UK) patent, 1926). These applications now also account
for substantial quantities of urea-formaldehyde resins. During the 1950s, the
large-scale production of chipboard (particleboard) was established and has
resulted in a large outlet for urea-formaldehyde polymers. In the 1960s, urea-
formaldehyde foams were developed and used for thermal insulation prepared *in situ*. However, there has been growing concern about the health
hazards resulting from the slow release of formaldehyde from these foams and
use of the foams for cavity insulation has been banned in some countries.

15.2.2   Raw materials

(a)   Urea

All current processes for the manufacture of urea are based on the reaction of
ammonia and carbon dioxide to form ammonium carbamate which is
simultaneously dehydrated to urea:

$$2NH_3 \; + \; CO_2 \; \rightleftharpoons \; NH_2COONH_4$$
$$\rightleftharpoons NH_2CONH_2 \; + \; H_2O$$

The dehydration of ammonium carbamate is appreciable only at temperatures above the melting point (about 150°C) and this reaction can only
proceed if the combined partial pressure of ammonia and carbon dioxide
exceeds the dissociation pressure of the ammonium carbamate (about
10 MPa at 160°C and about 30 MPa at 200°C). Thus commercial processes
are operated in the liquid phase at 160–220°C and 18–35 MPa (180–350
atmospheres). Generally, a stoichiometric excess of ammonia is employed,
molar ratios of up to 6:1 being used. The dehydration of ammonium
carbamate to urea proceeds to about 50–65% in most processes. The reactor
effluent therefore consists of urea, water, ammonium carbamate and the
excess of ammonia. Various techniques are used for separating the components. In one process the effluent is let down in pressure and heated at about
155°C to decompose the carbamate into ammonia and carbon dioxide. The
gases are removed and cooled. All the carbon dioxide present reacts with the
stoichiometric amount of ammonia to re-form carbamate, which is then
dissolved in a small quantity of water and returned to the reactor. The
remaining ammonia is liquefied and recycled to the reactor. Fresh make-up
ammonia and carbon dioxide are also introduced into the reactor. Removal of
ammonium carbamate and ammonia from the reactor effluent leaves an
aqueous solution of urea. The solution is partially evaporated and then urea is
isolated by recrystallization. Ammonium carbamate is very corrosive and at

one time it was necessary to use silver-lined equipment but now satisfactory alloy steel plant is available. Urea is a white crystalline solid, m.p. 133°C.

(b) Formaldehyde

The manufacture of formaldehyde is described in section 9.3.2. For the preparation of urea-formaldehyde resins, formaldehyde is normally used as the aqueous solution, formalin. Paraformaldehyde is occasionally used.

15.2.3 Resin preparation

As indicated previously, urea-formaldehyde polymers find practical utilization mainly in the form of network polymers. The polymerization is normally carried out in two separate operations. The first operation involves the formation of a low molecular weight fusible, soluble resin and the second operation involves curing reactions which lead to the cross-linked product. Various types of resins are produced commercially but they may be classified broadly into unmodified and modified resins.

(a) Unmodified resins

Straight urea-formaldehyde resins are used principally in the preparation of moulding compositions and adhesives. The methods employed to produce these resins vary somewhat according to the end-use envisaged but the following procedure, which is for a wood adhesive resin, illustrates the general principles involved.

Formalin is made slightly alkaline (pH 8) by the addition of sodium hydroxide and then urea is added to give a urea to formaldehyde ratio of about 1:2 molar. The resulting solution is boiled under reflux for about 15 minutes, acidified (to pH 4) with formic acid and then boiled for a further 5–20 minutes until the required degree of reaction is attained. The product is neutralized with sodium hydroxide and then evaporated under reduced pressure until the required solids contents is reached (about 70% in a typical liquid adhesive).

The reactions which occur in processes such as that described above are of two kinds:

(i) *Formation of methylolureas*. The initial products formed in the preparation of urea-formaldehyde resins are methylolureas. Under the mildly alkaline conditions used in the first stage of resin preparation, both monomethylolurea and dimethylolurea are obtained in high yield and can be isolated as pure, crystalline compounds:

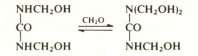

Although it has not been isolated, trimethylolurea is also formed under these conditions:

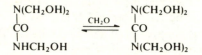

The presence of trimethylolurea in solutions containing more than two moles of formaldehyde to one mole of urea has been inferred from paper chromatography and NMR studies.

Tetramethylolurea allegedly has been detected in highly alkaline solutions of urea and formaldehyde:

$$
\begin{array}{ccc}
\underset{|}{N(CH_2OH)_2} & \xrightarrow{\quad CH_2O \quad} & \underset{|}{N(CH_2OH)_2} \\
CO & \rightleftharpoons & CO \\
| & & | \\
NHCH_2OH & & N(CH_2OH)_2
\end{array}
$$

However, tetramethylolurea does not appear to be formed under the conditions normally used in resin preparation. Thus after the initial reaction, a typical resin solution contains mono-, di- and trimethylolureas together with free urea and formaldehyde. The relative amounts of these components will depend on reaction conditions; for example, a solution which initially contained 2 moles/litre of urea and 4 moles/litre of formaldehyde at pH 6.5 and 35°C has the following molar composition at equilibrium [1]:

monomethylolurea	0.60	urea	0.05
dimethylolurea	1.03	formaldehyde	0.38
trimethylolurea	0.33		

The base-catalysed methylolation of urea probably proceeds by the following mechanism:

$$NH_2CONH_2 \ + \ OH^- \rightleftharpoons NH_2CO\bar{N}H \ + \ H_2O$$

$$NH_2CO\bar{N}H \ + \ CH_2O \rightleftharpoons NH_2CONHCH_2O^-$$

$$NH_2CONHCH_2O^- \ + \ H_2O \rightleftharpoons NH_2CONHCH_2OH \ + \ OH^-$$

(ii) *Condensation of methylolureas.* In the second stage of the resin preparation outlined above, reaction is continued under acidic conditions. It is believed that condensation occurs between methylol groups and amido hydrogen to form methylene links according to the following mechanism:

$$-CH_2OH \ + \ H^- \rightleftharpoons -CH_2\overset{+}{O}H_2 \rightleftharpoons -\overset{+}{C}H_2 \ + \ H_2O$$

$$-\overset{+}{C}H_2 \ + \ H_2N- \longrightarrow -CH_2-\overset{+}{N}H_2- \longrightarrow -CH_2-NH- \ + \ H^+$$

Since a typical resin solution includes several species which contain methylol and amido groups, various methylene compounds may result from this type of condensation, e.g.

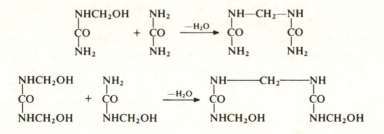

Products of the above kind still posses methylol and amido groups; thus condensation may continue to yield polymeric methylene compounds. Since a molar excess of formaldehyde is used to prepare resins, the polymers will be terminated predominantly by methylol groups. Further, since imido hydrogen is relatively unreactive towards methylol groups (as shown by the absence of mutual reaction between molecules of dimethylolurea [2]), the polymers will be essentially linear with little branching. The general structure of the methylene-containing polymers may therefore be represented as follows:

$$HOCH_2 \left[NH-CO-NH-CH_2 \right]_n NH-CO-NH-CH_2OH$$

Evidence for the above type of structure is provided by an analysis of products obtained by reaction of urea and fomaldehyde in acidic aqueous solution [3]. In this work, the urea residue ($-NH-CO-NH-$), methylene ($-CH_2-$) and methylol ($-CH_2OH$) contents were determined. The sum of the weights of these components was found to equal the weight of sample used, indicating that no other structural units are present. Further, an indication of molecular weight is given by the molar ratio of urea fragments to methylene groups; average molecular weights of 200–500 were computed by this method, corresponding to values of n in the above structure of about 1–5. The same molecular weight values have been obtained by gel-permeation chromatography [4].

(b) Modified resins

Unmodified urea-formaldehyde resins are not suitable for use in surface coating formulations because they are insoluble in common solvents, do not interact readily with other resins and are comparatively unstable. These limitations are substantially overcome when the resins are modified by alcohols. n-Butanol is the alcohol most widely used for this purpose and a typical process might be as follows.

Formalin is adjusted to pH 8 and urea is added to give a urea to formaldehyde ratio of about 1:2.5 molar. The resulting solution is boiled under reflux for 1 hour. Butanol (1.5–2.0 mole per mole of urea) is then added together with a little xylene. The latter forms, with butanol and water, a ternary azeotrope which on distillation yields a condensate separating into an upper organic layer and a lower aqueous layer. By discarding the lower layer and returning the upper layer to the reactor, water is progressively removed from the system. After a substantial proportion of the water has been removed, an acid catalyst (e.g. phosphoric acid or phthalic anhydride) is added and heating is continued. When the required degree of reaction is attained, the solution is neutralized and concentrated to the desired solids content.

In the procedure described above, the initial products are methylolureas; treatment with butanol results in the alkylation of a proportion of the methylol groups:

$$-NHCH_2OH \quad + \quad C_4H_9OH \longrightarrow -NHCH_2OC_4H_9 \quad + \quad H_2O$$

A typical resin contains 0.5–1.0 mole of combined butanol per mole of urea. On acidification and further heating polymerization occurs mainly through the remaining methylol groups as described in section 15.2.3(a).

Very commonly, butylated urea-formaldehyde resins are blended with alkyd resins (section 11.3) to give coatings with good flexibility and adhesion.

15.2.4 Cross-linking

The conversion of low molecular weight urea-formaldehyde resins into high molecular weight network polymers is usually accomplished by heating under acidic conditions. Acidic compounds such as phthalic anhydride are incorporated into moulding powders whilst ammonium chloride is commonly added to adhesive resins prior to use. In the latter case the chloride reacts with free formaldehyde to form hydrochloric acid (which serves as catalyst) and hexamethylenetetramine:

$$4NH_4Cl \quad + \quad 6CH_2O \longrightarrow (CH_2)_6N_4 \quad + \quad 4HCl \quad + \quad 6H_2O$$

The nature of the reactions which occur during cross-linking of urea-formaldehyde resins has not been firmly established. It is generally supposed that in the early stages of the process imido groups in the chain react with free formaldehyde present in the solution to give pendant methylol groups. Thereafter the principal reactions are thought to be those between methylol groups (pendant or terminal) and imido hydrogen and those involving self-condensation of methylol groups, e.g.

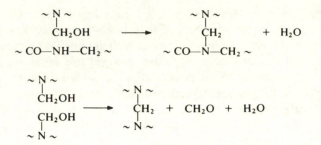

$$\sim N \sim \atop CH_2OH \quad \quad \longrightarrow \quad \sim N \sim \atop CH_2 \quad + \quad H_2O$$
$$\sim CO—NH—CH_2 \sim \quad \quad \sim CO—N—CH_2 \sim$$

Such reactions would lead to a network structure; a completely cross-linked structure containing no imido hydrogen atoms or methylol groups may be represented as follows:

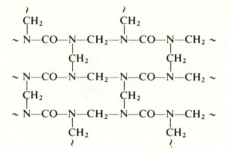

In practice, the hardened product is likely to have a structure far from the ideal, containing considerable numbers of unreacted imido and methylol groups.

The reactions involved in the cross-linking of butylated urea-formaldehyde resins may be supposed to be similar to those described above, with the added possibility of condensation between butoxy groups and imido hydrogen:

$$\sim N \sim \atop CH_2OC_4H_9 \quad \quad \longrightarrow \quad \sim N \sim \atop CH_2 \quad + \quad C_4H_9OH$$
$$\sim CO—NH—CH_2 \sim \quad \quad \sim CO—N—CH_2 \sim$$

Even after prolonged stoving, however, the hardened resins still retain a high proportion of butoxy groups. When butylated urea-formaldehyde resin/alkyd resin blends are stoved, there is the further possibility of ether interchange between the butoxy groups and free hydroxyl groups in the alkyd resin:

$$\sim N \sim \atop CH_2OC_4H_9 \quad \quad \longrightarrow \quad \sim N \sim \atop CH_2 \quad + \quad C_4H_9OH$$
$$OH \quad \quad \quad \quad \quad \quad O$$

An objection to the above proposals for the formation of network structures by the interaction of urea and formaldehyde is that they involve the participation of imido hydrogen atoms, which are known to be very unreactive towards methylol groups. In order to avoid this criticism, it has been suggested [5] that acidification of methylolureas yields methylene-imines which trimerize to give substituted trimethylenetriamine compounds which then polymerize to a network structure by interaction of methylol and amine groups:

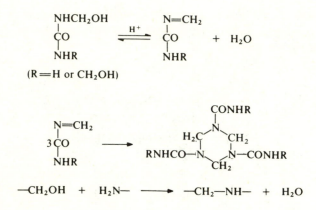

This mechanism now receives little support since most subsequent investigations indicate that open-chain structures and not cyclic structures are present in the early stages of the urea-formaldehyde reaction.

15.2.5 Properties of cross-linked polymers

When urea-formaldehyde resins have been cross-linked they are rigid, infusible and insoluble. Moulding compositions are generally filled with cellulose and some typical properties of general purpose moulding are given in Table 15.1. Urea-formaldehyde mouldings exhibit unusual surface hardness and, in contrast to phenolic mouldings, may be produced in a wide range of colours. Urea-formaldehyde polymers have relatively poor heat resistance and discolour and degrade if maintained continuously above about 80°C. Cured material exhibits good electrical insulation properties at low frequencies, having particularly good tracking resistance.

Cross-linked urea-formaldehyde polymers are very resistant to most organic reagents. They are, however, attacked by acids and alkalis and show relatively high water absorption.

Table 15.1 Typical values for various properties of urea-formaldehyde and melamine-formaldehyde mouldings

	Urea–formaldehyde	Melamine-formaldehyde		
	Cellulose filled	Cellulose filled	Glass filled	Mineral filled
Specific Gravity	1.5	1.5	2.0	1.8
Tensile strength				
(MPa)	52–79	55–83	41–69	28–41
(lbf/in^2)	7500–11 500	8000–12 000	6000–10 000	4000–6000
Cross-breaking				
strength (MPa)	76–120	90–145	62–97	41–76
(lbf/in^2)	11 000–17 000	13 000–21 000	9000–14 000	6000–11 000
Impact strength, Izod				
(J/m)	14–18	13–19	32–1000	–
(ft lbf/in)	0.26–0.34	0.24–0.36	0.60–19	–

15.3 MELAMINE-FORMALDEHYDE POLYMERS

15.3.1 Development

Melamine was first isolated by Leibig in 1834 from the mixture obtained by heating ammonium thiocyanate. A technically feasible route to melamine was developed in 1935 by Ciba AG (Switzerland) and at the same time Henkel patented the production of resins from melamine and formaldehyde. In general, melamine-formaldehyde polymers resemble urea-formaldehyde polymers but they have improved resistance to heat and water. The two materials have therefore found application in similar areas, melamine-formaldehyde resins now being widely used in the production of moulding powders, laminates, adhesives, surface coatings and textile finishes.

15.3.2 Raw materials

(a) Melamine

The standard route to melamine is from urea (section 15.2.2(a)). Urea is heated in the presence of ammonia at 250–350°C and 4–20 MPa. The reaction probably involves the simultaneous dehydration and hydration of urea to form cyanamide and ammonium carbamate; trimerization of the cyanamide then leads to melamine:

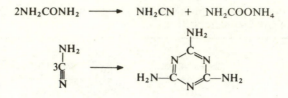

$$2NH_2CONH_2 \longrightarrow NH_2CN + NH_2COONH_4$$

Thus only 50% of the urea used gives melamine in one step and ammonium carbamate has to be separated and converted to urea for recycling. Despite this limitation, the urea route is the most economical of currently available routes.

At one time, melamine was mainly obtained from dicyandiamide, produced from calcium carbide as follows:

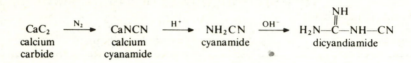

When dicyandiamide is heated above its melting point a vigorous reaction takes place to give a mixture which contains melamine:

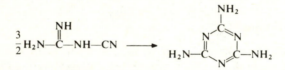

In order to reduce the formation of unwanted products, the conversion of dicyandiamide to melamine is carried out in the presence of liquid ammonia, typically at about 250°C and 4 MPa. Pure melamine is isolated from the reaction mixture by crystallization from water. This route is no longer of commercial importance.

Melamine is a colourless crystalline solid, m.p. 354°C. Although melamine is usually represented by the triazine structure shown above, X-ray analysis indicates that all the C–N bond lengths are approximately equal; thus solid melamine is probably a resonance hybrid:

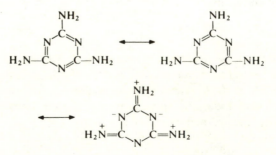

In aqueous solution, there is evidence that melamine exists in various tautomeric forms, e.g.

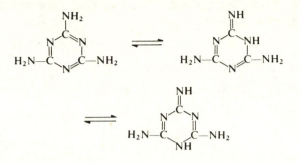

(b) Formaldehyde

The manufacture of formaldehyde is described in section 9.3.2. For the preparation of melamine-formaldehyde resins, formaldehyde is generally used as the aqueous solution, formalin. Paraformaldehyde is occasionally used.

15.3.3   Resin preparation

Melamine-formaldehyde polymers find practical utilization mainly in the form of network polymers. As with urea-formaldehyde polymers, polymerization is normally carried out in two operations, namely resin formation and curing. Both unmodified and modified resins are of commercial importance.

(a) Unmodified resins

Straight melamine-formaldehyde resins are used principally in the preparation of moulding compositions, laminates and textile finishes. The methods employed to produce these resins vary somewhat according to the end-use envisaged but the following procedure, which is for a laminating resin, illustrates the general principles involved.

Formalin is made slightly alkaline (pH 7.5–8.5) with aqueous sodium carbonate and then melamine is added to give a melamine to formaldehyde ratio of about 1:3 molar. The mixture is then heated at 80°C for 1–2 hours until the required degree of reaction is attained (conveniently judged by water tolerance). The resulting syrup is stabilized by the addition of borax (pH buffer) and is then used without further processing.

The reactions which occur in processes such as that described above are analogous to those which occur between urea and formaldehyde and are of two kinds:

(i) *Formation of methylolmelamines*. Under slightly alkaline conditions, melamine reacts with formaldehyde to give methylol derivatives with up to six methylol groups per molecule. The following examples illustrate the structures of these compounds:

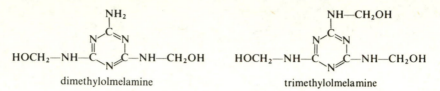

dimethylolmelamine   trimethylolmelamine

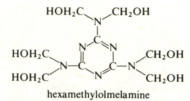

hexamethylolmelamine

The presence of all possible methylolmelamines in reaction mixtures prepared by using excess of formaldehyde (greater than 2:1 molar) has been demonstrated by paper chromatography [6]. However, only hexamethylolmelamine has been isolated in solid form.

(ii) *Condensation of methylolmelamines*. On heating, methylolmelamines condense to form resinous products which become increasingly hydrophobic until eventually a gel is formed. The rate of resinification is strongly dependent on pH. It is at a minimum at pH 10–10.5; an increase or decrease from this level results in a considerable increase in the reaction rate. The mechanism of resinification has not been investigated extensively but it may be envisaged that methylol groups can undergo reaction with amino, imino and other methylol groups:

(a) $R-NH-CH_2OH + H_2N-R \longrightarrow R-NH-CH_2-NH-R + H_2O$

(b) $R-NH-CH_2OH + R-NH-CH_2OH \longrightarrow$

$$R-NH-CH_2$$
$$R-N-CH_2OH + H_2O$$

(c) $R-NH-CH_2OH + HOCH_2-NH-R \longrightarrow$

$$R-NH-CH_2-O-CH_2-NH-R + H_2O$$

(d) $R-NH-CH_2OH + HOCH_2-NH-R \longrightarrow$

$$R-NH-CH_2-NH-R + CH_2O + H_2O$$

where R = melamine residue

In accordance with these proposals, it has been found that when methylolmelamines are heated both formaldehyde and water are evolved;

however, the relative importance of the various possible reactions has not been ascertained with any certainty. When hexamethylolmelamine (in which reactions (a) and (b) cannot occur) is heated, the product has an elemental analysis which corresponds to the presence of ether links (reaction (c)) and the virtual absence of methylene links (reaction (d)) [7]. It therefore seems likely that ether links are present in resins prepared using lesser amounts of formaldehyde. From analysis, it has been calculated that a typical melamine-formaldehyde resin contains an average of approximately three melamine residues per molecule.

(b) Modified resins

Methylolmelamines react with alcohols under acidic conditions to give ethers and both methylated and butylated products find commercial use.

In a typical process for the preparation of a methylated product, formalin is adjusted to pH 8 and melamine is added to give a melamine to formaldehyde ratio of 1:3.3 molar. The mixture is heated at 70°C for about 15 minutes until the melamine has completely dissolved. The solution is then either spray-dried or allowed to crystallize to give a solid composed essentially of monomeric methylolmelamines. This product is heated with about twice its weight of methanol and oxalic acid (0.5%) at 70°C until a clear solution is obtained. The solution is adjusted to pH 9 and concentrated to 80% solids under reduced pressure. The major component of this solution is the trimethyl ether of trimethylolmelamine:

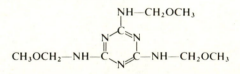

The methylated product is used for textile finishing, being preferred to straight methylolmelamines on account of the greater stability of the aqueous solutions.

Butylated melamine-formaldehyde resins are preferred for use in surface coating formulations because of their compatibility with hydrocarbon solvents and with other film-forming materials. Such resins may be prepared in a manner similar to that described previously for butylated urea-formaldehyde resins (section 15.2.3(b)). Generally, a melamine to formaldehyde ratio of 1:4–6 molar is used with a melamine to butanol (n- or iso) ratio of 1:4–8. About half of the available methylol groups become etherified in a typical process. Some of the remaining methylol groups then interact with one another to give a resin of molecular weight of about 800–1500.

Butylated melamine-formaldehyde resins give rather brittle films when used alone and are, therefore, generally blended with other film-forming materials which act as plasticizers. Alkyd resins (section 11.3) are most commonly used for this purpose.

15.3.4   Cross-linking

The conversion of low molecular weight melamine-formaldehyde resins into high molecular weight network polymers is usually accomplished by heating. The rate of cure may be increased by the addition of acidic compounds, but for most applications the rate of cure is adequate without an added catalyst. Cure of the resins proceeds through methylol groups, the reactions involved being the same as those which take place during the formation of resins from methylolmelamines (section 15.3.3(a)). When melamine-formaldehyde resins are heated at about 150°C water and formaldehyde are evolved. Since lesser quantities of formaldehyde are produced, it is generally supposed that ether links predominate over methylene links in the final product. A typical network structure might be represented as follows:

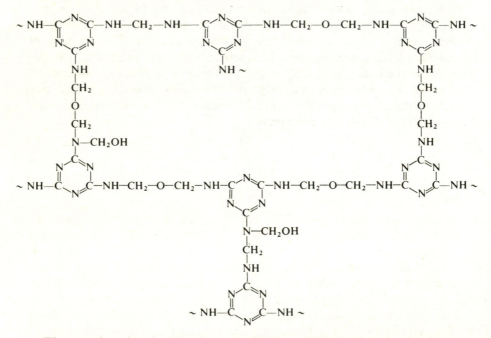

The reactions involved in the cross-linking of alkylated melamine-formaldehyde resins and their blends with alkyd resins may be supposed to be similar to those mentioned above. In addition, reactions analogous to those described for butylated urea-formaldehyde resins (section 15.2.4) may also take place.

15.3.5 Properties of cross-linked polymers

When melamine-formaldehyde resins have been cross-linked they are rigid, infusible and insoluble. Moulding compositions may contain various fillers and these have considerable influence on the properties of the mouldings (see Table 15.1). Melamine-formaldehyde mouldings have even greater surface hardness than urea-formaldehyde mouldings and, likewise, may be produced in a wide range of colours. The maximum temperature for continuous exposure without degradation is about 110°C for cellulose-filled mouldings but mineral-filled material withstands 200°C without serious loss of mechanical strength. Electrical insulating properties, which are similar to those of urea-formaldehyde mouldings under normal conditions are maintained better at elevated temperatures and in damp conditions.

Cross-linked melamine-formaldehyde polymers are very resistant to most organic reagents. They are generally attacked by acids and alkalis but the effect is much less than with urea-formaldehyde polymers. Melamine-formaldehyde mouldings are also much more resistant to staining by tea and coffee than urea-formaldehyde mouldings; largely because of this factor, the major use of melamine-formaldehyde moulding compositions is in the manufacture of tableware.

15.4 MELAMINE-PHENOL-FORMALDEHYDE POLYMERS

Resins prepared from melamine, phenol and formaldehyde have found some use in the preparation of moulding powders. These materials have properties which are intermediate between those of melamine-formaldehyde and phenol-formaldehyde moulding powders. In particular, mouldings have better dry-heat dimensional stability than those based on straight melamine-formaldehyde and they may be produced in a much wider range of colours than phenol-formaldehyde mouldings. Typical applications of melamine-phenol-formaldehyde materials include domestic mouldings such as iron handles where a combination of heat resistance and decorative appeal is required.

15.5 BENZOGUANAMINE-FORMALDEHYDE POLYMERS

Guanamines are closely related to melamine, one amino group of the latter being replaced by an alkyl or aryl group. They may be prepared by reaction of dicyandiamide with nitriles. Thus benzoguanamine, which is commercially available, is obtained from benzonitrile:

The reaction is carried out in liquid ammonia at 60–100°C in the presence of sodium or sodium hydroxide as catalyst.

Benzoguanamine closely resembles melamine in that reaction with formaldehyde gives methylol derivatives and then resinous condensates. Butylated benzoguanamine-formaldehyde resins have found some use in surface coatings. In a typical preparation, benzoguanamine (1 mole) is heated with formalin (3 mole formaldehyde) at pH 8.3 and 70–80°C until the guanamine has dissolved. n-Butanol (9 mole) and a small amount of hydrochloric acid are added and the azeotrope distilled off until removal of water is complete. Like their melamine counterparts, butylated benzoguanamine-formaldehyde resins are most commonly used in admixture with alkyd resins. The resulting coatings are very similar to those based on melamine but they show improved resistance to alkalis [8].

15.6   THIOUREA-FORMALDEHYDE POLYMERS

Thiourea is manufactured by heating ammonium thiocyanate at 140–145°C for about 4 hours; equilibrium is established when about 25% of the thiocyanate is converted to thiourea:

$$NH_4CNS \rightleftharpoons CS(NH_2)_2$$

Thiourea may also be prepared by the interaction of cyanamide and hydrogen sulphide:

$$NH_2CN + H_2S \longrightarrow CS(NH_2)_2$$

Thiourea closely resembles urea in that reaction with formaldehyde gives methylol derivatives and then resinous condensates which on continued heating yield network structures. Thiourea-formaldehyde resins are slower curing than urea-formaldehyde resins and the hardened products are more brittle and more water-resistant. At one time thiourea-formaldehyde resins were added to urea-formaldehyde resins to give mouldings and laminates with improved water-resistance. These mixed resins have now been largely superseded by melamine-formaldehyde resins which give products with better resistance to heat.

15.7   ANILINE-FORMALDEHYDE POLYMERS

Aniline and formaldehyde form resinous products under various conditions by means of rather ill-defined reactions. In a typical commercial process, aniline (1 mole) is treated with hydrochloric acid and then with formalin (1.2 mole formaldehyde). The resulting orange-red solution is fed into aqueous sodium hydroxide to precipitate a yellow material. This product has limited thermoplasticity but may be used to mould simple shapes. The material has good electrical insulation properties but has been largely superseded by polymers which are easier to process.

In the presence of acid, the amino group of aniline is protected and formaldehyde is directed to the *p*-position to give the hydrochloride of *p*-aminobenzyl alcohol. Treatment with alkali gives the free base which polymerizes, possibly as follows:

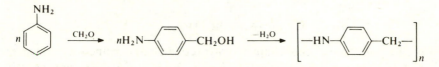

In the presence of an excess of formaldehyde, methylene bridges may be formed to give a slightly cross-linked product:

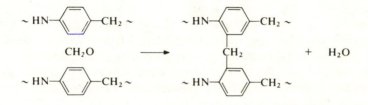

REFERENCES

1. De Jong, J. I. and de Jonge, J. (1953) *Rec. trav. chim.*, **72**, 88.
2. De Jong, J. I. and de Jonge, J. (1953) *Rec. trav. chim.*, **72**, 139.
3. De Jong, J. I. and de Jonge, J. (1953) *Rec. trav. chim.*, **72**, 1027.
4. Tsuge, M., Miyabayashi, T. and Tanaka, S. (1974) *Bunseki Kagaku*, **23**(10), 1146.
5. Marvel, C. S. *et al.* (1946) *J. Am. Chem. Soc.*, **68**, 1681.
6. Koeda, K. (1954) *J. Chem. Soc. Japan, Pure Chem. Sec.*, **75**, 571.
7. Gams, A. (1943) *British Plastics*, **14**, 508.
8. Grimshaw, F. P., (1957) *J. Oil Colour Chemists' Assoc.*, **40**, 1060.

BIBLIOGRAPHY

Blais, J. F. (1959) *Amino Resins*, Reinhold Publishing Corporation, New York.
Vale, C. P. and Taylor, W. G. K. (1964) *Aminoplastics*, Iliffe Books Ltd, London.
Meyer, B. (1979) *Urea-Formaldehyde Resins*, Addison-Wesley, Reading.

16

POLYURETHANES

16.1 SCOPE

For the purposes of this chapter, polyurethanes are defined as polymers which contain urethane groups (–NH–CO–O–) in the main polymer chain. However, it is to be noted that in technologically useful polymers of this type the urethane group is not usually the principal group present; other groups such as ester, ether, amide and urea groups are generally contained in the polymer chain in appreciable number.

Several kinds of polyurethanes are of commercial significance and are conveniently classified into the following major types – flexible foams; rigid foams; integral foams; elastomers; surface coatings; and adhesives. These various types of material are considered separately in this chapter after an account of the development of polyurethanes, the raw materials involved and the general reactions of isocyanates.

16.2 DEVELOPMENT

The commercial development of polyurethanes dates from 1937 when O. Bayer (I. G. Farbenindustrie and later Farbenfabriken Bayer, Germany) found that reaction of diisocyanates and glycols gave polyurethanes with properties which made them of interest as plastics and fibres. Further intensive research into polyurethanes soon indicated that the materials showed promise as adhesives, rigid foams and surface coatings. These various applications reached moderate commercial importance in Germany during the Second World War, but were not recognized elsewhere at this time. After the war, Allied intelligence teams reported on these activities and interest in polyurethanes was established in the UK and USA although commercial development was slow. The German chemical industry quickly recovered from the effects of war; Bayer again became active in the polyurethane field and developed elastomers (1950) and flexible foams (1952). In the period 1952–54 this company developed its diisocyanate-polyester flexible foam system to a degree suitable for commercial use and also introduced novel machines for continuous production of the foams. By 1955 the mass-production of polyester-based foams was established in most industrial countries. In 1957 several American companies developed polyether-based foams. These products offered a wider range of properties than polyester-based foam at a significantly lower price and substantially broadened the foam market. The advent of flexible foams assured the large scale commercial use of polyurethanes and with the ready availability of raw materials, interest was renewed in other applications of polyurethanes. At the present time, about 50% of polyurethane output is in the form of flexible foam and 25% is as rigid foam. The remainder finds use in such applications as elastomers, surface coatings, adhesives and binders.

16.3 RAW MATERIALS

The urethane group results from the interaction of an isocyanate and a hydroxyl compound:

$$R-NCO \ + \ HO-R' \longrightarrow R-NH-CO-O-R'$$

It will be apparent that this reaction leads to polyurethanes when multifunctional reactants are used. When a diisocyanate and a diol react together a linear polyurethane is obtained whilst a diisocyanate and a polyhydric compound (polyol) lead to a cross-linked polymer. A cross-linked polyurethane could also be derived from a compound containing three or more isocyanate groups and a diol but this approach is of limited commercial importance.

Thus diisocyanates and diols and polyols are the principal raw materials used in the manufacture of polyurethanes. The more important of these reactants are described in this section.

16.3.1 Diisocyanates

Several reactions are known by which isocyanates are formed. However, there is only one method of preparation of commercial importance, namely phosgenation of primary amines:

$$R-NH_2 \ + \ COCl_2 \longrightarrow R-N=C=O \ + \ 2HCl$$

The production of phosgene is described in section 12.4.1(a).

(a) Tolylene diisocyanate

Toluene is the starting material for the production of tolylene diisocyanate (TDI) and, as shown in Fig. 16.1, the process may be varied to give products of differing isomer contents. The nitration of toluene (with a

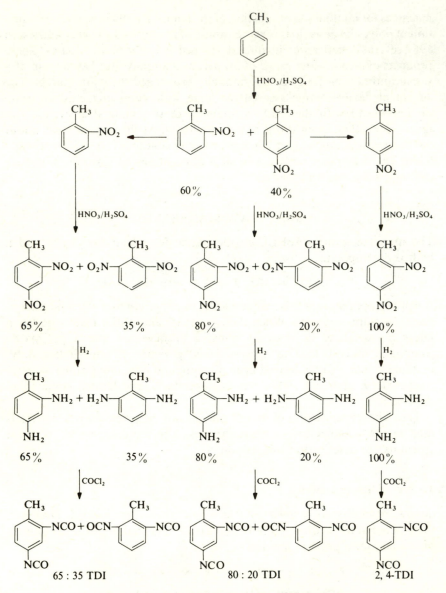

Fig. 16.1 Preparation of tolylene diisocyanate.

nitrating mixture containing 20% nitric acid, 60% sulphuric acid and 20% water at 30–45°C) gives a mixture of 2-nitrotoluene (about 60%) and 4-nitrotoluene (40%). If this mixture is nitrated further (with a mixture of 35% nitric acid and 65% sulphuric acid at 65–80°C) without separation, the product is a mixture of 2,4-dinitrotoluene (about 80%) and 2,6-dinitrotoluene (20%). If, on the other hand, the mixed mononitrates are separated (by distillation), then further nitration of the 2-nitrotoluene yields a mixture of 2,4-dinitrotoluene (about 65%) and 2,6-dinitrotoluene (35%) whilst further nitration of the 4-nitrotoluene gives only 2,4-dinitrotoluene.

Hydrogenation of the dinitrotoluenes in the presence of nickel, platinum or palladium catalysts (at about 5 MPa (50 atmospheres) and 90°C) leads to the corresponding diamines which on phosgenation give the corresponding diisocyanates. Thus it is possible to prepare '80:20' tolylene diisocyanate (80% 2,4-isomer, 20% 2,6-isomer), '65:35' tolylene diisocyanate (65% 2,4-isomer, 35% 2,6-isomer) and tolylene 2,4-diisocyanate; all of these materials are commercially available.

Phosgenation is usually carried out continuously as follows. The diamine feed is injected into a solution of phosgene in an inert solvent (e.g. *o*-dichlorobenzene) at 25–100°C (the cold stage) and then the product is passed through a heater at 150–160°C (the hot stage). Hydrogen chloride and unreacted phosgene are taken off and the diisocyanate is separated from the solvent by distillation.

The phosgenation process is generally considered to involve the formation of carbamoyl chloride in the cold stage and the decomposition of this product in the hot stage:

$$R\text{---}NH_2 \ + \ COCl_2 \ \longrightarrow \ R\text{---}NHCOCl \ + \ HCl$$

$$R\text{---}NHCOCl \ \longrightarrow \ R\text{---}NCO \ + \ HCl$$

Several intermediate steps, however, are involved. Various side-reactions may also occur during the phosgenation process, the chief of which is that between unused amine and isocyanate to form a urea:

$$R\text{---}NH_2 \ + \ R\text{---}NCO \ \longrightarrow \ R\text{---}NH\text{---}CO\text{---}NH\text{---}R$$

This reaction is minimized by the use of an excess of phosgene.

Overall, the above route to tolylene diisocyanate is not ideal for large scale operation, being complicated and involving hazardous materials at all stages. Nevertheless, it does not appear that any other process is operated commercially at the present time.

The tolylene diisocyanate isomer mixtures melt in the range 5–15°C and are therefore usually encountered as liquids; tolylene 2,4-diisocyanate is a solid, m.p. 22°C. Tolylene diisocyanates are respiratory irritants and require careful handling. Of the various products described above, the 80:20 isomer mixture is the cheapest and is widely used, particularly in the production of flexible foams. The other products find use when enhanced or reduced reactivity is desired; for steric reasons, the 4-isocyanato group is considerably more reactive than either the 2- or 6-isocyanato group.

(b)   Diphenylmethane diisocyanate

Diphenylmethane diisocyanate (MDI) is derived from aniline; the principal reactions involved are as follows:

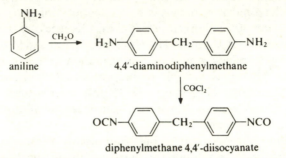

In the first step, aniline is treated with formaldehyde in the presence of concentrated hydrochloric acid as catalyst. The product is a mixture of amines, composed mainly of 4,4′-diaminodiphenylmethane together with lesser amounts of the 2,4′-isomer and various polyamines with up to about six amino groups per molecule; the polyamines have structures of the following type:

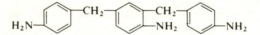

The precise composition of the mixture depends principally on the aniline to formaldehyde ratio used; increasing amounts of aniline favour the formation of diaminodiphenylmethanes. Sometimes the amine mixture is fractionated to give pure 4,4′-diaminodiphenylmethane which is then phosgenated to form diphenylmethane 4,4′-diisocyanate. More commonly, the amine mixture is phosgenated without separation to give 'polymeric' diphenylmethane diisocyanate which typically contains about 55% diphenylmethane diisocyanates (4,4′- and 2,4′-isomers), 25% triisocyanates and 20% higher polyisocyanates.

Diphenylmethane 4,4′-diisocyanate is a solid, m.p. 37–38°C whilst polymeric diphenylmethane diisocyanate is a liquid; both products have lower vapour pressure than tolylene diisocyanate and are therefore less toxic in use. Diphenylmethane 4,4′-diisocyanate is used in elastomer manufacture and polymeric diphenylmethane diisocyanate is extensively employed in the production of rigid foams.

(c)   Naphthylene 1,5-diisocyanate

Naphthylene 1,5-diisocyanate (NDI) is prepared from naphthalene as follows:

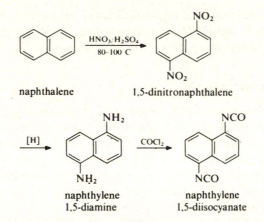

naphthalene   1,5-dinitronaphthalene

naphthylene
1,5-diamine

naphthylene
1,5-diisocyanate

Naphthylene 1,5-diisocyanate is a solid, m.p. 128°C. It has a lower vapour pressure than tolylene diisocyanate and is therefore less toxic in use; it does, however, have sensitizing properties. Naphthylene 1,5-diisocyanate is mainly used for the production of elastomers.

(d) Hexamethylene diisocyanate

Hexamethylene diisocyanate (HDI) is prepared by the phosgenation of hexamethylenediamine (section 10.2.2(b)):

$$H_2N-(CH_2)_6-NH_2 \xrightarrow{COCl_2} OCN-(CH_2)_6-NCO$$

Hexamethylene diisocyanate is a liquid with a volatility of the same order as that of tolylene diisocyanate. It is respiratory irritant and also has powerful effects on the skin and eyes. Hexamethylene diisocyanate was one of the first diisocyanates utilized for making polyurethanes, being used to prepare fibres and moulding compounds. These applications are no longer of importance but hexamethylene diisocyanate now finds use mainly in coatings which are more light stable than those based on aromatic isocyanates.

(e) Isophorone diisocyanate

Isophorone diisocyanate (IPDI) is prepared by the phosgenation of iso-phorone diamine (1-amino-3-aminomethyl-3,5,5-trimethylcyclohexane):

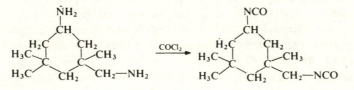

Isophorone diisocyanate is a liquid with a vapour pressure which is significantly lower than that of tolylene diisocyanate. Isophorone diisocyanate is used for the preparation of non-yellowing coatings.

16.3.2 Diols and polyols

The earliest polyurethanes were based on aliphatic diols (glycols) and, of the various glycols investigated, 1,4-butanediol was generally preferred for commercial operations. Since this time, however, production of polyurethanes has mainly involved polymeric hydroxyl compounds. The use of these materials permits the manufacture of a much wider range of products at relatively low cost. The polymeric hydroxyl compounds which have received most attention are polyesters and polyethers.

Polyethers are now used in about 90% of commercial polyurethanes whilst polyesters are used in most of the remainder.

(a) Polyethers

The polyethers most widely used for the production of polyurethanes are hydroxyl-terminated derivatives of propylene oxide and these polymers are described in section 9.4.3(a). Linear, glycol-initiated propylene oxide polymers and propylene oxide-ethylene oxide block copolymers find some use in the preparation of elastomers. Polyether triols of relatively high molecular weight (about 3000) are extensively used for the production of flexible foams whilst polyols of low molecular weight (about 500) are used for rigid foams and surface coatings. Poly(oxytetramethylene) glycols prepared from tetrahydrofuran (section 9.4.6) are used for the preparation of elastomers and 'spandex' fibres.

More recently, filled polyethers (so called 'polymer polyols') have been introduced. These contain dispersed organic filler such as acrylonitrile-styrene copolymer or polyurea, some of which is grafted on to the polyether chain. Filled polyethers are used principally for flexible foams of high resilience.

(b) Polyesters

The polyesters used in the preparation of polyurethanes generally have molecular weight in the range 1000–2000 and are liquids or low-melting solids; they are usually saturated. The essential feature of these polyesters is that they are hydroxyl-terminated and thus hydroxyl groups are available to participate in the urethane reaction. The polyesters are prepared by the reaction of dibasic acids (e.g. adipic and sebacic acids) with glycols (e.g. ethylene and diethylene glycols) and polyhydric alcohols (e.g. glycerol and trimethylolpropane). The reaction is carried out by procedures similar to those described in sections 11.2.3 and 11.11 for low molecular weight polyesters intended for use in laminating resins and as plasticizers. According to the reactants used, the products range from linear polymers to highly branched polymers; the structure of the polyester determines, of course, the nature of the polyurethane ultimately obtained. Compositions of typical polyesters used to prepare various types of polyurethanes are given in Table 16.1.

Hydroxyl-terminated polyesteramides are also used in the preparation of polyurethanes. A typical product may be made by reaction of adipic acid, ethylene glycol and ethanolamine:

$$HO—CH_2—CH_2—NH_2 \ + \ HOOC—(CH_2)_4—COOH \ + \ HO—CH_2—CH_2—OH$$

$$\xrightarrow{-H_2O} \sim O—CH_2—CH_2—NH—OC—(CH_2)_4—CO—O—CH_2—CH_2—O \sim$$

Polyurethanes based on polyesteramides were originally developed as elastomers but now find use primarily as leather adhesives and flexible coatings for rubber goods.

It may be noted here that castor oil is used as a hydroxyl-compound in the preparation of polyurethanes, particularly for coatings and adhesives. Castor oil is essentially a triol, being composed largely of glycerol triricinoleate (see Table 11.2).

Table 16.1 Compositions of typical polyesters used to prepare various types of polyurethanes

	Molar proportions			
Adipic acid	1.0	1.5	3.0	3.0
Sebacic acid	–	1.5	–	–
Ethylene glycol	0.75	–	–	–
Propylene glycol	0.35	–	–	–
1,3-Butanediol	–	–	3.0	–
Diethylene glycol	–	3.25	–	2.0
Glycerol	–	0.5	1.0	–
Trimethylolpropane	–	–	–	3.0
Structure	Linear	Slightly branched	Branched	Highly branched
Application of polyurethane	Elastomers	Flexible foams	Coatings	Rigid foams

16.4 ISOCYANATE REACTIONS

Isocyanates are very reactive materials and undergo a great many reactions. In this section, those reactions which have technological significance are discussed. The common reactions of isocyanates may be divided into two classes, namely addition reactions with compounds containing active hydrogen and self-addition.

The more important reactions involving active hydrogen compounds are shown below.

(i) *With alcohols*

$$R-NCO + R'-OH \longrightarrow R-NH-CO-O-R' \longrightarrow \text{reaction (v)}$$
$$\text{a urethane}$$

(ii) *With amines*

$$R-NCO + R'-NH_2 \longrightarrow R-NH-CO-NH-R' \longrightarrow \text{reaction (vi)}$$
$$\text{a urea}$$

(iii) *With carboxylic acids*

$$R-NCO + R'-COOH \longrightarrow [R-NH-CO-O-CO-R']$$
$$\text{anhydride}$$
$$\xrightarrow{-CO_2} R-NH-CO-R' \longrightarrow \text{reaction (vii)}$$
$$\text{amide}$$

(iv) *With water*

$$R-NCO + H_2O \longrightarrow [R-NH-COOH]$$
$$\text{a carbamic acid}$$
$$\xrightarrow{-CO_2} R-NH_2 \longrightarrow \text{reaction (ii)}$$

(v) *With urethanes*

$$R-NCO + R'-NH-CO-O-R'' \longrightarrow R-NH-CO-\overset{\displaystyle R'}{\underset{\displaystyle |}{N}}-CO-O-R''$$
$$\text{allophanate}$$

(vi) *With ureas*

$$R-NCO + R'-NH-CO-NH-R'' \longrightarrow R-NH-CO-\overset{\displaystyle R'}{\underset{\displaystyle |}{N}}-CO-NH-R''$$
$$\text{biuret}$$

(vii) *With amides*

$$R-NCO + R'-NH-CO-R'' \longrightarrow R-NH-CO-\overset{\displaystyle R'}{\underset{\displaystyle |}{N}}-CO-R''$$
$$\text{acylurea}$$

(viii) *With phenols*

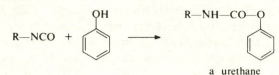

a urethane

The principal self-addition reactions of isocyanates are:

(ix) *Dimerization*

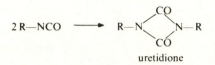

uretidione

(x) *Trimerization*

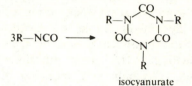

isocyanurate

(xi) *Carbodiimide formation*

$$2R—NCO \longrightarrow R—N=C=N—R \ + \ CO_2$$

It may be noted that the anionic polymerization of isocyanates to high molecular weight polymers is mentioned in section 1.4.2(c) but this process is not of commercial importance.

The mechanisms of the reactions between isocyanates and various active hydrogen compounds are probably broadly similar, but most experimental investigations in this field have involved the isocyanate-alcohol reaction. In the absence of an added catalyst, this reaction is believed to proceed by the following mechanism, in which the alcohol itself acts catalytically [1]:

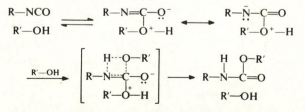

Electron-withdrawing groups on the isocyanate molecule and electron-donating groups on the active hydrogen molecule increase the rate of reaction. These effects indicate that the complex formed between the isocyanate and alcohol involves electron-pair donation and not hydrogen bonding.

Reactions between isocyanates and active hydrogen compounds are also extremely susceptible to catalysis and many commercial applications of isocyanates utilize catalysed reactions. The most widely used catalysts are tertiary amines and certain metal compounds, particularly tin compounds. The mechanisms through which these catalysts operate are probably similar to that shown above. Thus tertiary amine (R''_3N) catalysis is thought to proceed as follows:

$$R-NCO + R''_3N \rightleftharpoons R-N=\underset{\underset{R''_3N^+}{|}}{C}-\underset{..}{O}^- \longleftrightarrow R-\underset{\underset{R''_3N^+}{|}}{\overset{-}{N}}-C=O$$

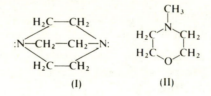

For a series of amine catalysts, the catalytic activity generally closely parallels the base strength of the amine except when steric effects become pronounced. The latter effects are illustrated in Table 16.2 in which it is seen that activity falls off as the size of the substituent groups in tertiary amines increases. In the case of triethylenediamine (I) there is little steric hindrance to complex formation and although this amine is a weak base it is a powerful catalyst.

$$\begin{array}{cc}
\begin{array}{c}
H_2C-CH_2 \\
:N\!\!\begin{array}{c}\\ -CH_2-CH_2-\end{array}\!\!N: \\
H_2C-CH_2
\end{array}
&
\begin{array}{c}
CH_3 \\
\overset{|}{N} \\
H_2C\overset{\diagup\ \ \diagdown}{}CH_2 \\
H_2C\underset{\diagdown\ \ \diagup}{}CH_2 \\
O
\end{array}
\\
(I) & (II)
\end{array}$$

Metal salt (MX_2) catalysts may operate through the following mechanism [3]:

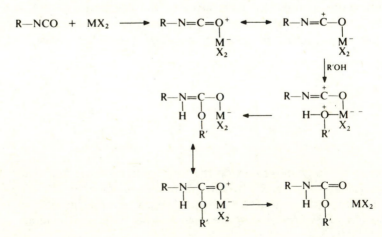

Table 16.2 Catalytic activity of tertiary amines in the phenyl isocyanate-butanol reaction [2]

Amine	Base strength (pK_a)	Relative activity
Trimethylamine	9.9	2.2
Ethyldimethylamine	10.2	1.6
Diethylmethylamine	10.4	1.0
Triethylamine	10.8	0.9
Triethylenediamine	8.2	3.3

As is discussed later, the relative rates at which various isocyanate reactions proceed is a matter of great practical significance, particularly in the preparation of foams. In general, the order of reactivity of active hydrogen compounds with isocyanates in uncatalysed systems is as follows (most reactive first):

Aliphatic amines; aromatic amines; primary alcohols; water; secondary alcohols; tertiary alcohols; phenols; carboxylic acids; ureas; amides; urethanes.

The self-addition reactions of isocyanates do not usually proceed as readily as reactions with active hydrogen compounds.

An additional point of importance is that the various isocyanate reactions are influenced to differing extents by different catalysts. This point is illustrated in Table 16.3. Further, tertiary amines are also effective catalysts for isocyanate self-addition reactions whereas metal compounds generally have less influence; tin compounds are particularly poor catalysts in these reactions.

Table 16.3 Catalytic activity in phenyl isocyanate reactions [4]

Catalyst	Relative rate of reaction with		
	n-Butanol	Water	Diphenylurea
None	1.0	1.1	2.2
N-Methylmorpholine (II)	40	25	10
Triethylamine	86	47	4
Tetramethyl-1,3-butanediamine	260	100	12
Triethylenediamine	1200	380	90
Tributyltin acetate	80 000	14 000	8000
Dibutyltin diacetate	600 000	100 000	12 000

16.5 FOAMS

Polyurethane foams are produced by forming a polyurethane polymer concurrently with a gas evolution process. Provided these two processes are
balanced, bubbles of gas are trapped in the polymer matrix as it is formed and
a cellular product results. The matching of the two reactions is essential for
the formation of satisfactory foams. If the evolution of gas is too rapid, the
foam initially rises well but then collapses because polymerization has not
proceeded sufficiently to give a matrix strong enough to retain the gas. If
polymerization is too fast, the foam does not rise adequately.

By selection of appropriate reactants, it is possible to prepare foams of
varying degrees of cross-linking. Slightly cross-linked products are flexible
whilst highly cross-linked products are rigid. Both flexible and rigid polyurethane foams are of commercial importance.

16.5.1 Flexible foams

(a) Preparation

The gas used in the production of flexible foams is usually carbon dioxide
formed by the interaction of isocyanate and water. In a system containing a
diisocyanate, a polyol and water, two principal reactions proceed simultaneously, namely:

diisocyanate + polyol → polyurethane
diisocyanate + water → carbon dioxide

As indicated above, a satisfactory foam is obtained only if these two reactions
are in step. Although choice of temperature and reactants permit a measure
of control over the reactions, much broader control may be exercised through
the use of catalysts. As will be seen, catalysts play a crucial role in commercial
operations.

Tolylene diisocyanate is usually the preferred isocyanate for the production of flexible polyurethane foams. As previously mentioned, flexible foams
are based either on polyesters or polyethers; polyether foams account for
about 90% of current flexible foam output. The two types of foams are
considered separately in this section.

Polyether foams. Flexible polyether foams are most commonly produced by
a 'one-shot' process. In this procedure, diisocyanate, polyol, water, catalysts
and surfactant are all mixed simultaneously. About 80% of such foam is
produced in slabstock form in large blocks (with cross-sections up to about
2.2 m wide and 1.25 m high) and 20% is moulded. Commonly, in the
production of slabstock the mixture is fed continuously into a moving trough

wherein, after a few seconds, the foam begins to rise. After 1–2 minutes the foam reaches its maximum height; it is then cut into blocks which are allowed to stand for several hours before being sliced into sheets of the required thickness. In the production of moulded foam, the mixture of reactants is fed into moulds which then become filled with the foam.

A typical formulation for flexible polyether foam would be as follows:

Polyether triol	100	parts by weight
80:20 Tolylene diisocyanate	40	
Water	3.0	
Triethylenediamine	0.5	
Stannous octoate	0.3	
Silicone block copolymer	1.0	

As mentioned in section 9.4.3(a) the majority of hydroxyl groups in a polyether triol are secondary groups and are comparatively unreactive towards isocyanates. It is therefore necessary to select a catalyst which favours the formation of urethane links relatively more than the formation of gas by the reaction of isocyanate and water. Tin compounds (e.g. stannous octoate and dibutyltin dilaurate) are particularly effective in this respect (cf. Table 16.3) and are very widely used. In addition to the primary isocyanate-polyol and isocyanate-water reactions, several secondary reactions occur during the preparation of foam. As shown in section 16.4, the final product may contain allophanate, biuret, isocyanurate and uretidione links. It will be appreciated that in a polymeric system, which is based on a diisocyanate, all of these links (except uretidione) represent points of branching or cross-linking. These secondary reactions are particularly favoured by tertiary amines (e.g. triethylenediamine and 4-dimethylaminopyridine) and these catalysts therefore contribute to the final cross-linking of the foam and hence to the achievement of, for example, a low compression set. Mixtures of tin compounds and tertiary amines are more effective than would be anticipated from the sum of the separate catalytic effects, i.e. they are synergistic. By varying the composition of the mixture of tin compound and tertiary amine it is possible to catalyse to a variable extent polymer build-up, gas formation and cross-linking with significant effects on foam properties. The final ingredient in the above formulation is a silicone-polyalkylene oxide block copolymer which acts as a foam stabilizer during the critical early stages of foam formation when the polymer is still weak.

Before the discovery of the catalytic activity of tin compounds, polyether foams were prepared by the 'pre-polymer' technique. In this two-stage process, the low reactivity of the polyether secondary hydroxyl groups is compensated for by carrying out the reaction between isocyanate and diol *before* the addition of water. The polyether is treated with excess of isocyanate to give an isocyanate-terminated pre-polymer which is stable in the absence of moisture. Addition of water, tertiary amine and surfactant to the pre-polymer results in a flexible foam of good quality. The pre-polymer process has been largely displaced by the more economical one-shot process but is still occasionally used since it offers more scope in the design of compounds.

A related technique is the 'quasi pre-polymer' (or 'semi pre-polymer') process in which the polyol is treated with a *large* excess of isocyanate to give a polymer of low molecular weight. This product is then treated with additional polyol, water and catalyst to produce a foam. The advantage of this procedure is that the initial polymer, being of low molecular weight, is less viscous than the product obtained in the pre-polymer process described previously and is easier to handle.

Slabstock flexible polyether foam is used mainly for furniture cushioning, mattresses and vehicular seating. The largest application of moulded flexible polyether foam is in automobile seating.

Polyester foams. Flexible polyester foams are nearly always prepared by a one-shot process; most commonly they are produced as slabstock by a continuous method exactly comparable to that described previously for polyether foams. A typical formulation might be as follows:

Polyester (slightly branched)	100	parts by weight
80:20 Tolylene diisocyanate	37	
Water	4.0	
Tertiary amine	1.5	
Silicone block copolymer	2.0	

The majority of hydroxyl groups in the polyesters used for foam making are primary groups and are comparatively reactive towards isocyanates. It is therefore sufficient to include only simple tertiary amine catalysts in the formulation to obtain a satisfactory foam. The reactions which occur during the foaming process are the same as those described for polyether-based foams. There is the added possibility of reaction between any carboxylic groups which the polyester may contain and the isocyanate to give amide and acylurea links. Indeed, the earliest foams were made from polyesters with high acid numbers, the isocyanate-carboxylic acid reaction being used as the gas-generating reaction.

(b)   Properties

Flexible polyurethane foams are open-cell structures which are usually produced with densities in the range 24–48 kg/m^3 (1.5–3 lb/ft^3). The major interest in flexible foams is for upholstery applications and thus the load-compression characteristics are of importance. Typical load-deflection curves

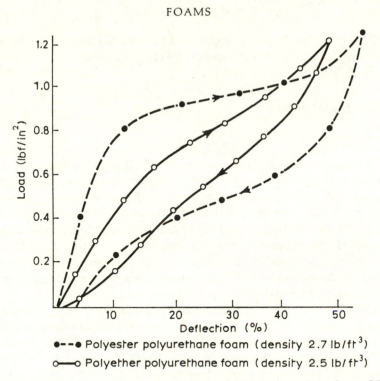

●--● Polyester polyurethane foam (density 2.7 lb/ft^3)
o—o Polyether polyurethane foam (density 2.5 lb/ft^3)

Fig. 16.2 Typical load-deflection curves for flexible polyurethane foams [5].

for polyether and polyester foams are shown in Fig. 16.2. The most obvious difference between polyether and polyester foams is the lower resilience of the polyester materials. This feature has led to a preference for polyether foams in cushioning applications. Compared to polyether foams, polyester foams have higher tensile strength, elongation at break and hardness; consequently polyester foams are preferred in such applications as textile laminates and coat shoulder pads.

Polyurethane foams are resistant to a wide range of solvents. In this respect, polyester foams are generally superior to polyether foams, particularly in resistance to dry cleaning solvents; this is a further reason for preferring polyester foams in textile applications. Polyurethane foams are subject to degradation by aqueous acids and alkalis and steam. Ester, amide and urethane groups represent sites for hydrolytic attack. Since the ether group is not readily attacked, polyether foams are generally more resistant to hydrolysis than polyester foams.

Flexible polyurethane foams have good dry heat stability, withstanding temperatures up to 150°C for long periods without serious loss of strength. Above this temperature the formation of biuret and allophanate cross-links is reversed and there is a reduction in load bearing properties. Urethane links are stable up to about 200°C. Exposure to ultraviolet light causes discoloration but no apparent deterioration in physical properties.

16.5.2   Rigid foams

(a)   *Preparation*

The principle underlying the production of rigid polyurethane foams is fundamentally the same as that used for flexible foams, namely a reacting isocyanate/polyol mixture is simultaneously expanded by gas generation. The essential difference between the two products lies in their degree of cross-linking; whereas flexible foams are lightly cross-linked, rigid foams are highly cross-linked. As indicated previously (section 16.3.2), this high degree of cross-linking is achieved by using relatively low molecular weight polyols for coupling with the isocyanate. Rigid foams, like flexible foams, are now mainly based on polyethers rather than polyesters.

Most commonly, rigid polyether foams are prepared by a one-shot process. Production is both in blocks by large factory-installed machines and *in situ* by small portable equipment. A difference between the manufacture of rigid and flexible foams is in the blowing agent generally used. Flexible foams are invariably blown by carbon dioxide (sometimes in conjunction with a small proportion of a halocarbon when particularly low density foam is required). Rigid foams can be blown by carbon dioxide generated by the isocyanate-water reaction but usually low-boiling halogenated alkanes are employed. Of these, trichlorofluoromethane (b.p. 24°C) is by far the most widely used. This inert liquid takes no part in the chemical reactions but is volatilized by the exothermic isocyanate-polyol reaction and expands the foam; the trichloro-fluoromethane gas remains in the closed cells of the rigid foam. Since this gas has very low thermal conductivity, the foams are outstandingly good insulators.

A typical basic formulation for rigid polyether foam might be as follows:

Polyether polyol	100 parts by weight
Polymeric diphenylmethane diiso-cyanate	Stoichiometric + 5%
Trichlorofluoromethane	50
Triethylenediamine	0.5
Silicone block copolymer	1.0
Glycerol	10 (cross-linking agent)

It will be noted that in the production of rigid foam it is not necessary to use a complex catalyst system. Commonly tertiary amines are used alone to catalyse the isocyanate-polyol reaction; metal catalysts are not widely used.

Diphenylmethane diisocyanate is the preferred isocyanate for rigid foams on account of its low volatility but tolylene diisocyanate may also be used. In the latter case, it is usual to utilize the quasi pre-polymer technique (section 16.5.1(a)) since the reactivity of the isocyanate makes it difficult to use in a one-shot process.

The chemical reactions which occur during the formation of rigid foams correspond to those described for flexible foams. However, since water is generally absent (although a small amount may be added to the formulation in order to boost the volatilization of the halocarbon with the isocyanate-water exotherm), biuret links are not produced.

In a modification of the usual process for the production of rigid foam, a high proportion of isocyanurate links are incorporated into the structure to give a so-called 'polyisocyanurate' foam. In this technique, the polyether is treated with an excess of isocyanate in the presence of an isocyanurate catalyst (e.g. an alkali metal alcoholate or carboxylate). Polyisocyanurate foams have significantly lower combustibility than conventional polyurethane foams.

It may be noted here that whilst most attention has been directed towards flexible and rigid polyurethane foams, intermediate products known as *semi-rigid foams* find use in the manufacture of car crash-pads and packaging. These products may be obtained from polyether triols of intermediate molecular weight or moderately branched polyesters.

(b) Properties

Rigid polyurethane foams are closed-cell structures which are usually produced with a density of about 32 kg/m^3 (2 lb/ft^3). Since the major interest in rigid foams has been for thermal insulation, the thermal conductivity of the foams is a physical property of some importance. About 97% of the volume of a typical foam is occupied by gas, the nature of which therefore has an appreciable effect on the thermal conductivity. This point is illustrated by Table 16.4 which shows the conductivities of various gases and foams. Newly

Table 16.4 Thermal conductivities of various gases and rigid polyurethane foams at 0°C [6]

	Btu in/(ft^2 hr °F)	W/(m K)
Air	0.17	0.024
Carbon dioxide	0.10	0.014
Trichlorofluoromethane	0.054	0.0078
Carbon dioxide-blown foam (initial)	0.15	0.022
Carbon dioxide-blown foam (final)	0.24	0.035
Trichlorofluoromethane-blown foam (final)	0.16	0.023

made carbon dioxide-blown foam has low conductivity but the carbon dioxide rapidly diffuses through the polymer and is replaced by air; the conductivity therefore increases until it reaches the value for air-filled foam. Trichlorofluoromethane diffuses very slowly and an equilibrium appears to be reached when the air to fluorocarbon in the cells is about 1:1; hence trichlorofluoromethane-blown foam has low ultimate thermal conductivity.

The chemical properties of rigid polyurethane foams are similar to those of flexible foams (section 16.5.1(b)).

Rigid polyurethane foams are used for the insulation of appliances such as freezers and refrigerators and for the production of building insulation panels.

16.5.3   Integral foams

Integral foams (also called integral skin or self-skinning foams) are continuous structures with a cellular core and a solid skin which are produced in one moulding operation. The skin provides both protection and finish to the core. In the production of such foams, a reaction mixture is injected into a cooled mould so that the mould is overpacked (i.e. contains more mixture than would be necessary to fill the mould cavity by free foam expansion). The surface layer of the reacting mass is cooled by contact with the mould surface and compressed by the expanding foam core. Thus the blowing agent in this surface layer cannot vaporize before gelation occurs and a solid, unexpanded skin is formed.

Both flexible and rigid integral foams are made commercially using one-shot systems. Formulations are similar to those discussed previously, with systems of high reactivity being preferred in order to minimize demould times. However, integral foams are always blown with low-boiling halogenated alkanes such as trichlorofluoromethane. Carbon dioxide cannot be used since it does not condense at the mould surface under the reaction conditions. By adjusting formulations and moulding conditions it is possible to obtain a spectrum of products ranging from highly foamed material with density of about 0.20 g/cm^3 (12 lb/ft^3) to solid material with density 1.2 g/cm^3 (75 lb/ft^3).

The production of integral foams requires the metering and fast mixing of reactive components and the injection of the resulting mixture into a closed mould. This process is known as reaction injection moulding (RIM). It may be noted that reaction injection moulding is applied mainly to polyurethanes but the process can be used with other systems, e.g. polyamides (see section 10.2.3(d)).

Flexible integral foams find wide application in the automotive industry for components such as bumpers, headrests and steering wheels whilst rigid integral foams are used in building components such as window frames and in furniture.

16.6 ELASTOMERS

Solid polyurethane elastomers (as distinct from flexible foams) may be divided into three categories, namely cast, millable and thermoplastic elastomers.

16.6.1 Cast elastomers

In the casting technique a liquid reaction mixture comprising low molecular weight material is poured into a heated mould, wherein the material is converted to a solid, high molecular weight elastomeric product.

The most obvious method of preparing polyurethane elastomers of this type is to cast a mixture of diisocyanate and slightly branched polyester or polyether. Elastomers of this kind, based on slightly branched poly(ethylene adipate) and tolylene diisocyanate are extensively used in printing rollers. In a typical process, a mixture of the polyester and isocyanate is degassed by heating at 70°C under reduced pressure. The mixture is then poured into a cylindrical mould (in which a roller spindle is located) and heated at 110°C for 2–3 hours. Although elastomers of this type possess properties (such as resilience and solvent resistance) which make them particularly useful in printing rollers and sealants, they are characterized by rather poor mechanical strength which restricts wider utilization. The reason for this inferiority is probably because the branch units in the polyester restrict intermolecular attraction. This limitation is overcome when the elastomer is cast from a mixture consisting of a *linear* polyester or polyether, diisocyanate and glycol or diamine. In this process, the hydroxyl-terminated polymer (e.g. poly(ethylene adipate) or poly(oxytetramethylene) glycol) is treated with an excess of diisocyanate (e.g. naphthylene 1,5-diisocyanate, tolylene diisocyanate or diphenylmethane 4,4'-diisocyanate). The product is a pre-polymer which is, in effect, a mixture of isocyanate-terminated polymer and unreacted diisocyanate. The pre-polymer is then mixed with either a glycol (e.g. 1,4-butanediol or 1,6-hexanediol) or diamine (which is usually a deactivated amine such as 3,3'-dichloro-4,4'-diaminodiphenylmethane (MOCA)). The mixture is poured into a heated mould where it quickly sets. After about 30 mins the casting is removed from the mould and cured at about 110°C for 24 hours.

The reactions which occur when a glycol is used in the above operations are not fully understood, but it is thought that firstly the glycol reacts with the isocyanate-terminated pre-polymer to give an extended polymer containing glycol-urethane links. The amount of glycol used is slightly less than that required to react with all the isocyanate groups in the pre-polymer and so the extended polymer is isocyanate terminated; the situation may be illustrated as follows:

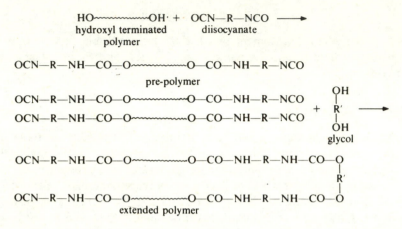

The glycol-urethane segments in the extended polymer have a strong tendency to molecular association through polar forces, accounting for the rapid set which is observed. The extended polymer is generally supposed to cross-link through reaction of terminal isocyanate groups with urethane groups in the polymer chain with the formation of allophanate links:

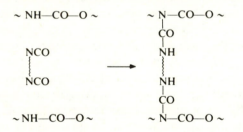

Cross-linking may also occur through trimerization of terminal isocyanate groups to isocyanurate:

A criticism of these proposals for the formation of cross-links is that the reactions shown do not usually occur readily below about 130°C except in the presence of a catalyst.

The reactions involved when a diamine is used in the preparation of a cast elastomer are more apparent. The amine and pre-polymer react to give an extended polymer containing urea groups, which then form biuret cross-links by reaction with terminal isocyanate groups:

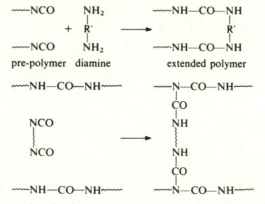

pre-polymer diamine   extended polymer

As mentioned previously, isocyanates react much more rapidly with ureas than urethanes and the above reaction occurs readily under the conditions used for curing.

In general, diamine-cured elastomers have similar properties to glycol-cured materials; commonly, diamines are preferred for use with tolylene diisocyanate based pre-polymers which set slowly with glycols. Some typical properties of various polyurethane elastomers are shown in Table 16.5, though it is to be noted that a wide range of properties can be obtained by appropriate formulation. In general, cast polyurethane elastomers are characterized by high tensile strength, abrasion resistance and tear strength; these properties have led to such uses as solid tyres, bearings and rollers.

The chemical properties of cast polyurethane elastomers are similar to those of flexible foams (section 16.5.1(b)). Of particular importance is the outstanding resistance to aliphatic hydrocarbon fuels and oils and to oxygen and ozone. Aromatic and chlorinated solvents cause some swelling.

16.6.2 Millable elastomers

Millable polyurethane elastomers are elastomers to which the conventional techniques of mill compounding and vulcanization can be applied. In this approach, stable (hydroxyl-terminated) polymers are prepared by the reaction of linear polyesters (commonly adipates) or polyethers (commonly poly(oxytetramethylene) glycol) with diisocyanates. These polymers are rubber-like gums which may be compounded on two-roll mills with other ingredients. Vulcanization may be effected by several types of reagents, but isocyanates, sulphur systems and peroxides are the most widely used.

Isocyanate-curing polyurethane elastomers incorporate groups containing active hydrogen atoms which serve as sites for cross-linking. The most

Table 16.5 Typical properties of various classes of polyurethane elastomers

Type	CAST			MILLABLE			THERMOPLASTIC
	Polyester/NDI	Polyether/TDI	Polyester	Polyether	Polyester		Polyester/MDI/1,4-butanediol
Curing agent	1,4-butanediol	MOCA	TDI-dimer	Sulphur	Dicumyl peroxide		None
Tensile strength							
(MPa)	29	25	23	36	29		40
(lbf/in^2)	4250	3600	3400	5150	4160		5840
Modulus at 300%							
extension(MPa)	6.9	4.3	—	17	21		8.6
(lbf/in^2)	1000	625	—	2475	3110		1240
Elongation at break (%)	650	800	650	540	430		540
Hardness (Shore A)	80	80	84	64	79		88

suitable reactive groups are urea and amide groups. Urea groups may be introduced by preparing the polymer from polyester or polyether, diisocyanate and water, diamine or aminoalcohol. Amide groups may be introduced into the polymer by using a polyesteramide as the hydroxyl-terminated base polymer. Reaction of the urea and amide groups with a diisocyanate leads to the formation of biuret and acylurea cross-links respectively:

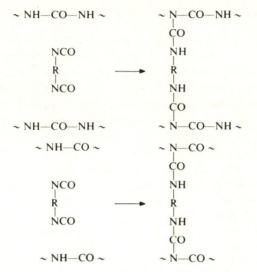

The most widely used isocyanate curing agent is the dimer of tolylene 2,4-diisocyanate; at the curing temperature (about 150°C) this dissociates into free isocyanate which effects cure:

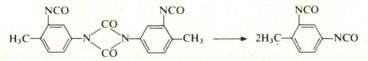

Sulphur- and peroxide-curing polyurethane elastomers usually incorporate olefinic groups as sites for cross-linking. A very suitable reactive group is the allyl ether group which is conveniently introduced by preparing the polymer from polyester or polyether, diisocyanate and a mono-allyl ether of a triol ($H_2C=CH-CH_2-O-R(OH)_2$). Cure is effected either by conventional sulphur systems (namely sulphur, accelerator and activator) or by peroxides (e.g. dicumyl peroxide). These curing agents are effective in the cure of other elastomers and their modes of action are discussed in Chapter 20. Alternatively, peroxide-curing polyurethanes may be prepared by incorporating active methylene groups, conveniently by using diphenylmethane diisocyanate as the chain-extending diisocyanate.

In general, the vulcanizates obtained by the use of the various curing systems described above are quite similar; isocyanates give products with somewhat higher mechanical strength whilst peroxide-cured materials have better heat resistance. Some properties of typical vulcanizates are shown in Table 16.5. The millable elastomers give vulcanizates similar to the cast products described previously but generally the cast materials have rather higher mechanical strength.

16.6.3 Thermoplastic elastomers

As discussed previously, thermoplastic elastomers are materials which have the functional properties of conventional vulcanized rubbers but which may be processed as normal thermoplastics (see section 2.9).

Effects of the above kind are shown by polyurethanes containing segments of polyurea or poly(glycol-urethane), but commercial development has been largely confined to the latter. Such polymers are prepared by the reaction of linear hydroxyl-terminated polyesters (usually adipates) or polyethers (commonly poly(oxytetramethylene) glycol), diisocyanates (usually diphenylmethane 4,4'-diisocyanate) and glycols (e.g. ethylene glycol and 1,4-butanediol). In order to obtain a linear high molecular weight polyurethane it is essential for the quantity of diisocyanate used to be equivalent to the combined polymer and glycol components. The product is a block copolymer made up of alternating polyester or polyether blocks (soft segments) and polyurethane blocks (hard segments):

HO⎯⎯⎯⎯⎯OH   +   OCN—R—NCO   +   HO—R'—OH ⎯⎯→
hydroxyl-terminated   diisocyanate   glycol
     polymer

—O⎯⎯⎯⎯⎯O—[—CO—NH—R—NH—CO—O—R'—O—]$_n$—CO—NH—R—NH—CO—

The properties of polyurethane thermoplastic elastomers are quite similar to those of cast elastomers (see Table 16.5) but the absence of primary cross-links results in higher compression set and more pronounced loss of strength with increasing temperature. Also, although the thermoplastic elastomers are insoluble in most common solvents they are soluble in highly polar solvents such as dimethylacetamide. A major use of polyurethane thermoplastic elastomers is in the construction of ski boots.

Polyurethane thermoplastic elastomers may also be produced in fibrous form to give elastic fibres (*spandex fibres*). Such fibres are made either by solution spinning or by reaction spinning. In the first process, a hydroxyl-terminated polyester (e.g. an adipate) or polyether (e.g. poly(oxytetramethylene) glycol) is treated with an excess of diisocyanate (e.g. diphenylmethane diisocyanate) to give an isocyanate-terminated pre-polymer similar to those used for cast elastomers (section 16.6.1). The pre-polymer is dissolved in a strongly polar solvent (e.g. dimethylformamide) and treated with an aliphatic diamine or hydrazine to effect chain extension; with hydrazine the following reaction occurs:

$$\sim NCO \; + \; H_2N\!-\!NH_2 \; + \; OCN\!\sim \; \longrightarrow \; \sim NH\!-\!CO\!-\!NH\!-\!NH\!-\!CO\!-\!NH\!\sim$$

The solution is then either spun into a coagulating bath or into a hot atmosphere (where the solvent evaporates) to produce a fibre. In reaction spinning, a liquid isocyanate-terminated pre-polymer of the type described above is spun into an aqueous solution of an aliphatic diamine (e.g. ethylene diamine). Rapid reaction occurs on the surface of the filament to give high molecular weight polyurethane. The filament is then treated with water at 70°C to complete reaction in the interior.

Spandex fibres have replaced natural rubber elastomeric fibres in various applications such as foundation garments on account of their greater strength, improved ageing characteristics and easier dyeability. They are also used in stretch fabrics.

16.7 SURFACE COATINGS

Several polyurethane-type products are utilized in coating applications. Coating compositions based on isocyanates can be divided into two broad categories, namely one-component systems (which cure by reaction with oxygen of the air, moisture or heat) and two-component systems (which cure through interaction of two materials on mixing).

16.7.1 One-component systems

Three common types of one-component systems may be distinguished, namely air-curing, moisture-curing and heat-curing systems.

(a) Air-curing systems

Air-curing systems are based on reaction products of diisocyanates, polyols and drying oils which cure by reaction of the unsaturated component with atmospheric oxygen. These products are commonly known as *urethane oils* and they are, in effect, alkyd resins (section 11.3) in which phthalic anhydride (and derived ester linkages) has been replaced by a diisocyanate (and derived urethane linkages). Urethane oils are produced in two stages. In the first stage a partial ester (i.e. an ester containing hydroxyl groups) is prepared either by esterification of a free fatty acid with a polyol or by alcoholysis of an oil with a polyol, using processes and reactants similar to those described for alkyd resins. In the second stage, the partial ester is treated with a diisocyanate (e.g. tolylene diisocyanate or isophorone diisocyanate at 40–100°C and solvent is added as required. The isocyanate reacts with the hydroxyl groups in the partial ester to give the oil-modified polyurethane.

Urethane oils are generally similar to alkyd resins of comparable type and oil length but are quicker drying. They give films with superior mar and abrasion resistance and resistance to solvents and chemicals. Urethane oils find application in wood varnishes and quick-drying enamels.

(b) Moisture-curing systems

Moisture-curing systems are based on isocyanate-terminated branched pre-polymers which cure by interaction with atmospheric moisture. The pre-polymers are prepared by reaction of an excess of diisocyanate (e.g. tolylene diisocyanate, diphenylmethane diisocyanate or hexamethylene diisocyanate) with polyol (polyester, polyether or castor oil). It is important that the residual free diisocyanate content should be low to reduce toxic vapour hazards. When the pre-polymer is exposed to moisture, the following reactions occur and cure is effected:

$$\sim NCO \;+\; H_2O \longrightarrow \sim NH_2 \;+\; CO_2$$

$$\sim NH_2 \;+\; OCN\sim \longrightarrow \sim NH\!-\!CO\!-\!NH\sim$$

The liberated carbon dioxide diffuses out of the film. Moisture-curing pre-polymers give hard films with good solvent and chemical resistance and their main use is for clear lacquers for floors, boats and cladding. Pigmented materials are difficult to prepare because pigments normally contain appreciable levels of moisture which lead to poor shelf life of the paint.

(c) Heat-curing systems

Heat-curing systems are based on a combination of hydroxyl-terminated polyesters or polyethers and blocked polyfunctional isocyanates. When the temperature is raised, the blocking agent is removed and the free isocyanate groups react with the hydroxyl groups to effect cure. The usual blocking agent is phenol and a typical blocked isocyanate is prepared by reaction of trimethylolpropane (1 mole), tolylene diisocyanate (3 mole) and phenol (3 mole):

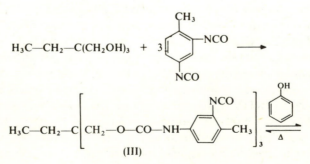

(III)

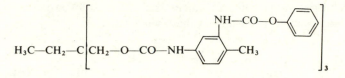

Free isocyanate groups are regenerated by heating at about 150°C. It will be noted that the above technique avoids the liberation of monomeric (toxic) tolylene diisocyanate during cure. Systems of this type are used for coating magnet wire.

16.7.2 Two-component systems

Two types of two-component systems are in use, namely isocyanate-polyol systems and the pre-polymer-polyol systems.

(a) Isocyanate-polyol systems

In these systems, isocyanates and hydroxyl-terminated polymers are mixed immediately prior to use. After the components are mixed curing takes place by direct reaction of isocyanate groups with hydroxyl groups. An obvious requirement of systems utilizing *in situ* polyurethane formation is freedom from hazard during the application. It is therefore necessary to use isocyanates of low volatility (e.g. diphenylmethane diisocyanate) or non-volatile derivatives of volatile diisocyanates such as tolylene diisocyanate. Derivatives which may be used for this purpose include polyol adducts such as that obtained from trimethylol propane and tolylene diisocyanate (III) and polymerized tolylene diisocyanate, which comprises a mixture of polyisocyanurates with free isocyanate groups; a typical component of such a mixture might be as follows:

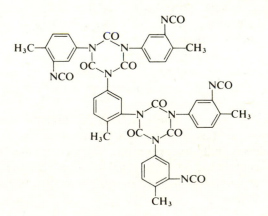

An aliphatic derivative which is widely used for surface coatings is obtained by treating hexamethylene diisocyanate with water:

$$3OCN-(CH_2)_6-NCO \ + \ H_2O \ \xrightarrow{-CO_2} \ OCN-(CH_2)_6-N \begin{array}{l} CO-NH-(CH_2)_6-NCO \\ \\ CO-NH-(CH_2)_6-NCO \end{array}$$

Both polyesters and polyethers may be used as the second component. A limitation of isocyanate-polyol systems is that in all practical systems there is interaction between isocyanate and water picked up from the atmosphere and substrate. This side-reaction, the extent of which depends on conditions prevailing at the time of application, can significantly affect film quality and reproducibility. This loss of isocyanate may be compensated for by using an initial amount of isocyanate greater than that required solely for the urethane reaction; however, the extra isocyanate is liable to give too much cross-linking, making the film hard and brittle. Two-component systems are readily pigmented and find use in such applications as floor paints.

(b)   Pre-polymer – polyol systems

As mentioned above, the technique of minimizing the variability of isocyanate-polyol systems by using an excess of the isocyanate has limitations. An alternative approach is to employ isocyanate-terminated pre-polymers (of the type described in section 16.7.1(b)) to cross-link the polyol. Since these pre-polymers are themselves film-forming materials, the use of a stoichiometric excess in a two-component system is not detrimental to the final film. This technique is applicable to the production of both flexible and hard coatings; the former are suitable finishes for leather and rubber whilst the latter find use as sanding sealers.

16.8   ADHESIVES

Several polyurethane-type products are used as adhesives. Three general categories may be distinguished, namely isocyanate-polyol systems, soluble elastomers and polyisocyanates.

Isocyanate-polyol adhesives are two-component systems similar to those used for surface coatings (section 16.7.2(a)) and may be used for bonding wood, plastics and metals. Adhesives of this type are low in viscosity and tack even when the solvent has evaporated.

Polyurethanes similar to those utilized as millable elastomers (section 16.6.2) may be used as adhesives provided the molecular weight is low enough to permit solubility in appropriate organic solvents. Isocyanates are generally the preferred cross-linking agents for these adhesives since they are effective at low temperatures. Such adhesives give flexible bonds and are used in the shoe industry.

Polyisocyanates, particularly triphenylmethane 4,4′,4″-triisocyanate, are used for the bonding of rubber to rubber, metals, glass and synthetic fibres. They may be used in rubber-containing cements or, in the case of triphenylmethane triisocyanate, alone.

The adhesive properties of the foregoing materials may be attributed to the polar nature of the polymers involved. Further, the isocyanates present in the compositions may react with any active hydrogen present in the substrate or with the films of water which are often present on the surfaces of such materials as ceramics, glass and metals. Such reactions result in good keying of the adhesive and permit the attainment of high bond strengths.

REFERENCES

1. Baker, J. W. *et al.* (1947) *J. Chem. Soc.*, 713; (1949) 9, 19, 24, 27.
2. O'Mant, D. M. and Twitchett, H. J. (1968) (via Johnson, P. C. in *Advances in Polyurethane Technology*, (eds J. M. Busit and H. Gudgeon), John Wiley and Sons Inc., New York, p. 10.
3. Britain, J. W. and Gemeinhardt, P. G. (1960) *J. Appl. Polymer Sci.*, **4**, 207.
4. Hostettler, F. and Cox, E. F. (1960) *Ind. Eng. Chem.*, **52**, 609.
5. Shell Chemicals U.K. Ltd, *Introduction to Flexible Polyurethane Foams*, Industrial Chemicals Technical Bulletin, ICO : 61 : 12.
6. Buist, J. M. *et al.* (1968) in *Advances in Polyurethane Technology*, (eds J. M. Buist and H. Gudgeon), John Wiley and Sons Inc., New York, p. 211.

BIBLIOGRAPHY

Saunders, J. H. and Frisch, K. C. (1962, 1964) *Polyurethanes: Chemistry and Technology* (Part I: Chemistry, Part II: Technology), Interscience Publishers, New York.

Healy, T. T. (ed.) (1963) *Polyurethane Foams*, Iliffe Books Ltd, London.

Phillips, L. N. and Parker, D. B. V. (1964) *Polyurethanes: Chemistry, Technology and Properties*, Iliffe Books Ltd, London.

Dombrow, B. A. (1965) *Polyurethanes*, 2nd edn., Reinhold Publishing Corporation, New York.

Ferrigno, T. H. (1967) *Rigid Plastic Foams*, Reinhold Publishing Corporation, New York.

Buist, J. M. and Gudgeon, H. (eds) (1968) *Advances in Polyurethane Technology*, John Wiley and Sons Inc., New York.

Saunders, J. H. (1969) in *Polymer Chemistry of Synthetic Elastomers*, Part II, (eds J. P. Kennedy and E. G. M. Törnqvist), Interscience Publishers, New York, ch. 8.

Wright, P. and Cumming, A. P. C. (1969) *Solid Polyurethane Elastomers*, Maclaren and Sons, London.

Buist, J. M. (ed.) (1978) *Developments in Polyurethanes*, Applied Science Publishers, London.

Hepburn, C. (1982) *Polyurethane Elastomers*, Applied Science Publishers, London.

Woods, G. (1982) *Flexible Polyurethane Foams*, Applied Science Publishers, London.

Oertel, G. (ed.) (1985) *Polyurethane Handbook*, Hanser Publishers, Munich.

SILICONES

17.1 SCOPE

For the purposes of this chapter, silicones are defined as polymers comprising alternate silicon and oxygen atoms in which the silicon atoms are joined to organic groups. The following types of structure come within this definition:

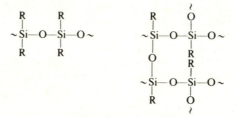

and both linear and network silicones find technological use. It is convenient to classify the silicones which are of commercial interest into three groups, namely fluids, elastomers and resins and these various types of materials are considered separately in this chapter.

17.2 DEVELOPMENT

The possibility of the existence of organosilicon compounds was first noted in 1840 by Dumas. In 1863, Friedel and Crafts prepared the first organosilicon compound, tetraethylsilane, by the reaction of diethylzinc with silicon tetrachloride:

$$2Zn(C_2H_5)_2 \ + \ SiCl_4 \longrightarrow Si(C_2H_5)_4 \ + \ 2ZnCl_2$$

Nevertheless, it is generally agreed that the foundations of organosilicon chemistry were really laid by Kipping (University College, Nottingham) (UK) during the period 1899–1944. However, his investigations were concerned almost exclusively with non-polymeric compounds and it is unlikely that Kipping foresaw any commercial application of his work. Indeed, the resinous materials which were frequently encountered were regarded merely as troublesome.

Commercial interest in silicon polymers developed in the 1930s during searches for heat-resistant electrical insulating materials. Research was carried out by the Corning Glass Works (USA) and the General Electric Co. (USA) and silicone resins were developed. It also became evident that these polymers had potential use in many other fields. The manufacture of silicones was started by the Dow Corning Corp. (USA) in 1943 and by the General Electric Co. (USA) in 1946.

Silicones are now well established as valuable materials. However, they are comparatively expensive; they are speciality products and, in terms of tonnage, are manufactured on a relatively modest scale.

17.3   NOMENCLATURE

Before discussing the detailed chemistry of silicones, it may be useful to review briefly the methods of naming some relevant silicon compounds.

Compounds with the general formula Si_nH_{2n+2} have the generic name *silanes*. SiH_4 is *silane*, $H_3Si–SiH_3$ is *disilane*, $H_3Si–SiH_2–SiH_3$ is *trisilane* and so on. Substituted silanes are named after the parent compound, as illustrated by the following examples:

$(CH_3)_2SiH_2$	dimethylsilane
CH_3SiCl_3	methyltrichlorosilane
$C_2H_5SiH_2—SiH_3$	ethyldisilane

Hydroxyl derivatives of silanes in which the hydroxyl groups are attached to silicon are *silanols* and are named by adding the suffices *-ol, -diol, -triol, etc.* to the name of the parent compound, e.g.

H_3SiOH	silanol
$H_2Si(OH)_2$	silanediol
$(HO)_3Si—Si(OH)_3$	disilanehexol

Compounds with the structure $H_3Si–(OSiH_2)_n–OSiH_3$ have the generic name *siloxanes*. According to the number of silicon atoms present, such compounds are *disiloxane, trisiloxane* and so on. Substituted siloxanes are named after the parent compound, e.g.

$Cl_3Si—O—SiCl_3$	hexachlorodisiloxane
$H_3Si—O—SiH_2(OH)$	disiloxanol

It may be noted that silicones are more properly referred to as *polyorganosiloxanes*. (The term 'silicone' was originated by Kipping to describe compounds with empirical formula R_2SiO; these were originally considered to be analogous to ketones, R_2CO.)

17.4 RAW MATERIALS

The basis of commercial production of silicones is that chlorosilanes readily react with water to give silanols which are unstable and condense to form siloxanes. Dichlorosilanes lead to linear silicones:

$$Cl—\underset{\underset{R'}{|}}{\overset{\overset{R}{|}}{Si}}—Cl \xrightarrow{H_2O} HO—\underset{\underset{R'}{|}}{\overset{\overset{R}{|}}{Si}}—OH$$

$$HO—\underset{\underset{R'}{|}}{\overset{\overset{R}{|}}{Si}}—O{\text -}H \;+\; HO—\underset{\underset{R'}{|}}{\overset{\overset{R}{|}}{S}}—O{\text -}H \;+\; HO—\underset{\underset{R'}{|}}{\overset{\overset{R}{|}}{Si}}—O{\text -}H \xrightarrow{-H_2O} \left[\underset{\underset{R'}{|}}{\overset{\overset{R}{|}}{Si}}—O—\right]_n$$

Polymers of this kind form the basis of silicone fluids and elastomers. Trichlorosilanes give branched and cross-linked silicones:

$$Cl—\underset{\underset{Cl}{|}}{\overset{\overset{R}{|}}{Si}}—Cl \xrightarrow{H_2O} HO—\underset{\underset{OH}{|}}{\overset{\overset{R}{|}}{Si}}—OH \xrightarrow{-H_2O} \sim\!\underset{\underset{O}{|}}{\overset{\overset{R}{|}}{Si}}—O—\underset{\underset{R}{|}}{\overset{\overset{O}{|}}{Si}}—O\!\sim$$

$$\sim\!\underset{\underset{R}{|}}{\overset{\overset{R}{|}}{Si}}—O—\underset{\underset{O}{|}}{\overset{\overset{R}{|}}{Si}}—O\!\sim$$

Structures of this type form the basis of silicone resins.

Commercial processes for the preparation of chlorosilanes start either from silicon or silicon tetrachloride, both obtained from silica:

$$SiO_2 \;+\; 2C \longrightarrow Si \;+\; 2CO$$
$$Si \;+\; 2Cl_2 \longrightarrow SiCl_4$$

Various methods of preparing chlorosilanes are of commercial importance and are described below.

17.4.1 Grignard process

In this process, organomagnesium compounds are used to transfer organo groups to silicon:

$$RMgX \;+\; XSi{\Big\langle} \longrightarrow RSi{\Big\langle} \;+\; MgX_2$$

The synthesis is normally carried out in two steps. In the first step, the Grignard reagent is prepared by adding an alkyl halide (generally chloride) or aryl halide (generally bromide) to a suspension of magnesium in an ether (usually diethyl ether), e.g.

$$RCl \;+\; Mg \longrightarrow R\,MgCl$$

In the second step, the resulting ethereal solution is treated with silicon tetrachloride (which is the preferred silicon compound on the grounds of

availability, although other halides and silicic acid esters may also be used).
An exothermic reaction occurs and cooling is essential. Magnesium chloride
is precipitated; this is filtered off and then the solvent is removed to yield a
mixture of various organosilicon compounds. These products arise through
the following sequence of reactions:

$$RMgCl \; + \; SiCl_4 \longrightarrow RSiCl_3 \; + \; MgCl_2$$

$$RMgCl \; + \; RSiCl_3 \longrightarrow R_2SiCl_2 \; + \; MgCl_2$$

$$RMgCl \; + \; R_2SiCl_2 \longrightarrow R_3SiCl \; + \; MgCl_2$$

$$RMgCl \; + \; R_3SiCl \longrightarrow R_4Si \; + \; MgCl_2$$

Although this method always results in a mixture of silanes, the relative
quantities of the components may be regulated to some extent by the amount
of Grignard reagent used and by reaction conditions. Normally, the
dialkyl(aryl)dichlorosilane predominates since steric effects oppose the introduction of further organic groups. The various products are separated by
distillation; this is frequently a difficult operation owing to the closeness in
boiling points of the compounds present. (Cf. Table 17.1.)

The Grignard process is very versatile, being applicable to a wide range of
materials. It is suitable for the preparation of both alkyl- and arylsilanes and
also 'mixed' compounds; the latter possibility is illustrated by the following
example:

$$CH_3SiCl_3 \; + \; C_6H_5MgBr \longrightarrow Cl-\underset{\underset{C_6H_5}{|}}{\overset{\overset{CH_3}{|}}{Si}}-Cl \; + \; MgBrCl$$

The process may also be used to prepare organo-H-chlorosilanes since
silicon-hydrogen bonds are not normally affected by Grignard reagents. In
this case, trichlorosilane is a convenient starting material:

$$RMgCl \; + \; SiHCl_3 \longrightarrow RSiHCl_2 \; + \; MgCl_2$$

$$RMgCl \; + \; RSiHCl_2 \longrightarrow R_2SiHCl \; + \; MgCl_2$$

$$RMgCl \; + \; R_2SiHCl \longrightarrow R_3SiH \; + \; MgCl_2$$

The Grignard process was the first process to be used for the large-scale
production of chlorosilanes but has now been largely displaced by the direct
process, which is more economical and easier to operate. However, the
method continues to be important for the preparation of special organosil-
anes for which the direct process cannot be used.

17.4.2 Direct process

In this process, elemental silicon is converted directly into chlorosilanes by
reaction with alkyl or aryl chlorides. The simplest reaction of this type may be

described by the equation:

$$2RCl \ + \ Si \longrightarrow R_2SiCl_2$$

In practice, other reactions occur simultaneously and a mixture of products is obtained. (See later.)

In a typical procedure, methyl chloride is passed through a mixture of silicon and copper (catalyst) at 250–280°C. The best results are obtained when the silicon and copper are in intimate contact. This may be achieved by superficially alloying the two metals by sintering a finely ground mixture at about 1000°C in hydrogen or by heating a mixture of silicon and cuprous chloride at 200–400°C. The composition of a typical reaction product obtained in the direct process is shown in Table 17.1, together with the boiling points of the various components. The composition of the product cannot easily be regulated by manipulation of reaction conditions. Separation of the mixture is effected by very careful distillation.

The reactions which occur in the direct process are complex and not well understood. The following free radical mechanism involving methylcopper has been suggested [2]:

$$2Cu \ + \ CH_3Cl \longrightarrow CuCH_3 \ + \ CuCl$$

$$CuCH_3 \longrightarrow Cu \ + \ \cdot CH_3$$

$$\cdot CH_3 \ + \ {\geqslant}Si \longrightarrow {\geqslant}Si{-}CH_3$$

$$CuCl \ + \ {\geqslant}Si \longrightarrow {\geqslant}Si{-}Cl \ + \ Cu$$

Table 17.1 Composition of crude product from the direct process (methyl chloride and silicon) [1]

Compound	Boiling point (°C)	Content (%)
Dimethyldichlorosilane, $(CH_3)_2SiCl_2$	70	75
Methyltrichlorosilane, CH_3SiCl_3	66	10
Trimethylchlorosilane, $(CH_3)_3SiCl$	58	4
Methyldichlorosilane, CH_3SiHCl_2	41	6
Silicon tetrachloride, $SiCl_4$	58	Small amounts
Tetramethylsilane, $(CH_3)_4Si$	26	Small amounts
Trichlorosilane, $SiHCl_3$	32	Small amounts
High-boiling residue (disilanes)	100–200	Small amounts

The last two reactions may occur at previously substituted silicon atoms and the process continues until the silicon is tetrasubstituted. The product is thus a mixture of silicon compounds with general formula $(CH_3)_nSiCl_{4-n}$ (where $n = 0$–4).

It has also been suggested that the reactions involved in the direct process are polar in nature. (See Reference 3 for a summary of this work.) In this scheme it is supposed that the silicon and copper interact to give a material containing bonds which constitute reaction sites. The formation of dimethyldichlorosilane may be represented as follows:

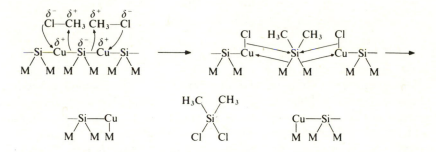

where M represents the body of the silicon

Evidence for this proposal is that the intermetallic phases existing in the silicon-copper system display differing activities toward methyl chloride. Against the radical mechanism, it has been found that although methyl chloride does react with copper to give cuprous chloride, no methylcopper can be detected.

Compared to the Grignard process, the direct process is not very versatile. In practice, it is satisfactory for the manufacture of only methyl- and phenylchlorosilanes. (For the latter, silver is the preferred catalyst and a reaction temperature of about 400°C is used.) When the process is used for other chlorosilanes, yields are unacceptably low. However, methylchlorosilanes are the chlorosilanes most widely used in the manufacture of commercial silicones and the direct process has become the dominant process.

17.4.3 Olefin addition process

In this process, compounds containing Si–H bonds are added to olefins:

$$\equiv Si-H + R_2C=CR_2 \longrightarrow R_2HC-CR_2-Si\equiv$$

The process may also be applied to alkynes to give unsaturated silanes:

$$\equiv\!\!\overset{\diagup}{Si}\!\!-\!\!H \ + \ RC\!\!\equiv\!\!CR \ \longrightarrow \ RHC\!\!=\!\!CR\!\!-\!\!\overset{\diagup}{Si}\!\!\!\diagdown$$

Reaction may be carried out either in the absence of catalysts in the temperature range 200–400°C or in the presence of catalysts (e.g. organic peroxides) at about 40–75°C.

The process involves a free radical mechanism, as illustrated by the reaction between trichlorosilane and ethylene to give ethyltrichlorosilane:

$$I\cdot \ + \ SiHCl_3 \longrightarrow IH \ + \ \cdot SiCl_3$$

$$HC_2\!\!=\!\!CH_2 \ + \ \cdot SiCl_3 \longrightarrow H_2\overset{\cdot}{C}\!\!-\!\!CH_2\!\!-\!\!SiCl_3$$

$$SiHCl_3 \ + \ H_2\overset{\cdot}{C}\!\!-\!\!CH_2\!\!-\!\!SiCl_3 \longrightarrow \ \cdot SiCl_3 \ + \ H_3C\!\!-\!\!CH_2\!\!-\!\!SiCl_3$$

where $I\cdot$ = initiator radical

Compared to the Grignard and direct processes, the addition method has the advantage of yielding one principal product rather than a mixture. The process is also economically attractive since olefins are low-cost materials and suitable Si–H containing compounds (trichlorosilane and methyldichlorosilane) are by-products of the direct process. The addition process is versatile but cannot, of course, be used to prepare methylchlorosilanes. It is particularly suitable for the production of silanes with organofunctional groups; useful reactions of this type include the following:

$$CH_3SiHCl_2 \ + \ H_2C\!\!=\!\!CH\!\!-\!\!CN \ \longrightarrow \ \overset{\displaystyle CH_3}{\underset{\displaystyle CH_2-CH_2-CN}{Cl-\underset{|}{\overset{|}{Si}}-Cl}}$$

β- cyanoethylmethyldichlorosilane

$$CH_3SiHCl_2 \ + \ H_2C\!\!=\!\!CH\!\!-\!\!CF_3 \ \longrightarrow \ \overset{\displaystyle CH_3}{\underset{\displaystyle CH_2-CH_2-CF_3}{Cl-\underset{|}{\overset{|}{Si}}-Cl}}$$

γ- trifluoropropylmethyldichlorosilane

17.4.4 Redistribution

Under appropriate conditions various substituents on the silicon atom can be exchanged for one another. This provides a convenient method of converting by-product chlorosilanes into more useful materials. A typical example, which is of practical interest, is the redistribution of a mixture of trimethylchlorosilane and methyltrichlorosilane to form dimethyldichlorosilane:

$$(CH_3)_3SiCl \ + \ CH_3SiCl_3 \rightleftharpoons 2(CH_3)_2SiCl_2$$

The reaction is carried out at 200–400°C in the presence of aluminium chloride.

17.5 POLYMERIZATION

As indicated in the previous section, the basis of the production of silicones is the hydrolysis of chlorosilanes to silanols which condense spontaneously to siloxanes. Both the functionality of the chlorosilane and the conditions used for hydrolysis have a decisive influence on the structure of the siloxane which is obtained. These factors are illustrated by consideration of the hydrolysis of the following organohalosilanes:

(i) *Trimethylchlorosilane*. In this case, the only hydrolysis product is the dimer, hexamethyldisiloxane:

$$(CH_3)_3SiCl \xrightarrow[-HCl]{H_2O} (CH_3)_3SiOH \xrightarrow{-H_2O} (CH_3)_3Si-O-Si(CH_3)_3$$

As is noted later, the dimer may be used to control molecular weight in the preparation of silicones.

(ii) *Dimethyldichlorosilane*. Hydrolysis of this compound leads to a mixture of linear and cyclic polymers:

$$(CH_3)_2SiCl_2 \xrightarrow[-HCl]{H_2O}$$

$$HO-\underset{\underset{CH_3}{|}}{\overset{\overset{CH_3}{|}}{Si}}-O-\left[\underset{\underset{CH_3}{|}}{\overset{\overset{CH_3}{|}}{Si}}-O\right]_n\underset{\underset{CH_3}{|}}{\overset{\overset{CH_3}{|}}{Si}}-OH$$

$$\left[\underset{\underset{CH_3}{|}}{\overset{\overset{CH_3}{|}}{Si}}-O\right]_n \quad (n=3,4,5,\ldots)$$

The relative amounts of the two types of polymers are determined by reaction conditions. Hydrolysis with water alone yields 50–80% linear polydimethyl-siloxane-α,ω-diols and 50–20% polydimethylcyclosiloxanes. Hydrolysis with 50–85% sulphuric acid gives mostly high molecular weight linear polymers with only small amounts of cyclosiloxanes. Conversely, the hydrolysis of dimethyldichlorosilane with water in the presence of immiscible solvents (e.g. toluene, xylene and diethyl ether) results in the preferential formation of lower polycyclosiloxanes. Such solvents, in which the organochlorosilane is readily soluble, lead to a reduction in concentration of dimethyldichlorosil-ane in the aqueous phase and thus intramolecular condensation is favoured over intermolecular condensation. Further, the hydrolysis products are also soluble in the organic solvent and the cyclic compounds are protected from the action of the aqueous acid. (See later.)

(iii) *Methyltrichlorosilane.* Hydrolysis of this compound with water alone gives highly cross-linked gel-like or powdery polymers.

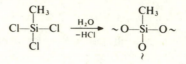

The presence of inert solvents promotes intramolecular condensation and the formation of ring compounds.

The condensation of silanols to polymeric products is catalysed by both acids and bases. Since the hydrolysis of chlorosilanes results in the formation of hydrochloric acid, condensation occurs rapidly in practical systems. The mechanism of the reaction is probably as follows:

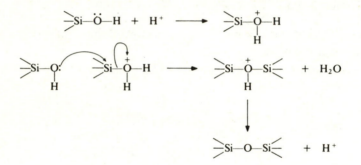

Base-catalysed condensation probably involves the following steps:

$$\geq Si-O-H \ + \ OH^- \ \longrightarrow \ \geq Si-O^- \ + \ H_2O$$

$$\geq Si-O^- \ + \ \geq Si-OH \ \longrightarrow \ \geq Si-O-Si \leq \ + \ OH^-$$

The condensation of silanols is also catalysed by metal compounds such as cobalt naphthenate and stannous octoate. Such compounds are used in connection with room-temperature vulcanizable elastomers (section 17.6.2(f)) and resins (section 17.6.3). Their mode of action does not appear to have been investigated in any detail.

It is also possible to obtain high molecular weight linear polysiloxanes from low molecular weight cyclosiloxanes. In this case, siloxane bonds are cleaved and then reformed by intermolecular reaction. Polymerization of this type is also catalysed by acids and bases. When the reaction is catalysed by acid (HX), bond scission may occur as follows:

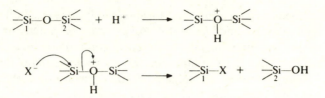

The following condensation reactions may then lead to new siloxane bonds:

$$\equiv \!Si\!-\!OH \ + \ X\!-\!Si\!\!\equiv \ \longrightarrow \ \equiv \!Si\!-\!O\!-\!Si\!\!\equiv \ + \ HX$$

$$\equiv \!Si\!-\!OH \ + \ HO\!-\!Si\!\!\equiv \ \longrightarrow \ \equiv \!Si\!-\!O\!-\!Si\!\!\equiv \ + \ H_2O$$

Alternatively, reaction may proceed as follows:

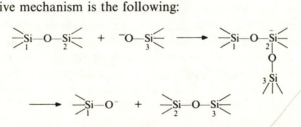

When the reaction is catalysed by base, the following sequence may be envisaged:

$$\equiv \!Si\!-\!O\!-\!Si\!\!\equiv \ + \ OH^- \ \longrightarrow \ \equiv \!Si\!-\!O\!-\!\!\underset{OH}{Si}\!\!\equiv \ \longrightarrow \ \equiv \!Si\!-\!O^- \ + \ HO\!-\!Si\!\!\equiv$$

$$\equiv \!Si\!-\!O^- \ + \ HO\!-\!Si\!\!\equiv \ \longrightarrow \ \equiv \!Si\!-\!O\!-\!Si\!\!\equiv \ + \ HO^-$$

An alternative mechanism is the following:

The conversion of cyclosiloxanes to high molecular weight linear siloxanes is utilized in the preparation of silicone elastomers (section 17.6.2(a)).

The cleavage and re-formation of siloxane bonds by the reactions described above also form the basis of the process of *equilibration*. In this process mixtures of siloxanes (both linear and cyclic) of differing molecular sizes are brought into molecular equilibrium with the result that a more homogeneous molecular weight distribution is produced. Equilibration may be effected by heating with either acid or base catalysts and is utilized in the preparation of silicone fluids (section 17.6.1) It may be noted that the addition of a siloxane containing trisubstituted silicon atoms (e.g. hexamethyldisiloxane) to a mixture to be equilibrated results in a lowering of the average molecular weight attained. The polymer chains formed become terminated by triorganosiloxy units. This technique is used as a means of preparing stable polymers of desired molecular weight for use in fluids.

17.6   PREPARATION OF SILICONE PRODUCTS

The silicones which are of commercial interest may be classified into three groups, namely fluids, elastomers and resins. The manufacture of these materials is based on the general reactions described in the preceding section. In the case of elastomers, the linear polymer which is produced initially is subjected to cross-linking reactions in order to develop elastic properties. Similarly, the practical utilization of resins depends on their conversion to highly cross-linked structures. The preparation of these various technical products is considered in this section.

17.6.1   Fluids

Silicone fluids consist essentially of linear polymers of rather low molecular weight, namely about 4000–25 000. Most commonly the polymers are polydimethylsiloxanes. In a typical process, dimethyldichlorosilane is continuously hydrolysed by mixing with dilute hydrochloric acid (the concentration of which is held at about 20%). The product separates as an oil and is removed; at this stage it consists of a mixture of long and short chain siloxane polymers and various cyclic siloxanes (principally the tetramer, octamethylcyclotetrasiloxane). The mixture is then equilibrated (see section 17.5) by heating with a small amount of aqueous sulphuric acid or sodium hydroxide at about 150°C for several hours. At the equilibration stage a 'chain stopper' such as hexamethyldisiloxane is added to give a product with the desired viscosity. The fluid is separated from the aqueous layer and remaining traces of catalyst are neutralized. The fluid is then filtered, dried and heated under reduced pressure to remove volatile material.

For use as fluids with enhanced thermal stability, silicones containing both methyl and phenyl groups are manufactured. Generally, the phenyl groups make up 10–45% of the total number of substituent groups present. Such silicones are obtained by hydrolysis of mixtures of methyl- and phenylchlorosilanes.

Fluids for use in textile treatment incorporate reactive groups so that they may be cross-linked to give a permanent finish. Commonly, these fluids contain Si–H bonds (introduced by including methyldichlorosilane in the polymerization system) and cross-linking occurs on heating with alkali:

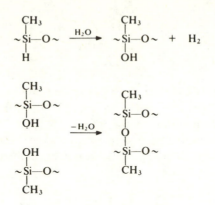

17.6.2   Elastomers

Silicone elastomers are based on linear polymers which are analogous to the fluids but which have higher molecular weight. As with other elastomeric materials, it is necessary to cross-link the linear polymers in order to obtain characteristic elastic properties. General purpose elastomers are based on polydimethylsiloxanes but special purpose materials which contain a small proportion of groups other than methyl are also available. These various products are described below.

(a)   Dimethyl silicone elastomers

Dimethyl silicone gums (i.e. unvulcanized material) consist of linear polymers of very high molecular weight, namely 300 000–700 000. Such polymers cannot satisfactorily be prepared by the direct hydrolysis of dimethyldichlorosilane owing to the difficulty of obtaining the silane in a sufficiently pure state. Traces of monofunctional impurities result in a lowering of molecular weight whilst trifunctional impurities lead to branching. However, it is possible to obtain very pure octamethylcyclotetrasiloxane, from which satisfactory gums can be prepared (see section 17.5).

In a typical process, dimethyldichlorosilane is dissolved in ether and the solution is mixed with an excess of water. Cyclic polymers make up about 50% of the product; the tetramer, octamethylcyclotetrasiloxane (I) is the main cyclic compound. This tetramer has b.p. 175°C (cf. Table 17.1) and is readily purified by distillation. The tetramer is then polymerized by heating at 150–200°C with a trace of sodium hydroxide and a very small amount of monofunctional material to control the molecular weight. The product is a highly viscous gum with no elastic properties.

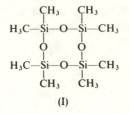

(I)

Dimethyl silicone gums are normally cured by heating with organic peroxides such as benzoyl peroxide and 2,4-dichlorobenzoyl peroxide (section 11.2.4) at temperatures in the range 110–175°C. Thermal decomposition of the peroxides gives rise to free radicals (see section 1.4.2(a)) which abstract hydrogen from methyl groups; combination then results in ethylene cross-links:

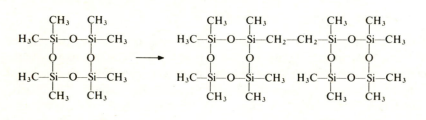

Support for this mechanism comes from the finding that treatment of octamethylcyclotetrasiloxane with benzyol peroxide gives 1,2-bis-(heptamethylcyclotetrasiloxanyl)ethane [4]:

$$
\begin{array}{ccc}
\text{CH}_3 \quad \text{CH}_3 & & \text{CH}_3 \quad \text{CH}_3 \qquad\qquad \text{CH}_3 \quad \text{CH}_3 \\
\text{H}_3\text{C—Si—O—Si—CH}_3 & \longrightarrow & \text{H}_3\text{C—Si—O—Si—CH}_2\text{—CH}_2\text{—Si—O—Si—CH}_3 \\
\text{O} \qquad\quad \text{O} & & \text{O} \qquad\quad \text{O} \qquad\qquad \text{O} \qquad\quad \text{O} \\
\text{H}_3\text{C—Si—O—Si—CH}_3 & & \text{H}_3\text{C—Si—O—Si—CH}_3 \quad \text{H}_3\text{C—Si—O—Si—CH}_3 \\
\text{CH}_3 \quad \text{CH}_3 & & \text{CH}_3 \quad \text{CH}_3 \qquad\qquad \text{CH}_3 \quad \text{CH}_3
\end{array}
$$

(b)   Vinyl silicone elastomers

Vinyl silicone gums are very similar to the dimethyl gums described above except for the incorporation of a small number of methylvinylsiloxane groups (about 0.1% molar). The vinyl groups are particularly reactive toward peroxides and the gums are readily vulcanized by less active peroxides such as di-*tert*-butyl peroxide. Since the presence of so few vinyl groups is sufficient to give an adequate level of cross-linking, it is unlikely that vinyl-to-vinyl linking occurs. It has been suggested that 1,2-propylene and trimethylene links are formed [5]:

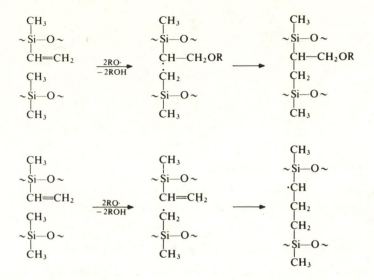

Compared to dimethyl silicone elastomers, vinyl silicone elastomers have much reduced compression set. A further point is that although silicone elastomers are mostly reinforced with silica, special purpose electrically conducting elastomers are filled with carbon black. Carbon black does not inhibit cure by less active peroxides and may be used with vinyl silicone elastomers. This filler does, however, inhibit vulcanization by benzoyl peroxide (and its derivatives) and therefore cannot be added to dimethyl silicone gums.

(c)   Phenyl silicone elastomers

In these elastomers, 5–15% of the methyl groups of polydimethylsiloxane are replaced by phenyl groups to give materials with especially good low temperature properties. The relatively large phenyl groups inhibit crystallization and reduce the tendency of the polymer to stiffen as the temperature is lowered. Thus dimethyl silicone elastomers lose their elasticity at about $-50°C$ whereas phenyl silicones retain their elasticity at temperatures as low as $-100°C$.

(d)   Nitrile silicone elastomers

These elastomers contain cyanoalkyl groups along the polymer chain. Such groups are introduced by the incorporation of β-cyanoethylmethylsiloxane (II) or γ-cyanopropylmethylsiloxane units (III) into the polymer. The elastomers have enhanced oil resistance and low temperature properties but have not achieved much commercial importance.

$$
\begin{array}{ccc}
\begin{array}{c} CH_3 \\ | \\ -Si-O- \\ | \\ CH_2 \\ | \\ CH_2 \\ | \\ CN \\ (II) \end{array}
&
\begin{array}{c} CH_3 \\ | \\ -Si-O- \\ | \\ CH_2 \\ | \\ CH_2 \\ | \\ CH_3 \\ | \\ CN \\ (III) \end{array}
&
\begin{array}{c} CH_3 \\ | \\ -Si-O- \\ | \\ CH_2 \\ | \\ CH_2 \\ | \\ CF_3 \\ (IV) \end{array}
\end{array}
$$

(e) Fluorosilicone elastomers

These elastomers contain γ-trifluoropropylmethylsiloxane units (IV) and have improved solvent resistance (see also section 7.15.3).

(f) Room temperature vulcanizing silicone elastomers

Like other types of silicone elastomers, the room temperature vulcanizing (RTV) elastomers are based essentially on polydimethylsiloxanes. Their distinguishing feature is that the siloxane chains of the basic gums are terminated by silanol groups which are reactive and readily participate in cross-linking reactions. Also, the gums generally have molecular weights in the range 10 000–100 000, which is much lower than that used for heat-cured materials. In fact, the lowest molecular weight polymers are free-flowing liquids which can be fabricated by casting techniques.

Low molecular weight, hydroxyl-terminated polydimethylsiloxanes may be prepared by hydrolysing dimethyldichlorosilane in the presence of small quantities of a base, e.g. calcium carbonate. The base is added continuously and neutralizes the hydrochloric acid formed and thus protects the silanol groups from further condensation.

The gums may be converted to rubber-like products by reaction with an alkoxysilane such as a tetraalkoxysilane, trialkoxysilane or polyalkoxysiloxane. By a suitable choice of catalyst, cure may be effected at room temperature in times ranging from 10 minutes to 24 hours; stannous octoate and dibutyltin dilaurate are particularly satisfactory. The reactions which take place with tetraethoxysilane illustrate the cross-linking process:

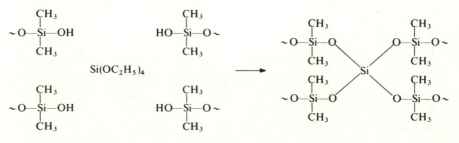

Alternatively, the silanol-terminated gums may be cured by the use of polysiloxanes containing silanic hydrogen:

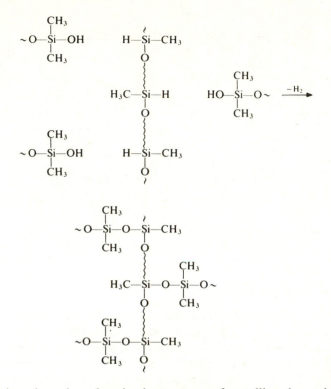

This reaction also takes place in the presence of metallic salt catalysts. The products obtained in this way have good strength but since hydrogen is evolved during cure this type of system is generally used for treatment of substrates such as paper, where only thin films are involved.

The room temperature vulcanizing elastomers described above are 'two-pack' systems, i.e. they are supplied as two separate components which must be mixed together prior to use. The first part contains the silanol-terminated gum and the second part contains the catalyst; the cross-linking agent may be present in either part. The need for mixing two components is avoided by the use of 'one-pack' systems. In these, the silanol-terminated gum, cross-linking agent and catalyst are supplied mixed together but curing does not commence until the mixture is exposed to atmospheric moisture. The cross-linking agents most frequently used in one-pack systems are tri- or tetra-acetoxysilanes. When, for example, methyltriacetoxysilane is mixed with a silanol-terminated silicone, reaction occurs to give a stable acetoxy-terminated polymer (V) (which corresponds to the gum supplied to the fabricator). When this product is exposed to moisture, some of the acetoxy groups are hydrolysed to silanol groups which then condense with remaining acetoxy groups to give a cross-linked material:

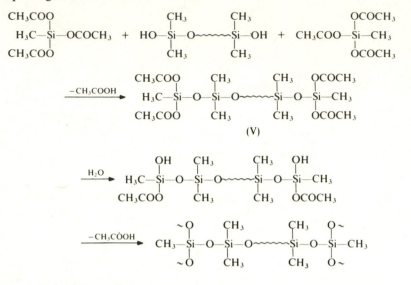

(V)

(g) Liquid silicone elastomers

The characteristic feature of these elastomers is that they are based on liquid materials which can be fabricated by injection moulding. The basic components are low viscosity materials, supplied as a two-pack system, which are blended before injection. The mixture is then cured by heating in the mould at 150–200°C. Liquid silicone elastomers are based on polydimethylsiloxanes containing vinyl groups which are cross-linked by compounds containing silanic hydrogen. In the presence of a platinum catalyst, the following addition reaction occurs:

$$H_3C-\underset{\underset{?}{O}}{\overset{?}{Si}}-CH=CH_2 \;+\; H-Si\underset{\diagdown}{\diagup} \;\longrightarrow\; H_3C-\underset{\underset{?}{O}}{\overset{?}{Si}}-CH_2-CH_2-Si\underset{\diagdown}{\diagup}$$

Liquid silicone elastomers should not be confused with two-pack room temperature vulcanizing elastomers. The former cure only on heating. Blended liquid silicone components can be stored for at least 24 hours at room temperature; the shelf-life of the separate components is approximately 1 year.

(h) Borosilicones

The inclusion of a small number of B–O–Si bonds in the polydimethyl-siloxane structure leads to significant effects. Two products are of some commercial interest, namely 'fusible' elastomers and 'bouncing putty'.

In the preparation of fusible elastomers, silanol-terminated polydimethyl-siloxanes with molecular weights in the range 1000–10 000 are treated with boric acid to give gums with molecular weights 350 000–500 000. These gums contain about 1 boron atom per 300 silicon atoms as a result of the following condensation:

$$\text{>B—OH} + \text{HO—Si<} \longrightarrow \text{>B—O—Si<} + H_2O$$

Alternatively, alkoxy-terminated siloxanes may be treated with acetoxyboron compounds; e.g.

$$\text{>B—OCOCH}_3 + \text{C}_2\text{H}_5\text{O—Si<} \longrightarrow \text{>B—O—Si<} + \text{CH}_3\text{COOC}_2\text{H}_5$$

The gums can be cross-linked by means of peroxides to give non-tacky rubbers which adhere strongly to each other to form homogeneous masses when left in contact under slight pressure at room temperature.

Bouncing putty is prepared by heating a dimethyl silicone with ferric chloride and boric oxide or other boron compounds and adding suitable fillers and softeners to the resultant gum. The material is readily shaped by kneading and may be drawn into threads on application of moderate tension. When dropped on a hard surface, the material shows high elastic rebound; it shatters, however, when given a sharp blow.

The unusual characteristics of borosilicones are generally supposed to stem from weak intermolecular bonds formed by attraction of electron pairs from oxygen to boron:

$$\begin{array}{c} \equiv\text{Si—O—Si<} \\ \big| \\ \text{—O—B—O—} \\ \big| \end{array}$$

17.6.3 Resins

Silicone resins consist of branched polymers, production of which is based on the hydrolysis of trichlorosilanes. If pure trichlorosilanes are hydrolysed, the products are highly cross-linked and intractable and are unsuitable for normal applications. In order to reduce the degree of cross-linking, it is usual to subject a blend of tri- and dichlorosilanes to hydrolysis. A convenient measure of the functionality of a blend is given by the R/Si value, which is the ratio of the numbers of organic groups and silicon atoms. (Thus pure dimethyldichlorosilane and methyltrichlorosilane have R/Si values of 2 and 1 respectively.) For the preparation of commercial resins, R/Si values in the range 1.2–1.6 are usual. Most commercial silicone resins contain both methyl and phenyl groups. The introduction of phenyl groups into the methyl-siloxane network results in improved heat resistance, flexibility and compatibility with pigments although pure phenyl silicone resins give products which are too weak for most applications. In methyl phenyl resins, the methyl and phenyl groups may be attached to either the same or different silicon atoms. Thus resins are prepared by the hydrolysis of blends of chlorosilanes chosen from methyltrichlorosilane, phenyltrichlorosilane, dimethyldichlorosilane, diphenyldichlorosilane and methylphenyldichlorosilane.

In a typical process, the chlorosilane blend is dissolved in a solvent such as toluene or xylene and then stirred with water. When the blend contains mainly methylchlorosilanes, reaction is very rapid and exothermic and cooling may be necessary to prevent gelation. When substantial quantities of phenylchlorosilanes are present, however, it is often necessary to raise the temperature to 70–75°C to ensure complete hydrolysis. At the end of the reaction, the mixture is allowed to separate into two layers. The organic layer is washed free from hydrochloric acid and then some of the solvent is distilled off to leave a solution with a solids content of about 80%. At this stage, only about 90% of the silanol groups initially present have condensed to form siloxane links; the resin therefore consists of silanol-terminated linear, branched and cross-linked polymers and cyclic polymers of low average molecular weight. The resin solution is then 'bodied' by heating at about 150°C in the presence of a catalyst such as zinc octoate. During this treatment, the average molecular weight of the resin increases through condensation of some of the remaining silanol groups. The batch is then cooled to prevent further reaction.

Silicone resins are normally supplied for use as solutions, prepared as described above. The final conversion of the partially polymerized soluble material into a fully cross-linked product is carried out *in situ*. In the cross-linking process, remaining silanol groups are condensed by heating in the presence of a catalyst, e.g. zinc octoate, cobalt naphthenate or triethanol-amine.

(a)   Modified resins

Silicone resins may be combined with other resinous materials to give modified resins which find use in surface coatings. Combination of the two materials may be achieved either by blending or copolymerization. Silicone resins with a high phenyl content are compatible to some extent with various resins such as alkyd, esterified epoxy and phenol-, urea- and melamine-formaldehyde resins and cold blending may be carried out. However, the limited compatibility of the components restricts the scope of the products which may be obtained. Copolymerization has been applied mainly to silicone-alkyd resins, which are prepared by heating a low molecular weight silicone resin containing a high proportion of free silanol groups with an alkyd resin. Interaction between silanol groups in the silicone and free hydroxyl groups in the alkyd gives a copolymer.

17.7 PROPERTIES

In this section, the general characteristics of silicones are reviewed and then some important properties of the various technical products are considered in more detail.

The outstanding property of silicones is their thermal stability. In general, the polymers can be heated in air to about 200°C without appreciable change in properties. This stability is attributable to the relatively high bond strengths found in the polymers (see Table 17.2). It is interesting to note that, in contrast to carbon, the ability of silicon to form catenated compounds is limited, although a few high molecular weight polysilanes have been prepared [6]. The low energy of the Si–Si bond indicates that such chains are less thermodynamically stable than the corresponding carbon chains.

Although the Si–O bond has good thermal stability, its relatively high ionic character (51%) renders it easily cleaved by strong acids and bases (as in equilibration reactions) (see section 17.5).

The presence of electropositive silicon atoms in the main chain might be expected to lead to dipole interactions with neighbouring chains, resulting in brittle, glass-like behaviour. However, the unusually large Si–O–Si bond angle permits very free rotation about the Si–O bond and the organic substituents occupy a large effective volume. Thus close packing is not possible and chain interactions are diminished.

A further characteristic property of all silicones is that they are water repellent due to the sheath of hydrophobic organic substituents which encases the polymer backbone. Related to this property is the 'non-stick' behaviour of silicone surfaces.

Table 17.2 Bond energies of various carbon and silicon bonds

Bond	Bond energy (kJ/mol)
Si—O	444
C—H	414
C—O	360
C—C	348
Si—C	318
Si—H	318
Si—Si	222

17.7.1 Fluids

In common with other silicone products, the fluids have outstanding heat resistance. Methyl silicone fluids are stable in air up to 150°C for long periods; in the absence of air this temperature is raised to 250°C. The presence of phenyl groups in the polymer enhances heat stability and the aforementioned temperatures are increased by about 100 deg. C. The changes that occur in air and in an inert atmosphere are different. In air, cross-linking occurs and the fluid viscosity increases. The process is initiated by oxygen attack on methyl groups and some of the reactions which may be involved are shown below [7]:

$$\equiv Si-CH_3 + O_2 \longrightarrow \equiv Si-\dot{C}H_2 + \cdot OOH$$

$$\equiv Si-CH_3 + O_2 \longrightarrow [\equiv Si-CH_2OOH] \longrightarrow \equiv Si\cdot + \cdot OH + CH_2O$$

$$\equiv Si-\dot{C}H_2 + O_2 \longrightarrow \equiv Si-CH_2OO\cdot \longrightarrow \equiv SiO\cdot + CH_2O$$

$$\equiv Si\cdot + \equiv SiO\cdot \longrightarrow \equiv Si-O-Si\equiv$$

The oxidative stability of fluids may be improved by the incorporation of an antioxidant such as an iron soap. When silicone fluids are heated in the absence of air, chain scission occurs and the viscosity falls.

An important characteristic of silicone fluids is the relatively small variation of viscosity with temperature (provided, of course, the temperature is not sufficiently high to cause degradation). This property is illustrated in Table 17.3, wherein the viscosities of a silicone fluid and a petroleum oil are compared. The small change in viscosity of silicone fluids stems from the low interaction between chains.

Silicone fluids are soluble in aliphatic, aromatic and chlorinated hydrocarbons as is to be expected from their low solubility parameter. They are

Table 17.3 Viscosities of a silicone fluid and a petroleum oil [8]

Temperature (°C)	Viscosity ($m^2/s \times 10^6$ (centistokes))	
	Silicone fluid	Petroleum oil
99	40	10
38	100	100
−18	350	11 000
−37	660	230 000
−57	1560	—

insoluble in liquids of higher solubility parameter such as acetone, ethanol, glycerol and water. The fluids are resistant to many inorganic reagents but are attacked by strong acids and alkalis.

Because of their thermal stability and low viscosity-temperature coefficient, silicone fluids find use as hydraulic fluids, lubricants and greases. Other applications include textile finishes, mould release agents and anti-foaming agents where water repellency, non-stick characteristics and surface activity respectively are utilized.

17.7.2 Vulcanized elastomers

The most important characteristic of silicone elastomer vulcanizates is their maintenance of physical properties over a large temperature range. General purpose material is serviceable over the approximate range -50 to $200°C$ but both ends of the range may be extended by the use of special purpose compounds. When elastomers are exposed to air at temperatures above $200°C$, cross-linking and embrittlement slowly occur; in the absence of air chain scission and softening predominate (cf. fluids).

The mechanical properties of silicone elastomers at room temperature are inferior to those of other elastomers. For example, natural rubber vulcanizates normally have tensile strengths in the range 21–28 MPa (3000–4000 lbf/in^2) whilst the corresponding range for general purpose heat-cured silicones is only 3.5–7.0 MPa (500–1000 lbf/in^2) (room temperature-cured material is usually inferior). This low strength results from the small intermolecular attraction between silicone chains and from the highly coiled nature of the chains which results in a small end-to-end length and, consequently, limited chain entanglement. However, the mechanical properties of silicone elastomers at elevated temperatures are greatly superior to those of other elastomers (except fluoro elastomers). For example, in a heat ageing test at $250°C$ a silicone elastomer had useful tensile strength and elongation after 2 weeks whereas a natural rubber vulcanizate had lost all rubbery properties after 4 hours [9].

Silicone elastomers have good electrical insulating properties, especially if the decomposition products of the vulcanizing agent are removed by heating the vulcanizate after cure.

Silicone elastomers are unaffected by oxygen and ozone on natural ageing. Because of their water-repellency, silicone elastomers are not seriously affected by aqueous solutions of most chemical reagents. They are attacked by strong acids and alkalis. General purpose elastomers are swollen by aliphatic, aromatic and chlorinated hydrocarbons but much improved resistance is shown by nitrile- and fluorine-containing silicones. In general, room temperature-cured materials have lower solvent resistance.

The major use of conventional silicone elastomers is in aircraft components where high and low temperature properties and electrical insulation characteristics are utilized, e.g. gaskets, sealing strips, ducting and cable insulation. Room-temperature vulcanizing silicone elastomers are used for the encapsulation of electronic and electrical equipment and for caulking. Liquid silicone elastomers are used for injection moulding of such components as seals, diaphragms and O-rings. They are ideal for moulding intricate shapes and are particularly suited to high-speed, high-volume operations.

17.7.3   Cross-linked resins

The principal uses of silicone resins are in surface coatings and glass-fibre laminates. Thus it is appropriate to consider both film and bulk properties.

The use of silicones in surface coatings depends largely on their heat resistance and water repellency. In general, the film properties (e.g. flow and scratch-resistance) are inferior to those of conventional polymers such as alkyds; thus blends or copolymers of the two classes of material are frequently used to give systems with better film properties. However, as the silicone content of a film decreases so does heat resistance. This feature is illustrated in Table 17.4 which gives comparative maximum service temperatures of various types of coating. The heat resistance and colour retention of coatings based on silicone-alkyd copolymers are generally superior to those given by the corresponding blends.

Cross-linked silicone films are not generally seriously affected by aqueous solutions of chemical reagents, including strong acids. Strong alkalis, however, cause degradation. It may be noted that resistance to steam is rather poor; this is because silicone films have high water vapour permeability and the penetration of water vapour to the substrate substantially reduces the adhesion of the film. Silicone films are attacked by aliphatic, aromatic and chlorinated hydrocarbons; resistance to fats and oils is good.

Silicone coatings are applied, typically, to electrical, heating and food-handling equipment. In the last use, the non-stick and non-toxic characteristics of silicones are important.

Table 17.4   Maximum service temperature of various types of coating [10]

Type of coating	Maximum service temperature (°C)
Alkyd	150
Silicone-alkyd blend	200
Silicone-alkyd copolymer	200
Silicone	250

The second major use of silicone resins is for the production of glass-fibre laminates. These laminates are mainly distinguished by their excellent electrical insulating properties which show little change on prolonged heating at 250°C. The laminates are extensively used in electrical equipment operating at high temperatures, e.g. slot wedges, spacers and terminal boards in electrical motors. The mechanical properties of the laminates are somewhat inferior to those of laminates based on other resins such as polyester, epoxy and melamine-formaldehyde resins.

REFERENCES

1. Watt, J. A. C. (1970) *Chem. Brit.*, **6**, 519.
2. Hurd, D. T. and Rochow, E. G. (1945) *J. Am. Chem. Soc.*, **67**, 1057.
3. Noll, W. (1968) *Chemistry and Technology of Silicones*, Academic Press, New York, p. 31.
4. Nitzche, S. and Wick, M. (1957) *Kunstoffe*, **47**, 431; U.K. Patent 751,325.
5. Dunham, M. L. *et al.* (1957) *Ind. Eng. Chem.*, **49**, 1373.
6. David, L. D. (1987) *Chem. Brit.*, **23**, 553.
7. Nielsen, J. M. (1968) in *Stabilization of Polymers and Stabilizer Processes*, (ed. R. F. Gould), American Chemical Society, Washington, ch. 8.
8. Wilcock, D. F. (1946) *Gen. Elect. Rev.*, **49**, No. 11, 14.
9. Konkle, G. M. *et al.* (1956) *Rubber Age*, **79**, 445.
10. McGregor, R. R. (1954) *Silicones and their Uses*, McGraw-Hill Publishing Company Ltd., London.

BIBLIOGRAPHY

Rochow, E. G. (1951) *An Introduction to the Chemistry of the Silicones*, 2nd edn., Chapman and Hall Ltd., London, (2nd ed.).

McGregor, R. R. (1954) *Silicones and their Uses*, McGraw-Hill Publishing Company Ltd., London.

Meals, R. N. and Lewis, F. M. (1959) *Silicones*, Reinhold Publishing Corporation, New York.

Fordham, G. (ed.) (1960) *Silicones*, George Newnes Limited, London.

Freeman, G. G. (1962) *Silicones*, Iliffe Books Ltd., London.

Noll, W. (1968) *Chemistry and Technology of Silicones*, Academic Press, New York.

Lewis, F. M. (1969) in *Polymer Chemistry of Synthetic Elastomers*, Part II, (ed. J. P. Kennedy, and E. G. M. Törnqvist), Interscience Publishers, New York, ch. 8.

Lynch, W. (1978) *Handbook of Silicone Rubber Fabrication*, Van Nostrand Reinhold Co., New York.

EPOXIES

18.1 SCOPE

For the purposes of this chapter, epoxies are defined as cross-linked polymers in which the cross-linking is derived from reactions of the epoxy group.

This definition excludes products based on the polymerization of epoxy compounds such as ethylene and propylene oxides. Although these products may be cross-linked as in, for example, elastomers (section 9.4.4) and foams (section 16.5), the cross-linking reactions do not involve epoxy groups.

It may be noted that there are various practices with respect to nomenclature in the literature. The epoxy group is also called the epoxide, oxirane or ethoxyline group. Both cross-linked polymers and low molecular weight resin precursors are variously known as epoxies or epoxides. In this chapter, the terms epoxy, epoxy resin and epoxy group are used to indicate cross-linked polymer, precursor and functional group respectively.

18.2 DEVELOPMENT

Epoxies were developed, independently, principally by Ciba AG. (Switzerland) (first patent to Castan, 1943) and the Devoe and Raynolds Co. (USA) (first patent to Greenlee, 1950). The former company concentrated mainly on adhesives whilst the latter company was more concerned with surface coatings.

Despite their relatively high cost, epoxies are now firmly established in a number of important industrial applications. The largest single use is for surface coatings, which account for about 50% of current epoxy resin output. Other applications include laminated circuit boards, carbon fibre composites, electronic component encapsulations and adhesives.

At the present time, 80–90% of commercial epoxy resins are prepared by the reaction of 2,2-bis(4'-hydroxyphenyl)propane (bisphenol A) and epichlorhydrin with the other types of products described in this chapter being produced in minor amounts.

18.3   BISPHENOL A-EPICHLORHYDRIN EPOXIES

18.3.1   Raw materials

The principal raw materials required for resin production are bisphenol A (2,2-bis(4'-hydroxyphenyl)propane) and epichlorhydrin (1-chloro-2,3-epoxy-propane).

(a)   Bisphenol A

Bisphenol A is so called since it is formed from phenol (2 mole) and acetone (1 mole):

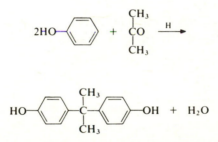

Although the reaction theoretically requires the molar ratio of reactants to be 2:1, an improved yield of bisphenol A is obtained if additional phenol is present; the optimum molar ratio is 4:1. In a typical process, the phenol and acetone are mixed and warmed to 50°C. Hydrogen chloride (catalyst) is passed into the mixture for about 8 hours, during which period the temperature is kept below 70°C to suppress the formation of isomeric products. Bisphenol A precipitates and is filtered off and washed with toluene to remove unreacted phenol (which is recovered). The product is then recrystallized from aqueous ethanol. Since epoxy resins are of low molecular weight and because colour is not normally particularly important, the purity of bisphenol A used in resin production is not critical. Material with a p,p'-isomer content of 95–98% is usually satisfactory; the principal impurities in such material are o,p'- and o,o'-isomers.

The formation of bisphenol A is thought to proceed as follows [1]:

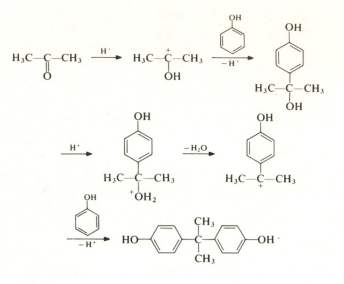

(b) *Epichlorhydrin*

The preparation of epichlorhydrin ($CH_2-CH-CH_2Cl$, with epoxide bridge) is outlined in section 11.3.2(a). Epichlorhydrin is a colourless liquid with an irritating odour, b.p. 115°C.

18.3.2 Resin preparation

In a typical process for the preparation of a liquid epoxy resin, a mixture of bisphenol A and epichlorhydrin (about 1:4 molar) is heated to about 60°C with stirring. Solid sodium hydroxide (2 mole per mole bisphenol A) is added slowly at such a rate that the reaction mixture remains neutral. The reaction is exothermic and cooling is applied to keep the temperature at 60°C. Excess of epichlorhydrin is then removed by distillation under reduced pressure. The residue consists of epoxy resin mixed with sodium chloride. The latter is filtered off, toluene having been added to the mixture in order to facilitate filtration. The toluene is removed by distillation under reduced pressure and then the resin is heated at 150°C/0.6 kPa to remove traces of volatile matter. This last step is important since the presence of volatiles may lead to bubble formation when the resin is subsequently used. Finally, the resin is clarified by passage through a fine filter.

In the preparation of solid epoxy resins the above process is slightly modified [2]. A mixture of bisphenol A and epichlorhydrin (the molar ratio of reactants used depends on the resin molecular weight required; see later) is heated to 100°C and aqueous sodium hydroxide is added slowly with vigorous stirring. When reaction is complete the agitator is stopped and a 'taffy' (which is an emulsion of about 30% water in resin) rises to the top of the reaction mixture. The lower layer of brine is removed; the resinous layer is coagulated and washed with hot water. The resin is heated at 150°C under reduced pressure to remove water, clarified by passage through a filter and then allowed to solidify. Solvent can be added at the washing stage but whilst this facilitates washing and filtration it is very difficult to remove subsequently all traces of the solvent. Alternatively, solid epoxy resins may be prepared by a two-step process in which a pre-formed liquid resin is heated with bisphenol A in the presence of a basic catalyst to effect chain extension. This method avoids the difficulty of washing sodium chloride from highly viscous material.

The reactions which are involved when epichlorhydrin reacts with a phenol (ROH) in the presence of sodium hydroxide are as follows:

(i) Formation of phenoxy anion

$$ROH + OH^- \longrightarrow RO^- + H_2O$$

(ii) Reaction of epoxy group of epichlorhydrin with phenoxy anion

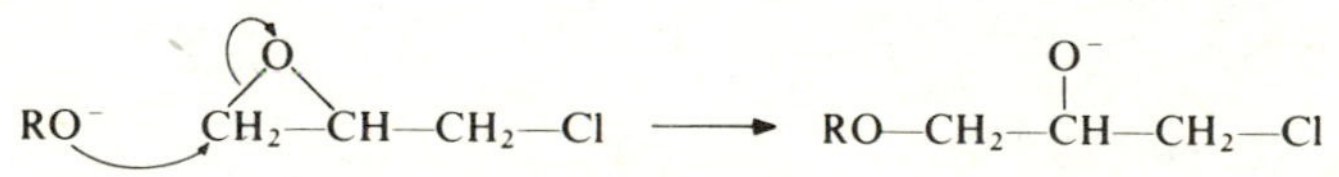

$$RO^- \quad CH_2-CH-CH_2-Cl \longrightarrow RO-CH_2-\overset{O^-}{\underset{|}{CH}}-CH_2-Cl$$

(iii) Elimination of chloride anion to form a glycidyl ether

$$RO-CH_2-\overset{O^-}{\underset{|}{CH}}-CH_2-Cl \longrightarrow RO-CH_2-CH-CH_2 + Cl^-$$

(iv) Reaction of epoxy group in glycidyl ether with phenoxy anion

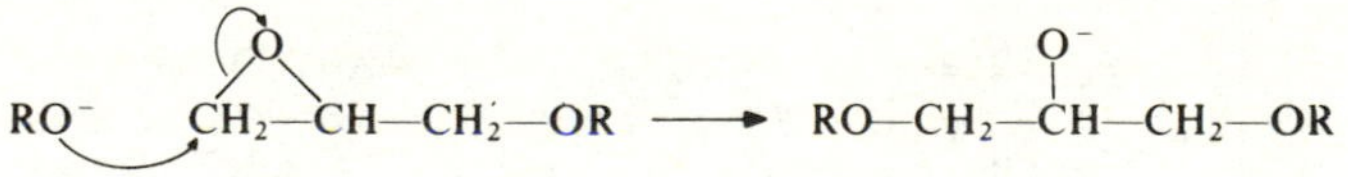

$$RO^- \quad CH_2-CH-CH_2-OR \longrightarrow RO-CH_2-\overset{O^-}{\underset{|}{CH}}-CH_2-OR$$

(v) Formation of hydroxyl group by protonation

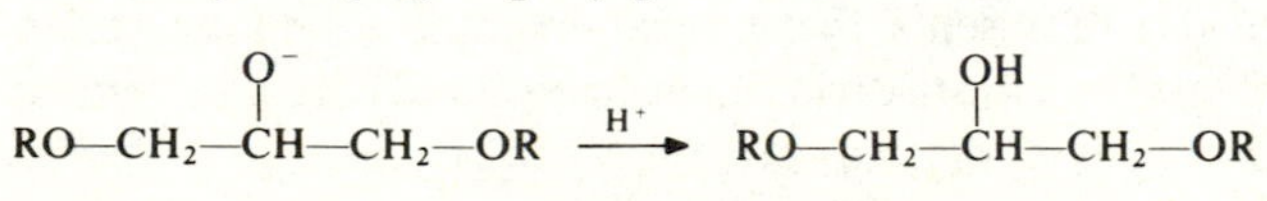

$$RO-CH_2-\overset{O^-}{\underset{|}{CH}}-CH_2-OR \xrightarrow{H^+} RO-CH_2-\overset{OH}{\underset{|}{CH}}-CH_2-OR$$

It is to assist protonation that the addition of sodium hydroxide is carried out slowly throughout the reaction. If protonation did not occur the anion could react with an epoxy group to give a branched product.

When epichlorhydrin reacts with a difunctional phenol such as bisphenol A the above reactions lead to linear polymers. In order for such polymers to be commercially useful as epoxy resins it is necessary that the polymers are terminated by epoxy groups (through which they may be subsequently cross-linked). This is achieved by carrying out the polymerization with a molar excess of epichlorhydrin. The formation of epoxy resins is illustrated by the following scheme (wherein the reaction mechanisms are as described previously):

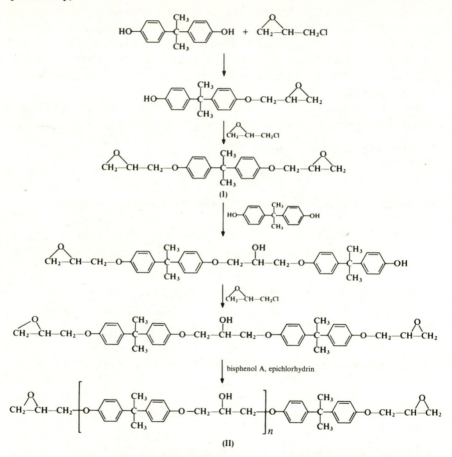

(I)

(II)

Structure (II) represents the general formula of epoxy resins based on bisphenol A and epichlorhydrin. Such compounds are also termed *diglycidyl ethers* since they contain two glycidyl ether groups,

$$-O-CH_2-CH-CH_2$$

per molecule. It may be noted that the simplest diglycidyl ether of this kind is represented by Structure (I) and that this may be regarded as the case where $n = 0$ in the general formula. The value of n is determined by the molar ratio of

Table 18.1 Effect of reactant ratio on molecular weight of epoxy resins [3]

Molar ratio epichlorhydrin/ bisphenol A	Molecular weight	Softening point (°C)
10.0:1	370	9
2.0:1	451	43
1.4:1	791	84
1.33:1	802	90
1.25:1	1133	100
1.2:1	1420	112

the reactants; the more nearly this ratio approaches unity the higher the molecular weight of the product. The effect of the ratio of starting materials on molecular weight is shown in Table 18.1. Commercial liquid epoxy resins based on bisphenol A and epichlorhydrin generally have average molecular weights of about 400. The molecular weight of the simplest diglycidyl ether (I) is 340 and it is therefore clear that the liquid resins are composed largely of this material. Solid resins have higher molecular weights; commercial solid resins are low melting solids and seldom have average molecular weights exceeding 4000 (this corresponds to an average value of n of about 13). Resins with molecular weights above this value are of limited use since their high viscosity and restricted solubility make subsequent processing difficult. It may also be noted that the higher molecular weight resins have pendant hydroxyl groups; the value of n corresponds to the number of hydroxyl groups per molecule. Broadly speaking, the lower molecular weight resins find use in such applications as adhesives, castings, encapsulations and laminates whilst the higher molecular weight resins are used in surface coatings (such resins often being modified through the pendant hydroxyl groups).

18.3.3   Cross-linking agents

The bisphenol A-epichlorhydrin resins produced by the methods described above cannot be cross-linked at a reasonable rate by heat alone; even heating at 200°C has little effect. In order to convert the resins into cross-linked structures it is necessary to add a curing agent. Most of the curing agents in common use can be classified into three groups, namely tertiary amines, polyfunctional amines and acid anhydrides; these types of curing agent are considered in this section. Polysulphides have also found some use; these materials are described in Chapter 19.

(a)   Tertiary amines

Examples of tertiary amines used as curing agents for epoxy resins include benzyldimethylamine (BDMA) (III), 2-(dimethylaminomethyl)phenol

(DMAMP) (IV), 2,4,6-tris(dimethylaminomethyl)phenol (TDMAMP) (V), triethanolamine (VI) and *N*-n-butylimidazole (VII).

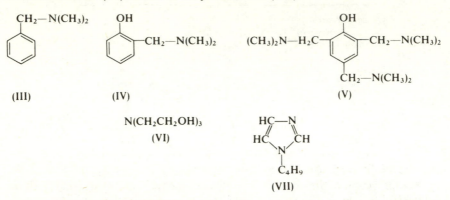

(III)   (IV)   (V)

N(CH$_2$CH$_2$OH)$_3$

(VI)

(VII)

The reaction between an epoxy resin and a tertiary amine (R$_3$N) is thought to proceed via the following steps [4, 5]:

(i) Formation of quaternary base

$$R_3N: \quad CH_2-CH\sim \quad \longrightarrow \quad R_3\overset{+}{N}-CH_2-\overset{O^-}{\underset{|}{CH}}\sim$$

(ii) Protonation of quaternary base with formation of an anion

$$R_3\overset{+}{N}-CH_2-\overset{O^-}{\underset{|}{CH}}\sim \; + \; R'OH \quad \longrightarrow \quad R_3\overset{+}{N}-CH_2-\overset{OH}{\underset{|}{CH}}\sim \; + \; R'O^-$$

$$\text{or} \; R_3\overset{+}{N}-CH_2-\overset{O^-}{\underset{|}{CH}}\sim \; + \; H_2O \quad \longrightarrow \quad R_3\overset{+}{N}-CH_2-\overset{OH}{\underset{|}{CH}}\sim \; + \; OH^-$$

The proton donor may either be the curing agent itself (if a phenolic tertiary amine is used) or water present as impurity.

(iii) Polymerization through epoxy groups (initiated by above-formed anion)

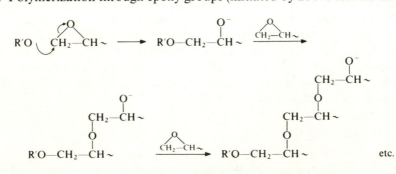

etc.

Evidence for this scheme is that chemically-bound nitrogen is found in the final product; the rate of reaction is increased by the presence of phenols; and anhydrous non-phenolic tertiary amines are not effective curing agents. Since bisphenol A-epichlorhydrin resins have epoxy groups at each end of the polymer the above scheme results in the formation of a cross-linked polymeric structure.

Tertiary amines are used alone as curing agents for epoxy resins mainly in adhesive and coating applications. Tertiary amine salts, such as the tri-2-ethylhexoate salt of tris(dimethylaminomethyl)phenol (VIII) have been introduced to provide tertiary amine curing agents which can be used in higher

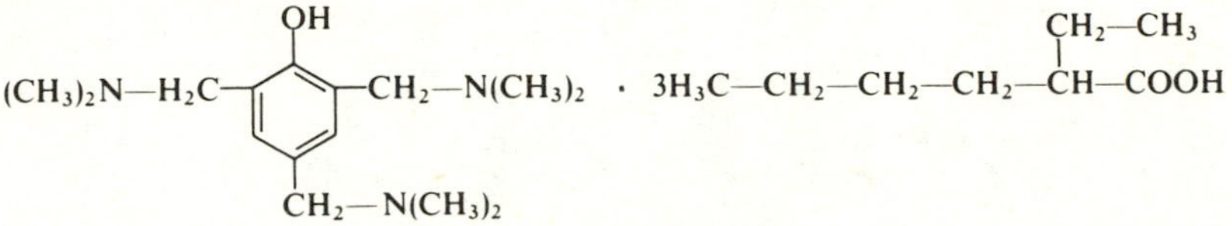

$$(CH_3)_2N-H_2C- \overset{\overset{\displaystyle OH}{|}}{\underset{\underset{\displaystyle CH_2-N(CH_3)_2}{|}}{\bigcirc}} -CH_2-N(CH_3)_2 \quad \cdot \quad 3H_3C-CH_2-CH_2-CH_2-\overset{\overset{\displaystyle CH_2-CH_3}{|}}{CH}-COOH$$

(VIII)

percentages (making their use easier by permitting greater weighing errors to be made before the characteristics of the resin-curing agent mix are significantly altered) and still provide relatively long pot lives. The curing mechanism involves two steps. The first and rate determining step is the removal of the fatty acid by esterification of some of the epoxy groups in the resin:

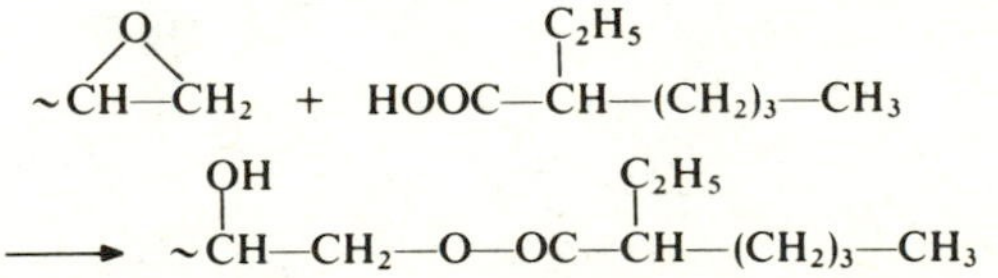

$$\sim \overset{\displaystyle O}{\overset{\displaystyle /\backslash}{CH-CH_2}} + HOOC-\overset{\overset{\displaystyle C_2H_5}{|}}{CH}-(CH_2)_3-CH_3$$

$$\longrightarrow \sim \overset{\overset{\displaystyle OH}{|}}{CH}-CH_2-O-OC-\overset{\overset{\displaystyle C_2H_5}{|}}{CH}-(CH_2)_3-CH_3$$

Then the liberated tertiary amine initiates reaction of the remaining epoxy groups as described previously.

Tertiary amines are commonly referred to as 'catalytic' curing agents since they induce the direct linkage of epoxy groups to one another. This is in contrast to other classes of curing agents which function by providing intervening groups through which epoxy groups are linked to one another (see later). Nonetheless, tertiary amines, as sole curing agents, are normally required in greater than 'catalytic' amounts, indicating that chain termination frequently occurs. However, it is as accelerators for anhydride curing agents that tertiary amines find their widest use and in this application small concentrations of amine are employed (see section 18.3.3(c)).

(b) Polyfunctional amines

Both aliphatic and aromatic compounds having at least three active hydrogen atoms present in primary and/or secondary amine groups are widely used as curing agents for epoxy resins. Examples of such polyfunctional amines are diethylenetriamine (DTA) (IX), triethylenetetramine (TET) (X), *m*-phenylenediamine (MPD) (XI), 4,4′-diaminodiphenylmethane (DDM) (XII) and 4,4′-diaminodiphenylsulphone (DDS) (XIII).

$H_2N—(CH_2)_2—NH—(CH_2)_2—NH_2$ $H_2N—(CH_2)_2—NH—(CH_2)_2—NH—(CH_2)_2—NH_2$
                (IX)   (X)

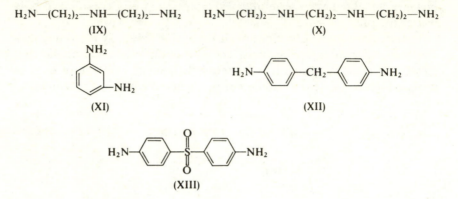

(XI)   (XII)

(XIII)

The reaction between an epoxy resin and a primary amine (RNH_2) may be written simply as:

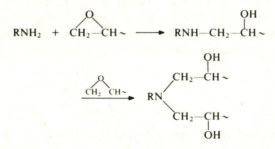

However, there is much evidence to suggest that the reaction requires the presence of a proton-donor (which may be either hydroxyl groups present in the resin or traces of water). It is supposed that these compounds aid the opening of the epoxy ring by hydrogen bonding the oxygen atom in the transition state, e.g.

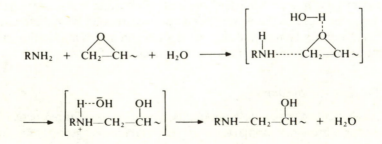

The secondary amine formed then undergoes further similar reaction, giving a tertiary amine. Any secondary amine groups originally present in the curing agent will, of course, react comparably. Evidence for the intervention of a proton donor is provided by studies on the reaction between the model compound, phenylglycidyl ether, and diethylamine. The reaction is greatly accelerated by the presence of phenol, isopropanol or water [6]. Also, it has been found that whilst anhydrous diethylamine and ethylene oxide do not react, interaction proceeds readily in the presence of catalytic amounts of water, methanol or ethanol [7]. It has been noted that the final product of the reaction between an epoxy resin and primary or secondary amine is a tertiary amine and it is conceivable that the latter could initiate polymerization of epoxy groups in the manner described in section 18.3.3(a). However, this tertiary amine is probably so immobile as to be virtually unreactive. Also, the reaction between an epoxy resin and primary or secondary amine results in the formation of hydroxyl groups and a possible subsequent reaction is etherification by epoxy groups:

$$-OH \ + \ CH_2\overset{O}{-}CH\sim \ \longrightarrow \ -O-CH_2-\overset{OH}{CH}\sim$$

However, the tertiary amine does not appear able to promote this reaction to any great extent.

Reaction between a bisphenol A-epichlorhydrin epoxy resin and a polyfunctional amine thus results in a cross-linked polymeric structure. The process is illustrated by the reaction with triethylenetetramine (which has six reactive hydrogen atoms):

$$H_2N-(CH_2)_2-NH-(CH_2)_2-NH-(CH_2)_2-NH_2 \ + \ 6CH_2\overset{O}{-}CH\sim \ \longrightarrow$$

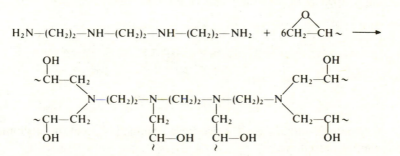

Generally speaking, aliphatic amines provide fast cures and are effective at room temperature whilst aromatic amines are somewhat less reactive and give products with higher heat distortion temperatures. Polyfunctional amines are widely used in adhesive, casting and laminating applications.

As a class, the amines suffer from the disadvantage of usually being toxic. They are often capable of causing severe irritation and may cause a serious rash or an asthmatic response in sensitized persons. In order to reduce toxicity, the polyfunctional amines are often used in the form of adducts. A further point is that unmodified amines are usually used in stoichiometric ratio (which typically amounts to about 12 parts or less per hundred of resin); this necessitates careful weighing which may not be feasible in some situations. The effect of using modified amines is to increase the weight of curing agent required which allows greater weighing errors to be made before the characteristics of the resin-curing agent mix are significantly altered. Examples of amine adducts which have found commercial application are:

(i) Glycidyl ether adducts, e.g. the product obtained by adding the diglycidyl ether of bisphenol A (I) to a large excess of diethylenetriamine and cooling and stirring vigorously to prevent gelation:

$$CH_2\text{---}CH\text{---}CH_2\text{---}O\text{---}R\text{---}O\text{---}CH_2\text{---}CH\text{---}CH_2 \;+\; 2 \;\; H_2N\text{---}(CH_2)_2\text{---}NH\text{---}(CH_2)_2\text{---}NH_2$$

$$\longrightarrow \quad
\begin{array}{l}
O\text{---}CH_2\text{---}\overset{\displaystyle OH}{CH}\text{---}CH_2\text{---}HN\text{---}(CH_2)_2\text{---}NH\text{---}(CH_2)_2\text{---}NH_2\\[4pt]
R\\[4pt]
O\text{---}CH_2\text{---}\overset{\displaystyle OH}{CH}\text{---}CH_2\text{---}HN\text{---}(CH_2)_2\text{---}NH\text{---}(CH_2)_2\text{---}NH_2
\end{array}$$

Where R =

(ii) Ethylene oxide and propylene oxide adducts, e.g. the product obtained by treating propylene oxide with excess of ethylene diamine:

$$H_3C\text{---}CH\text{---}CH_2 \;+\; H_2N\text{---}CH_2\text{---}CH_2\text{---}NH_2 \longrightarrow$$

$$H_3C\text{---}\overset{\displaystyle OH}{CH}\text{---}CH_2\text{---}HN\text{---}CH_2\text{---}CH_2\text{---}NH_2$$

(iii) Acrylonitrile adducts, e.g. the products obtained by treating acrylonitrile with diethylenetriamine:

$$NC\text{---}CH{=}CH_2 \;+\; H_2N\text{---}(CH_2)_2\text{---}NH\text{---}(CH_2)_2\text{---}NH_2 \longrightarrow$$

$$NC\text{---}CH_2\text{---}CH_2\text{---}HN\text{---}(CH_2)_2\text{---}NH\text{---}(CH_2)_2\text{---}NH_2 \xrightarrow{\;H_2C=CH\text{---}CN\;}$$
(XIV)

$$NC\text{---}CH_2\text{---}CH_2\text{---}HN\text{---}(CH_2)_2\text{---}NH\text{---}(CH_2)_2\text{---}NH\text{---}CH_2\text{---}CH_2\text{---}CN$$
(XV)

Commercial curing agents are mixtures of the two addition compounds (XIV and XV). Since accelerating hydroxyl groups are not present in these adducts (in contrast to (i) and (ii) above) they represent curing agents of reduced activity and so give longer pot lives.

Polyfunctional amines which are commonly used as curing agents for epoxy resins but which are of a rather different kind to those described above are the so-called 'fatty polyamides.' These polymers, which are described in section 10.4 are of low molecular weight (2000–5000) and are prepared by treating dimerized and trimerized fatty acids with ethylenediamine or diethylenetriamine. Fatty polyamides are used to cure epoxy resins where a more flexible product is required, particularly in adhesive and coating applications.

Finally, in this discussion of polyfunctional amines, mention may be made of amines which are used as curing agents but which have less than three hydrogen atoms available for reaction with epoxy groups. Examples of this type of amine are diethanolamine (DEA) (XVI), piperidine (XVII) and dimethylaminopropylamine (DMAPA) (XVIII). These curing agents operate by means of a two-part reaction. Firstly, the active hydrogen atoms of the primary and secondary amine groups are utilized in the manner which has been described previously for polyfunctional amines. Thereafter, since in these cases the resulting tertiary amine is sufficiently reactive to initiate polymerization of epoxy groups, the reaction proceeds in the manner described in section 18.3.3(a).

$$NH(CH_2CH_2OH)_2$$

(XVI)

$$\begin{array}{c} CH_2 \\ H_2C \quad CH_2 \\ | \quad\quad | \\ H_2C \quad CH_2 \\ NH \end{array}$$

(XVII)

$$(CH_3)_2N—CH_2—CH_2—CH_2—NH_2$$

(XVIII)

(c)   Acid anhydrides

Cyclic acid anhydrides are widely employed as curing agents for epoxy resins. Both mono- and dianhydrides are used. Examples of anhydrides which find application are maleic anhydride (MA) (XIX), dodecenylsuccinic anhydride (DDSA) (XX), hexahydrophthalic anhydride (HPA) (XXI), phthalic anhydride (PA) (XXII), pyromellitic dianhydride (PMDA) (XXIII), nadic methyl anhydride (NMA) (a mixture of maleic anhydride adducts of methylcyclopentadienes; cf. section 11.2.2(c)) (XXIV) and chlorendic (HET) anhydride (XXV; see also section 11.2.2(c)).

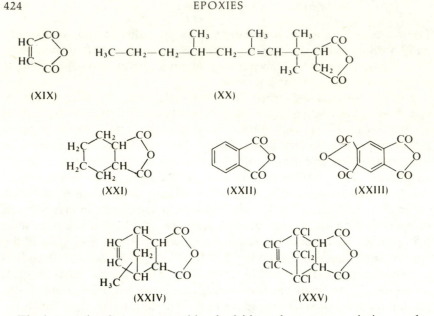

(XIX)   (XX)

(XXI)   (XXII)   (XXIII)

(XXIV)   (XXV)

The interaction between an acid anhydride and an epoxy resin is complex. Several different reactions are involved, the relative extent of which may be affected by prevailing conditions. In general, two main types of reaction occur, namely opening of the anhydride ring with the formation of carboxyl groups and opening of the epoxy ring. The various reactions which may occur are shown below, using phthalic anhydride as example:

(i)   Reactions of the anhydride group

The anhydride ring may be opened to produce one or two carboxyl groups by reaction with (a) water (traces of which may be present in the system) or (b) hydroxyl groups (which may be present as pendant groups in the original resin or which may be produced by reaction (iia) below):

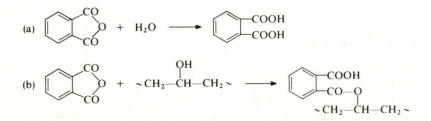

(ii)   Reactions of the epoxy group

The epoxy ring may be opened by reaction with (a) carboxyl groups (formed by reactions (ia, b) above) or (b) hydroxyl groups (which may be present as pendant groups in the original resin or which may be produced by reaction (iia) below), e.g.

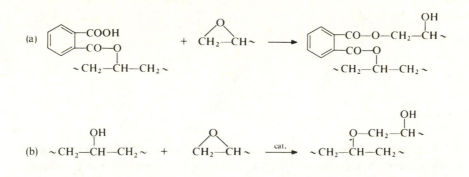

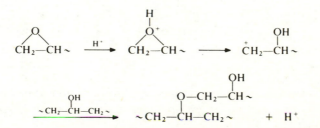

The uncatalysed reaction between aliphatic hydroxyl groups and epoxy groups is very slow, as is illustrated by the fact that epoxy resins may be heated to 200°C with little effect. However, the reaction is catalysed by proton donors [4]. In the present case the carboxyl group functions as a catalyst, possibly through a carbenium ion intermediate:

In an investigation [4] in which the model compound, phenylglycidyl ether, was heated with phthalic anhydride at 200°C, epoxy groups disappeared about twice as fast as anhydride groups. This observation indicates that reactions (iia)and (iib) proceed at about the same rate so that roughly equal numbers of ester and ether links are formed.

It is possible that the above scheme does not always apply. For example, when a low molecular weight bisphenol A-epichlorhydrin resin was cured with hexahydrophthalic anhydride or nadic methyl anhydride there was no indication of ether formation during the reaction prior to the gel point. It has been suggested, on the basis of kinetic data, that these cases involve a termolecular transition state which results in the production of a diester [8]:

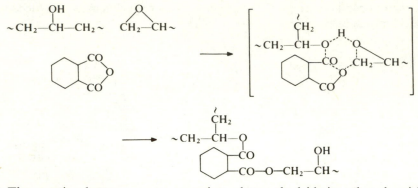

The reaction between an epoxy resin and an anhydride is rather sluggish and it is common practice to increase the rate by the addition of a catalyst, usually a tertiary amine. In this case, the tertiary amine appears to react preferentially with the anhydride to generate a carboxyl anion. This anion opens an epoxy ring to give an alkoxide ion which forms another carboxyl anion from a second anhydride molecule and so on, e.g.

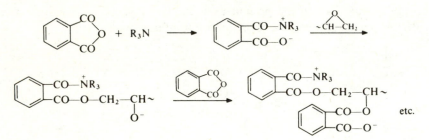

The selectivity of these reactions is indicated by the disappearance at an equal rate of anhydride and epoxy groups from a model system based on allyl glycidyl ether [9].

Compared to the polyfunctional amines, anhydrides are generally less skin sensitizing and give lower exotherms when used to cure epoxy resins. Anhydride-cured resins generally have better thermal stability, electrical insulation properties and chemical resistance (except to alkalis, which hydrolyse the ester groups). Phthalic anhydride is the cheapest anhydride curing agent but it suffers from the disadvantage of being rather difficult to mix with resin. In order to attain solubility it is necessary to heat the two components to about 120°C, at which temperature the pot life of the mix is limited and irritating fumes are evolved. Liquid anhydrides (e.g. dodecenylsuccinic anhydride and nadic methyl anhydride), low melting anhydrides (e.g. hexahydrophthalic anhydride) and eutectic mixtures greatly facilitate mixing procedures. Maleic anhydride is seldom used by itself as it produces brittle products; it is usually admixed with other anhydrides. Dodecenylsuccinic anhydride gives flexible products whilst chlorendic anhydride confers flame resistance. Pyromellitic dianhydride, by virtue of its increased functionality, produces tightly cross-linked products of high heat distortion temperatures.

Anhydride curing agents find use in most of the important applications of epoxy resins, particularly in castings and laminates.

18.3.4 Properties of cross-linked polymers

Since the characteristic group in epoxy resins largely disappears on cross-linking, it is difficult to make simple generalizations relating structure to properties. Furthermore, there is available a wide range of curing agents and as already indicated the choice of curing agent greatly affects the properties of the final product. Also the time and temperature of cure can have substantial effects on properties. Typical values of some properties of epoxies cured with various agents are shown in Table 18.2. In addition, it may be noted that the presence of fillers can also have profound effects on the properties of the cross-linked polymer. The cure of epoxy resins involves little molecular reorientation and no evolution of volatile by-products. Hence shrinkage is low and the cured products are relatively strain-free; they are thus usually tough. The main skeletons of the resins themselves are rather stable structures with good heat resistance but the overall stability of the cured products is very dependent on the nature of the cross-links present. Thus anhydride-cured systems are stable up to 200°C in air whereas amine-cured systems may start to decompose at 150°C. The electrical insulating properties of epoxies are very good although the presence of polar hydroxyl and ether groups results in comparatively high dielectric constant and power factor. These polar groups also lead to outstanding adhesion to many surfaces.

Being cross-linked, epoxies do not dissolve without decomposition but are swollen by liquids of similar solubility parameter (e.g. chlorinated hydrocarbons and alcohols). Resistance to acids and alkalis is largely determined by the curing agent used. Ether links (formed by the use of catalytic hardeners) are stable toward most inorganic and organic acids and alkalis whereas ester links (produced by anhydrides) are sensitive to strong alkalis and inorganic acids. The C–N bond (formed by amines) is generally stable toward inorganic acids and alkalis but not organic acids.

18.4 MODIFIED BISPHENOL A-EPICHLORHYDRIN EPOXIES

Whilst the straight bisphenol A-epichlorhydrin epoxies described above have found widespread use in such applications as adhesives, castings, encapsulations, composites and laminates they are used to a relatively small extent in surface coatings. In this important field, mainly modified bisphenol A-epichlorhydrin epoxies are used. Two principal types of modification are commercially practised, namely combination with other resins and esterification.

Table 18.2 Typical values for various properties of unfilled castings prepared from a liquid bisphenol A-based epoxy resin and various curing agents [10]

	Curing agent			
	Aliphatic amines (a)	Aromatic amines (b)	Fatty poly-amides (c)	Anhydrides (HPA or NMA) (d)
Tensile strength (MPa)	48–69	69–90	31–45	83–90
(lbf/in^2)	7000–10 000	10 000–13 000	4500–6500	12 000–13 000
Compressive strength (MPa)	83–100	120–130	48–62	120–130
(lbf/in^2)	12 000–15 000	17 000–19 000	7000–9000	17 000–19 000
Flexural strength (MPa)	83–100	120	48–62	120–130
(lbf/in^2)	12 000–15 000	17 000–18 000	7000–9000	18 000–19 000
Impact strength, Izod (J/m)	21–27	27–32	53–64	21
(ft lbf/in)	0.4–0.5	0.5–0.6	1.0–1.2	0.4
Heat deflection temperature (°C)	70–110	145–150	40–60	125–135

Cure schedules:
(a) 4–7 days at room temperature or gel at 23°C + 1–2 hours at 100°C.
(b) Gel at 80°C + 4 hours at 150°C.
(c) 7 days at room temperature or gel at 23°C + 1–2 hours at 100°C.
(d) 2 hours at 80°C + 5–15 hours at 125–150°C.

18.4.1 Resin-modified epoxies

Bisphenol A-epichlorhydrin resins may be blended with a variety of other resins which contain reactive groups. On curing, interaction occurs to give a cross-linked copolymer which exhibits characteristics of the two straight resins. Examples of resins used in conjunction with epoxy resins are:

(i) *Phenol-formaldehyde resins*. The phenolic resins most widely used are low molecular weight butylated resols (Chapter 14), which contain phenolic hydroxyl groups and etherified and unetherified methylol groups. The epoxy resins used have a molecular weight of 3000–4000 and therefore contain secondary hydroxyl groups. At stoving temperatures of 180–200°C, the phenolic hydroxyl groups react with epoxy groups and both types of methylol groups react with the hydroxyl groups of the epoxy resin, e.g.

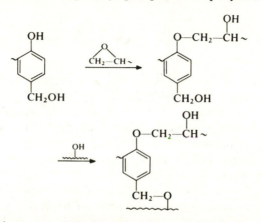

The net result is a cross-linked polymer network which exhibits the best chemical and heat resistance of all epoxy coating systems. Coatings of this type are used for corrosion-resistant pipes and containers.

(ii) *Amino resins*. The amino resins most widely employed are butylated urea-formaldehyde resins (section 15.2.3(b)) although melamine-formaldehyde resins may also be used. The epoxy resins generally used have a molecular weight of 3000–4000. On heating at 200°C (or 150°C in the presence of an acidic accelerator) the following reactions may occur:

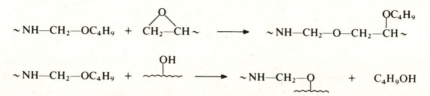

The resultant coatings have good colour, hardness and heat resistance and are used for industrial and domestic equipment and containers.

(iii) *Acrylic resins*. Low molecular weight epoxy resins are compatible with acrylic resins and give coatings widely used for refrigerators and washing machines. The reactions involved are discussed in section 6.7.2.

18.4.2   Esterified epoxies

The epoxy resins which are most commonly subjected to esterification have a molecular weight of about 1400; such resins can be esterified with carboxylic acids through their hydroxyl and epoxy groups:

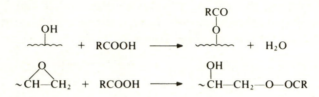

Catalysts such as alkali metal salts (e.g. sodium carbonate) are used to minimize etherification of epoxide groups by hydroxyl groups, since this reaction can lead to gelation. When fatty acids are used to esterify epoxy resins, products which are somewhat similar to alkyd resins (section 11.3) are obtained. The esterification of epoxy resins may be carried out in either the absence or presence of solvent using procedures entirely analogous to those described in section 11.3.4. Long drying oil or semi-drying oil epoxy esters are, in the presence of conventional driers, air drying. The films are more resistant to alkali than alkyd finishes (since the polymer backbone does not contain ester links) and show improved adhesion. Air-drying epoxy esters are widely used for the protection of plant where mildly corrosive conditions are encountered. Short or medium drying oil or semi-drying oil epoxy esters can be stoved to form films with good adhesion, flexibility and chemical resistance. They are used for metal primers and collapsible tube coatings. Short semi-drying or non-drying oil epoxy esters may be blended with melamine-formaldehyde resins and stoved to give films with superior hardness and chemical resistance. These systems are used for domestic appliance primers and drum coatings.

18.5   OTHER EPOXIES

As has been noted previously, most commercial epoxy resins are prepared by reaction of bisphenol A and epichlorhydrin. However, other types of resin are available and some of these are described below. These resins are conveniently divided into two groups, namely glycidyl ethers and non-glycidyl ethers.

18.5.1 Epoxies based on glycidyl ethers

It has been shown previously that resins containing the glycidyl ether group

$$-O-CH_2-\overset{\displaystyle O}{\overset{\displaystyle \triangle}{CH}}-CH_2$$

result from the reaction of epichlorhydrin and hydroxyl compounds. Whilst bisphenol A is the most commonly used hydroxyl compound, a few glycidyl ether resins which are based on other hydroxyl compounds are commercially available.

(a) Novolak epoxies

Novolak resins (see Chapter 14) are low molecular weight polymers consisting of phenolic nuclei linked in the o- and p- positions by methylene groups. They may be epoxidized through the phenolic hydroxyl groups by treatment with epichlorhydrin:

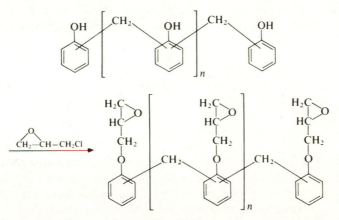

A typical commercial novolak epoxy resin has an average molecular weight of 650 and contains about 3.6 epoxy groups per molecule. Because of their multifunctionality, the novolak epoxy resins give, on curing, more tightly cross-linked products than the bisphenol A-based resins. This results in improved elevated temperature performance and chemical resistance. Although novolak epoxy resins have found some application in laminates and adhesives, their use has been restricted by their high viscosity and consequent handling difficulties.

(b) Polyglycol epoxies

Linear polyglycols such as poly(propylene glycol) (see section 9.4.3) may be epoxidized through the terminal hydroxyl groups to give diglycidyl ethers:

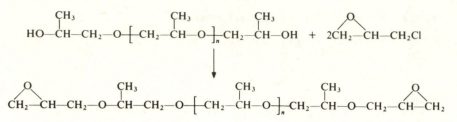

Commercial products are available where *n* varies from 1 to 6. When used alone these resins cure to soft products of low strength so they are normally used in blends with bisphenol A- or novolak-based resins. Generally, they are added to the extent of 10–30% when improved resilience is required without too large a loss in strength in such applications as adhesives and encapsulations.

(c)   Halogenated epoxies

Epoxies containing halogen may be prepared from halogenated hydroxyl compounds and resins are available based on tetrabromobisphenol A and tetrachlorobisphenol A (XXVI). Treatment of the tetrahalobisphenol A with

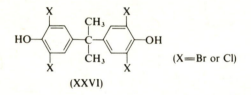

(X=Br or Cl)

(XXVI)

epichlorhydrin gives resins of the same general structure as the conventional resins based on bisphenol A except for the halogen atoms in the aromatic rings. The presence of halogen renders these resins flame retardant. The ability of the resins to retard or extinguish burning is due to the evolution of hydrogen halide upon decomposition at elevated temperatures. The brominated resins are more effective than the chlorinated resins and have become predominant commercially. The brominated resins are more stable than the chlorinated resins but, once begun, the evolution of hydrogen bromide is more rapid than that of hydrogen chloride and the system is more effectively blanketed. The superiority of the brominated resins is illustrated by the finding that it takes a bromine content of 13–15% by weight to make an unfilled cured epoxy resin system flame retardant while in an identical system 26–30% by weight chlorine is required to obtain the same result [11]. Brominated epoxy resins are generally used in blends with other epoxy resins to confer flame retardance with the maintenance of physical properties in such applications as adhesives and laminates.

18.5.2 Epoxies based on non-glycidyl ethers

Various types of non-glycidyl ether epoxy resins are commercially available and are briefly considered below.

(a) Cyclic aliphatic epoxies

Cyclic aliphatic epoxy resins which are commercially available include 3,4-epoxy-6-methylcyclohexylmethyl-3,4-epoxy-6-methylcyclohexane carboxylate (XXVII), 4-vinylcyclohexene dioxide (XXVIII) and dicyclopentadiene dioxide (XXIX).

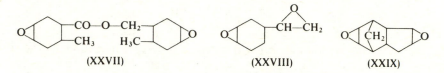

(XXVII)   (XXVIII)   (XXIX)

These resins may be cured by means similar to those described for the bisphenol A-epichlorhydrin resins but generally acid anhydrides are the preferred curing agents. Amines are less effective because of low reactivity and apparent aminolysis of ester linkages, if they are present (as in XXVII). Even as catalysts for anhydride curing agents, tertiary amines are not very effective with these resins. Thus a hydroxyl compound, such as ethylene glycol, is often added as initiator. This reacts quickly with the anhydride to give carboxyl groups which then react with epoxy groups in the manner described in section 18.3.3(c).

The cyclic aliphatic resins have a more compact structure than the bisphenol A-epichlorhydrin resins and thus give cured products in which the cross-links are closer together. This generally leads to higher heat distortion temperatures and increased brittleness. The products also have better tracking behaviour than the bisphenol A-epichlorhydrin epoxies. The latter decompose in the presence of a high temperature arc to produce carbon which acts as a conductor and leads to insulation failure. The cyclic aliphatic epoxies, on the other hand, oxidize to volatile products which do not cause tracking.Cyclic aliphatic epoxy resins have thus found use in such applications as heavy duty electrical castings and laminates.

(b) Acyclic aliphatic epoxies

Two types of acyclic aliphatic epoxy resins are commercially available, namely epoxidized oils and epoxidized diene polymers.

Drying and semi-drying oils such as linseed and soybean oils are unsaturated (see section 11.3.2(c)) and treatment with peracetic acid leads to epoxidation:

$$\text{\Large$\diagup$}C{=}C\text{\Large$\diagdown$} \ + \ H_3C{-}\overset{\displaystyle O}{\overset{\|}{C}}{-}O{-}OH \ \longrightarrow \ \text{\Large$\diagup$}C{\overset{O}{\diagdown}\diagup}C\text{\Large$\diagdown$} \ + \ H_3C{-}\overset{\displaystyle O}{\overset{\|}{C}}{-}OH$$

Epoxidized oils find use primarily as plasticizers/stabilizers for poly(vinyl chloride).

Typical of epoxidized diene polymers are the products obtained by treatment of polybutadiene (Chapter 20) with peracetic acid. Such materials contain a variety of structural units as illustrated in the following representation of a polymer segment:

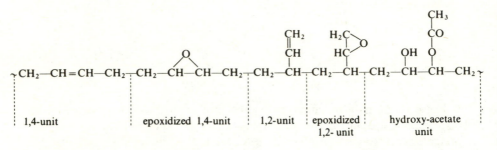

| 1,4-unit | epoxidized 1,4-unit | 1,2-unit | epoxidized 1,2- unit | hydroxy-acetate unit |

Epoxidized dienes are not very reactive towards amines but may be cross-linked with anhydride hardeners to give products with high heat distortion temperatures.

(c) Glycidyl amine epoxies

Various epoxy resins obtained by glycidylation of amines with epichlorhydrin are available. Two examples are N, N, N', N'-tetraglycidyl-4,4'-diaminodiphenylmethane (XXX) prepared from 4,4'-diaminodiphenylmethane and triglycidyl isocyanurate (XXXI) prepared from cyanuric acid. These multifunctional epoxy resins give highly cross-linked structures which have enhanced heat resistance and chemical resistance.

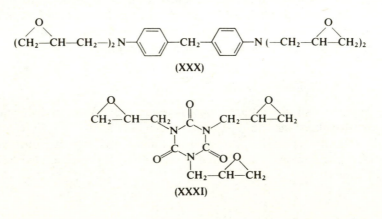

(XXX)

(XXXI)

REFERENCES

1. Leibnitz, E. and Naumann, K. (1958) *Ind. Chem. Tech.*, **3**, 5 (1951); ((1968) **RAPRA** Translation 1544).
2. Hutz, C. E. (1964) in *Manufacture of Plastics*, (ed. W. M. Smith), Reinhold Publishing Corporation, New York, ch. 13.
3. Shokal, E. C. *et al.* (1951) U. S. Patent 2575558; (1953) 2643239.
4. Shechter, L. and Wynstra, J. (1956) *Ind. Eng. Chem.*, **48**, 86.
5. Soldatos, A. C. and Burhans, A. S. (1967) *Ind. Eng. Chem. Product Research and Development*, **6**, 205.
6. Shechter, L. *et al.* (1956) *Ind. Eng. Chem.*, **48**, 94.
7. Horne, W. H. and Shriner, R. L. (1932) *J. Am. Chem. Soc.*, **54**, 2925.
8. Taniaka, Y. and Kakiuchi, H. (1963) *J. Appl. Polymer Sci.*, **7**, 1063; (1963) **7**, 2951.
9. Fischer, R. F. (1960) *Ind Eng. Chem.*, **52**, 321.
10. Potter, W. G. (1970) *Epoxide Resins*, Iliffe Books, London, ch. 4.
11. Meath, A. R. (1968) in *Epoxy Resin Technology*, (ed. P. F. Bruins), Interscience Publishers, New York, ch. 3.

BIBLIOGRAPHY

Skeist, I. (1958) *Epoxy Resins*, Reinhold Publishing Corporation, New York.

Lee, H. and Neville, K. (1967) *Handbook of Epoxy Resins*, McGraw-Hill Book Company, New York.

Bruins, P. F. (ed.) (1968) *Epoxy Resin Technology*, Interscience Publishers, New York.

Potter, W. G. (1970) *Epoxide Resins*, Iliffe Books, London.

Ranney, M. W. (1977) *Epoxy Resins and Products*, Noyes Data Corporation, Park Ridge.

DiStasio, J. I. (ed.) (1982) *Epoxy Resin Technology*, Noyes Data Corporation, Park Ridge.

Bauer, R. S. (ed.) (1983) *Epoxy Resin Chemistry II*, American Chemical Society, Washington.

POLYSULPHIDES

19.1 SCOPE

Polysulphides of commercial interest are alkyl polysulphides with the general structure

$$[-R-S_x-]_n$$

where x is commonly in the range 2–4. These polymers find limited use as elastomers.

19.2 DEVELOPMENT

The first polysulphide elastomer was discovered in 1924 by Patrick during an attempt to prepare ethylene glycol by hydrolysis of ethylene dichloride in the presence of sodium polysulphides. Commercial production was begun in 1929 by the Thiokol Chemical Corp. (USA). The current consumption of polysulphide elastomers, compared to that of other elastomers, is very small, use being restricted mainly to special applications involving resistance to oils and solvents.

19.3 RAW MATERIALS

The most widely used method for the preparation of polysulphide polymers is the reaction of sodium polysulphides with alkyl dichlorides:

$$n\text{Cl}-\text{R}-\text{Cl} \ + \ n\text{Na}_2\text{S}_x \longrightarrow [-\text{R}-\text{S}_x-]_n \ + \ 2n\text{NaCl}$$

19.3.1 Sodium polysulphides

Sodium polysulphides have the general formula Na_2S_x where x can be 2–6. All of these compounds have been prepared in the solid state, in which they are well defined. For the preparation of polymers, however, aqueous solutions of sodium polysulphides are used and such solutions contain equilibrium mixtures of anions ranging from S^{2-} to S_6^{2-}. The equilibration is effected by nucleophilic displacement interchange at points along the poly-sulphide chain, e.g.

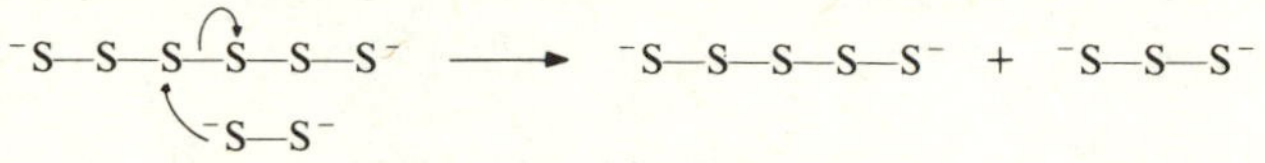

$$^-S-S-S-\overset{\frown}{S}-S-S^- \longrightarrow \ ^-S-S-S-S-S^- \ + \ ^-S-S-S^-$$
$$\underset{^-S-S^-}{\searrow}$$

The average number of sulphur atoms per polysulphide unit in a mixture is known as the *rank* of the system.

Sodium polysulphide solutions are most commonly prepared by treating sulphur with aqueous sodium hydroxide:

$$6NaOH \ + \ (2x+1)S \longrightarrow 2Na_2S_x \ + \ 3H_2O \ + \ Na_2SO_3$$

The following reactions also occur:

$$Na_2SO_3 \ + \ S \longrightarrow Na_2S_2O_3$$
$$S_x^{2-} \ + \ H_2O \rightleftharpoons HS_x^- \ + \ OH^-$$

A slight excess of sodium hydroxide is used to ensure complete reaction of all the sulphur in the system. Unreacted sodium hydroxide is then neutralized with sodium hydrosulphide:

$$NaOH \ + \ NaSH \longrightarrow Na_2S \ + \ H_2O$$

This last step is carried out in order to minimize hydrolysis of chloride in the subsequent preparation of polysulphide polymers.

The sodium polysulphide solutions used for the preparation of poly-sulphide elastomers generally have a rank of approximately 2; for elastomers with a high sulphur content which have especially good solvent resistance, a rank of about 4 is used. In general, the rank of a polymer is close to that of the sodium polysulphide solution from which it is formed.

It may be noted that chemical and physical evidence shows that the sulphur chains in inorganic and organic polysulphides are unquestionably linear and not branched as was once believed.

19.3.2 Alkyl dichlorides

Although a number of alkyl dichlorides can be used to prepare polysulphide elastomers, only a few have been utilized commercially. Ethylene dichloride was used originally but had the disadvantage of giving products with objectionable odour during processing. Ethylene dichloride now finds little use except for the preparation of elastomers with high sulphur content. At least 90% of current production is based on bis(2-chloroethyl) formal, obtained by the following route from ethylene oxide (section 9.4.1(a)):

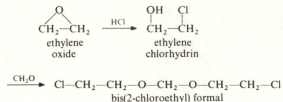

This dichloride gives elastomers with good all-round properties and little odour. In some cases, a small proportion (0.5–2% molar) of a trihalide (e.g. 1,2,3-trichloropropane) is included in the system to introduce branching and cross-linking.

19.4 PREPARATION

The base polysulphide polymers are usually prepared by a suspension process. Typically, the aqueous sodium polysulphide (about 20% solids) is heated to 80°C in a stirred reactor and an alkyl aryl sulphonate (surfactant) is added together with aqueous sodium hydroxide and magnesium chloride. (The latter reagents form magnesium hydroxide which serves as a nucleating agent.) This system leads to a dispersion which does not coagulate during polymerization but which settles readily when stirring is stopped, so that the polymer particles can be washed easily. The bis(2-chloroethyl)formal is added slowly over 2 hours with stirring; the reaction is exothermic and cooling is applied to keep the temperature at 90°C. The mixture is then heated for 1–2 hours to complete reaction. In order to assist in greater utilization of the alkyl halide, a stoichiometric deficiency (about 20%) is used in the process. Sometimes a further 20% excess of sodium polysulphide is added toward the end of the process to ensure complete reaction of the halide; this operation is known as 'toughening' since it gives a polymer of increased molecular weight. Finally, the resulting polymer dispersion is washed with hot water. Depending on the next step, various types of products are commercially available. Before these products are discussed it is convenient to consider the reactions involved in the polymerization process described above.

The basic reaction in the polymerization process is a nucleophilic displacement of halide by polysulphide anion, e.g. with disulphide:

$$^-S{-}S^- \longrightarrow R{-}Cl \longrightarrow \ ^-S{-}S{-}R{-}Cl \ + \ Cl^-$$

Since an excess of sodium polysulphide is present in the system, the above reaction leads to low molecular weight polymers which are predominately polysulphide-terminated:

$$nCl—R—Cl \;+\; (n+1)S_2^{2-} \;\longrightarrow\; {}^-S{-}{\Big[}S{-}R{-}S{\Big]}_n{-}S^- \;+\; 2nCl^-$$

Of course, this equation represents an ideal case; in practice, polysulphide anions of varying ranks are involved and the polymer contains a distribution of polysulphide groups. Some unreacted chloride terminals are present together with hydroxyl terminals, formed by hydrolysis:

$$\sim S{-}S{-}R{-}S{-}S{-}R{-}Cl \;+\; OH^- \;\longrightarrow\; \sim S{-}S{-}R{-}S{-}S{-}R{-}OH \;+\; Cl^-$$

Besides reacting with halide, as shown above, the polysulphide anions are continually interchanging with the polysulphide groups present in the polymeric product, e.g.

$$\sim S{-}S{-}R{-}S{-}S{-}R{-}S{-}S{-}R \sim \;\rightleftharpoons$$

$${}^-S{-}S^-$$

$$\sim S{-}S{-}R{-}S{-}S{-}S^- \;+\; {}^-S{-}R{-}S{-}S{-}R\sim \qquad \text{(i)}$$

It may be noted that if the interchange occurs near a terminal carrying a hydroxyl group a low molecular weight hydroxyl-fragment is formed:

$$\sim S{-}S{-}R{-}S{-}S{-}R{-}OH \;\rightleftharpoons\; \sim S{-}S{-}R{-}S{-}S{-}S^- \;+\; {}^-S{-}R{-}OH$$

$${}^-S{-}S^-$$

This fragment is water-soluble and passes into the aqueous phase. The significance of this point is referred to later.

Analogous interchange reactions may also occur between an anionic terminal of one polymer and polysulphide groups in another, e.g.

$$\sim S{-}S{-}R{-}S{-}S{-}R \sim \;\rightleftharpoons\; \sim S{-}S{-}R{-}S{-}S{-}R{-}S{-}S{-}R{-}S{-}S\sim$$
$$\sim S{-}S{-}R{-}S{-}S{-}R{-}S^- \qquad\qquad\qquad +\; {}^-S{-}R \sim$$

$$\sim S{-}S{-}R{-}S{-}S{-}R{-}OH \;\rightleftharpoons\; \sim S{-}S{-}R{-}S{-}S{-}R{-}S{-}S{-}R{-}S{-}S\sim$$
$$\sim S{-}S{-}R{-}S{-}S{-}R{-}S^- \qquad\qquad\qquad +\; {}^-S{-}R{-}OH$$

The final stage in the preparation of the polymer consists of several washing and decantation treatments. During these operations the soluble hydroxyl-terminated fragments are removed; the polymer becomes more completely mercaptide-terminated and increasingly able to participate in the equilibrium (i) shown above. Also, the washing process serves to remove inorganic polysulphide anions from the system and as a result the equilibrium (i) is driven to the left, i.e. the average molecular weight of the polymer is increased. If reaction with excess of polysulphide and the washing process do not yield a product with a high enough molecular weight, the polymer dispersion can be retreated with fresh polysulphide solution to completely convert any remaining chloride terminals and to further the solubilization of remaining hydroxyl groups, thereby allowing the equilibrium (i) to move to the desired position. In general practice, molecular weights of at least 500 000 are readily achieved. It may be noted that this polymerization differs from conventional condensation polymerization in that a stoichiometric excess of one of the reactants is used to obtain a high molecular weight product.

As mentioned earlier, various types of commercial products are prepared from the base polymer dispersion obtained by the process outlined above. These various products are described below.

19.4.1 Dispersions

For a few special applications such as solvent resistant coatings for concrete, the alkyl polysulphide polymer dispersion is used directly without further treatment.

19.4.2 Solid polymers

Two types of solid product are prepared commercially, viz hydroxyl-and thiol-terminated polymers. The significance of the nature of the end-groups lies in the means whereby the polymer may be cured (see section 19.5).

(i) *Hydroxyl-terminated polymers.* The aqueous polymer dispersion is co-agulated with sulphuric acid and the resultant crumb is washed and dried. As indicated earlier, the hydroxyl terminal groups in the polymer arise by hydrolysis of chloride terminals; a few unreacted chloride terminals may also be present.

(ii) *Thiol-terminated polymers.* The aqueous polymer dispersion is heated at about 80°C for 1 hour with a mixture of sodium hydrosulphide and sodium sulphite. The product is washed by decantation and then coagulated by acidification. In this treatment, the hydrosulphide splits the polymer at the polysulphide links by another interchange reaction. In the case of a di-sulphide link, fragments with mercaptide and thiothiol terminals are formed:

$$\sim S—S—R—S—S—R—S—S—R\sim \ + \ SH^- \rightleftharpoons$$
$$\sim S—S—R—S—SH \ + \ {}^-S—R—S—S—R\sim$$

If this were the only reaction involved, the washing process would drive the equilibrium to the left and the original polymer would be regenerated. However, the sodium sulphite strips sulphur from the thiothiol terminal to produce a thiol group:

$$\sim S—S—R—S—SH \ + \ SO_3^{2-} \longrightarrow \ \sim S—S—R—SH \ + \ S_2O_3^{2-}$$

Thus reversal of the equilibrium is prevented. At the same time the sodium sulphite also removes high rank sulphur present in polysulphide groups in the polymer backbone to give disulphide groups, e.g.

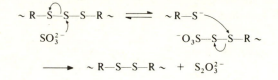

When the system is acidified, the mercaptide terminals formed in the splitting process are converted to thiol groups:

$$\sim R\!-\!S\!-\!S\!-\!R\!-\!S^- \;+\; H^+ \longrightarrow \; \sim R\!-\!S\!-\!S\!-\!R\!-\!SH$$

Thus the final product is essentially completely thiol-terminated. Clearly the molecular weight of the product is lower than that of the parent polymer but nevertheless, molecular weights of up to 50 000–100 000 can be obtained.

19.4.3   Liquid polymers

By increasing the proportion of sodium hydrosulphide and sodium sulphite in the splitting process described above for the preparation of solid thiol-terminated polymers it is possible to obtain low molecular weight polymers which are liquid. Commercial liquid polymers generally have molecular weights in the range 1000–4000. These polymers are, of course, also thiol-terminated. Liquid polymers make up about 80% of the current output of alkyl polysulphide polymers, being used for sealing and caulking.

19.5   CURING PROCESSES

As indicated above, the means whereby elastomeric properties are developed in polysulphide polymers are governed by the nature of the terminal groups in the base polymers.

19.5.1   Cure of hydroxyl-terminated polymers

The use of hydroxyl-terminated polysulphide polymers is limited to applications where outstanding solvent resistance is required. The reason for this is that the best solvent resistance is shown by polymers with a high sulphur content but these cannot be prepared with thiol end-groups by the hydrosulphide/sulphite treatment because sulphite removes high rank sulphur from the polymer backbone (see section 19.4.2). Thus since polysulphide polymers of high sulphur content cannot be subjected to this after-treatment, they are supplied as hydroxyl-terminated polymers of high molecular weight.

Because of their high molecular weight, the hydroxyl-terminated polymers are difficult to process and a 'chemical plasticizer' is usually added to facilitate milling. The plasticizer is a disulphide such as benzothiazyl disulphide (mercaptobenzothiazole disulphide, MBTS); interchanges of the following type lead to a reduction in molecular weight:

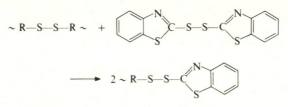

The material is then cured by heating with a metal oxide, generally zinc oxide, whereby the polymer is reformed:

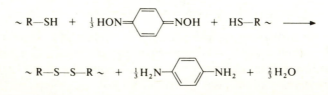

The metal oxide also probably effects a further increase in molecular weight by coupling hydroxyl terminals. It is to be noted that the final product is still an essentially linear polymer; there is no cross-linking and 'vulcanizates' have poor compression set resistance.

19.5.2 Cure of thiol-terminated polymers

Thiol-terminated polymers are by far the most important type of polysulphide polymer manufactured. They may be readily cured by oxidative coupling of the thiol terminals:

$$\sim SH \ + \ [O] \ + \ HS\sim \ \longrightarrow \ \sim S—S\sim \ + \ H_2O$$

A wide variety of reagents may be used to effect this reaction.

The standard curing system for the solid polymer is p-quinone dioxime (GMF), the reaction of which may be written as:

$$\sim R—SH \ + \ \tfrac{1}{3}HON{=}\!\!\left\langle\!\!\bigcirc\!\!\right\rangle\!\!{=}NOH \ + \ HS—R\sim \ \longrightarrow$$

$$\sim R—S—S—R\sim \ + \ \tfrac{1}{3}H_2N{-}\!\!\left\langle\!\!\bigcirc\!\!\right\rangle\!\!{-}NH_2 \ + \ \tfrac{2}{3}H_2O$$

For the liquid polymers, lead peroxide is the most widely used curing agent; the principal reaction is as follows:

$$\sim R—SH \ + \ PbO_2 \ + \ HS—R\sim \ \longrightarrow \ \sim R—S—S—R\sim \ + \ PbO \ + \ H_2O$$

An undesirable side-reaction which may occur is the formation of lead mercaptide links:

$$\sim R\!-\!SH \;+\; PbO \;+\; HS\!-\!R\sim \;\longrightarrow\; \sim R\!-\!S\!-\!Pb\!-\!S\!-\!R\sim \;+\; H_2O$$

These bonds readily participate in interchange reactions with polysulphide groups and result in products with poor stress relaxation characteristics (see later). The inclusion of sulphur in the system reduces the number of lead mercaptide links by the formation of lead sulphide:

$$\sim R\!-\!S\!-\!Pb\!-\!S\!-\!R\sim \;+\; S \;\longrightarrow\; \sim R\!-\!S\!-\!S\!-\!R\sim \;+\; PbS$$

An important difference between solid hydroxyl- and thiol-terminated polymers is that since the latter are of relatively low molecular weight it is possible to prepare base polymers which are slightly branched (by including a trihalide in the polymerization system) but still easily handled. When the branched material is cured a cross-linked product is obtained and this has improved compression set resistance. In contrast, if a trihalide is included in the polymerization system and the resulting polymer is not cleaved by a hydrosulphide/sulphite treatment, the product is too tough to be worked on a rubber mill.

Liquid thiol-terminated polymers are also frequently used in conjunction with epoxy resins (Chapter 18). Interaction of the two materials leads to chain extension:

$$\sim R\!-\!SH \;+\; \overset{O}{\overset{\triangle}{CH_2\!-\!CH}}\sim \;\longrightarrow\; \sim R\!-\!S\!-\!CH_2\!-\!\overset{OH}{\underset{|}{CH}}\sim$$

The reaction is accelerated by tertiary amines, which may function by either (i) forming mercaptide ions or (ii) opening the epoxy ring:

(i) $$R_3'N \;+\; \sim R\!-\!SH \;\rightleftharpoons\; R_3'\overset{+}{N}H \;+\; \sim R\!-\!S^-$$

$$\sim R\!-\!S^- \;+\; \overset{O}{\overset{\triangle}{CH_2\!-\!CH}}\sim \;\longrightarrow\; \sim R\!-\!S\!-\!CH_2\!-\!\overset{O^-}{\underset{|}{CH}}\sim$$

$$\xrightarrow{\;R_3'\overset{+}{N}H\;} \sim R\!-\!S\!-\!CH_2\!-\!\overset{OH}{\underset{|}{CH}}\sim \;+\; R_3'N$$

(ii) $$R_3'N \;+\; \overset{O}{\overset{\triangle}{CH_2\!-\!CH}}\sim \;\longrightarrow\; R_3'\overset{+}{N}\!-\!CH_2\!-\!\overset{O^-}{\underset{|}{CH}}\sim$$

$$\xrightarrow{\;\sim R\!-\!SH\;} \sim R\!-\!S\!-\!CH_2\!-\!\overset{OH}{\underset{|}{CH}}\sim \;+\; R_3'N$$

Tertiary amines also catalyse interaction between the hydroxyl groups formed and remaining epoxy groups and a cross-linked structure is built up. Epoxy-polysulphide compositions have good impact strength and find use in adhesives and castings.

19.6 PROPERTIES

As indicated earlier, polysulphide elastomers are distinguished by outstanding resistance to oils and solvents. The resistance improves as the
sulphur content of the polymer increases. Products of high sulphur content
are also less permeable to vapours and gases. Polysulphide elastomers are
also noted for oxygen and ozone resistance. The polymers are attacked by
strong oxidizing acids and strong alkalis.

A characteristic of polysulphide elastomers, especially those with high
sulphur content, is poor compression set resistance. This does not arise
through oxidative degradation, as is the case with most elastomers, but
through interchange reactions at the polysulphide bonds [1]. Interchange
reactions may proceed through both radical and ionic mechanisms. In the
free radical process, initiation depends on the homolytic cleavage of a
polysulphide link, e.g. a tetrasulphide link:

$$\sim R—S—S—S—S—R\sim \;\rightleftharpoons\; 2\sim R—S—S\cdot$$

Various displacement reactions, resulting in stress relaxation, may then
occur, e.g.

$$\sim R—S—S—S—S—R\sim \;\rightleftharpoons\; \sim R—S—S—S—R\sim \;+\; R—S—S—S\cdot$$
$$\sim R—S—S\cdot$$

The radical interchange process occurs at an appreciable rate at temperatures
above about 80°C.

It appears that radicals containing more than one sulphur atom are
resonance stabilized, perhaps by formation of a 'three-electron bond' [2], e.g.

$$\sim R—\ddot{S}—S\cdot \;\longleftrightarrow\; \sim R—S\!\cdots\!S$$

Obviously, radicals of the type $\sim R—S\cdot$ cannot be stabilized in this way. It
may therefore be predicted that the ease of homolytic cleavage of polysulphide links is in the order disulphide < trisulphide < tetrasulphide ≃ higher
sulphides, since the following radicals are involved:

$\sim R—S—S—R\sim \;\longrightarrow\; 2\sim R—S\cdot$ (neither fragment stabilized)

$\sim R—S—S—S—R\sim \;\longrightarrow\; \sim R—S—S\cdot \;+\; \sim R—S\cdot$ (one fragment stabilized)

$\sim R—S—S—S—S—R\sim \;\longrightarrow\; 2\sim R—S—S\cdot$ (both fragments stabilized)

In agreement with this prediction, the dissociation energies of the three types
of bond have been calculated to be 289, 193 and 155 kJ/mole respectively [3].
Thus the poor compression set resistance of elastomers with high sulphur
content results from the ease of initiation of interchange reactions.

Ionic interchange reactions may also contribute to stress relaxation and
these can occur at disulphide bonds as well as at higher polysulphide bonds.

Such reactions may be initiated by ionic impurities in the elastomer or by mercaptides formed by the use of metallic compounds as curing agents. In the latter case, the following reaction may be envisaged:

$$\sim R{-}S{-}S{-}R\sim \quad + \quad \sim R'{-}S^- \longrightarrow \sim R{-}S{-}S{-}R'\sim \quad + \quad \sim R{-}S^-$$

Ionic reactions of this type may occur even at room temperature.

Compared to many other elastomers, polysulphide elastomers have rather low tensile strength and abrasion resistance. Polysulphide elastomers have good low temperature flexibility; this feature is determined largely by the length of the chain between the polysulphide linkages and products based on bis(2-chloroethyl)formal are more satisfactory than those from ethylene dichloride. The sulphur content has an effect on the thermal stability of elastomers since, as discussed above, polysulphide links are more readily homolytically cleaved than disulphide links. The presence of mercaptide bonds also has an adverse effect on thermal stability by leading to a high weight loss on heat ageing. Volatilization is thought to occur by a 'backbite' interchange:

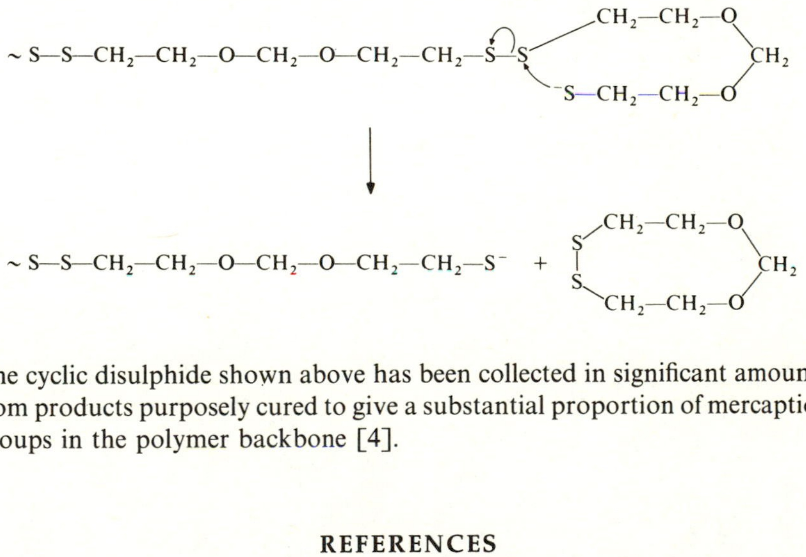

The cyclic disulphide shown above has been collected in significant amounts from products purposely cured to give a substantial proportion of mercaptide groups in the polymer backbone [4].

REFERENCES

1. Tobolsky, A. V. and MacKnight, W. J. (1965) *Polymeric Sulfur and Related Polymers*, Interscience Publishers, New York, ch. 4.
2. Fairbrother, F., Gee, G. and Merall, G. T. (1952) *J. Polymer Sci.*, **74**, 1023.
3. Pickering, T. L., Saunders, K. J. and Tobolsky, A. V. (1967) *J. Am. Chem. Soc.*, **89**, 2364; (1968) in *The Chemistry of Sulfides*, (ed. A. V. Tobolsky), Interscience Publishers, New York, p. 61.
4. Berenbaum, M. B. (1968) in *The Chemistry of Sulfides*, (ed. A. V. Tobolsky), Interscience Publishers, New York, p. 221.

BIBLIOGRAPHY

Fettes, E. M. (1961) in *Organic Sulfur Compounds*, Vol. 1, (ed. N. Kharasch), Pergamon Press, Oxford, ch. 24.

Berenbaum, M. B. (1964) in *Chemical Reactions of Polymers*, (ed. E. M. Fettes), Interscience Publishers, New York. ch. 7 and 8.

Tobolsky, A. V. and MacKnight, W. J. (1965) *Polymeric Sulfur and Related Polymers*, Interscience Publishers, New York.

Tobolsky, A. V. (ed.) (1968) *The Chemistry of Sulfides*, Interscience Publishers, New York.

Gobran, R. H. and Berenbaum, M. B. (1969) in *Polymer Chemistry of Synthetic Elastomers, Part II*, (eds. J. P. Kennedy and E. G. M. Tornqvist), Interscience Publishers, New York, ch. 8.

20

POLYDIENES

20.1 SCOPE

Polydienes constitute an extremely important group of polymers. This group comprises natural rubber (and related natural materials) and its derivatives together with the products of polymerization and copolymerization of conjugated dienes. These various materials make up the contents of this chapter. The importance of the polydienes lies in the fact that they encompass the bulk of the commercial elastomers currently in use.

20.2 NATURAL RUBBER

20.2.1 Development

When the Europeans first explored Central and South America, they found that the natives coagulated the exudation (latex) of certain trees to make balls which bounced. Also, the latex was spread on fabric and allowed to dry to give a product from which waterproof shoes and flexible bottles were made. In 1770, Priestley named the solid material 'rubber' after finding it could be used to rub out pencil marks; the name has remained although this application is a very minor one. It may be noted that the term 'rubber' is now commonly used also as a generic name for any material with rubber-like properties; to avoid confusion the latter materials may be called 'elastomers'.

There were several attempts early in the 19th century to produce commercial articles from natural rubber, but few of the products were successful. Two major problems were encountered. Firstly, only the solid rubber was available for manufacturing processes since the latex was not stable enough to be transported. The coagulated material was highly elastic and could not be shaped by moulding or extrusion; thus recourse had to be made to somewhat laborious solution methods of fabrication. In 1820, Hancock discovered that if rubber is masticated it becomes plastic or capable of flow and hence easier to fabricate. However, there remained the second problem which was that rubber products became hard and stiff in cold weather and soft and sticky in hot weather; the situation was further aggravated by the poor ageing properties shown by the products. Thus by 1830 the world consumption of natural rubber reached only about 100 tons. In 1839, Goodyear found that rubber which had been heated with sulphur retained its elasticity over a wider range of temperature than the raw material and had better resistance to solvents. Subsequently, Hancock also found that when plastic masticated rubber was heated with sulphur a strong, elastic material was regenerated. The interaction of rubber and sulphur was termed 'vulcanization' by Brockedon, a friend of Hancock. Although the discovery of Hancock was subsequent to that of Goodyear (indeed, it was initiated by examination of some of Goodyear's samples), Hancock obtained a British patent in 1843, one year before Goodyear received a USA patent for the same invention. The discoveries of mastication and vulcanization represent the foundations of the rubber industry.

The great improvement which vulcanization brought about in the properties of rubber led to an increased demand for rubber products such as raincoats, footwear, solid tyres for carriages and hose. The pneumatic tyre was patented by Dunlop in 1888 and the development of the bicycle and automobile resulted in greater requirements of rubber. This demand led to high prices and ruthless exploitation of all sources of wild rubber, production of which reached a maximum of about 100 000 tons per year in 1910. These developing conditions fostered investigations into establishing plantations. In 1876, Wickham collected a large number of seeds of *Hevea brasiliensis* from the Amazon Basin and sent them to the Royal Botanical Gardens at Kew whence seedlings were sent to Ceylon. Small trees were transferred from here a few years later to Malaya (Malaysia) and the Dutch East Indies (Indonesia). It took several years before the production of plantation rubber reached significant proportions; by 1910 it accounted for only about 10% of the total world output. From this date, growth was phenomenal and by 1920 the plantations produced 90% of the total output; wild rubber now has negligible commercial importance. At the present time, the world production of natural rubber is about 4 million tonnes per annum.

20.2.2 Production

On the plantations, the Hevea trees are 'tapped', i.e. the bark is cut and the latex which flows is collected in small cups. The latex is a stable dispersion of rubber in water, proteinous material being the dispersing agent; a typical analysis might be as follows:

Rubber hydrocarbon	35%
Non-rubber solids (proteins, lipids, quebrachitol, inorganic salts, etc)	5
Water	60

The bulked latex is then diluted to a solids content of about 15% and coagulated by the addition of acetic or formic acid. The coagulum is sheeted on rollers, washed and dried. Two types of drying process are used. In the first method, the sheets of rubber are dried in wood smoke at about 60°C for 4 days; the resultant brown product is known as *smoked sheet*. In the second method, the sheets are dried in air at about 40°C for 7 days; the product is *pale crepe*. Compared to smoked sheet, pale crepe is more expensive to produce (since the drying conditions are more critical and require better control) and is more prone to subsequent mould growth; however, it has the advantage of being light coloured.

About 10% of the latex produced is exported, as such, from the plantations, Commonly, field latex is centrifuged to give *concentrated latex* which has a solids content of about 60%. To preserve such latices over long periods of time, ammonia and other bactericides are added.

20.2.3 Structure

The principal component of the raw rubber of commerce is the single polymeric material known as *rubber hydrocarbon*. This is present to the extent of about 94%; the balance consists mainly of proteins, lipids, quebrachitol (monomethyl ether of hexahydroxycyclohexane) and inorganic salts. Pure rubber hydrocarbon may be obtained from raw rubber by fractional precipitation from dilute solutions. If, for example, methanol is added to a solution of raw rubber in toluene, the first fraction carries down most of the nitrogenous impurities and subsequent fractions consist of fairly pure rubber hydrocarbon; the other impurities remain in solution.

Although the minor constituents of rubber have some effect on the bulk properties, it is clearly the rubber hydrocarbon which determines the major features. Accordingly, the elucidation of the structure of rubber hydrocarbon has received the attention of many workers over the past 100 years or so. In 1826, Faraday established the empirical formula of rubber hydrocarbon as C_5H_8 and in 1860 Williams obtained isoprene (I) by dry distillation of rubber. In 1879, Bouchardat converted isoprene to a rubber-like material by treatment with hydrochloric acid and concluded that isoprene was the 'mother substance' of natural rubber. In 1888, Gladstone and Hibbert showed that rubber was unsaturated; under normal conditions, 2 atoms of bromine were added for each C_5H_8 unit.

Thus by the turn of the century it was generally accepted that rubber hydrocarbon could be represented as $(C_5H_8)_n$ but the way in which the units were joined was unknown. Important indications of the mode of linkage were obtained by Harries in 1904 during the course of extensive investigations into the action of ozone on olefins. It had been found that ozone reacts readily with unsaturated compounds to form products which, on hydrolysis, yield aldehydes and/or ketones, depending on the structure of the starting material, e.g.

$$RHC=CR'R'' \xrightarrow[H_2O]{O_3} RHC=O \;\; + \;\; R'R''C=O \;\; + \;\; H_2O_2$$

In practice, acids are also formed through oxidation of aldehyde by hydrogen peroxide. When Harries subjected rubber to ozonolysis, he obtained high yields of levulinic aldehyde (II) and levulinic acid (III). These results are readily explained if it is assumed that rubber hydrocarbon consists of isoprene units joined head-to-tail by 1,4-links (see also sections 1.4.2(a) and 1.6.3):

$$\underset{\text{(I)}}{H_2C{=}\overset{\overset{\displaystyle CH_3}{|}}{C}{-}CH{=}CH_2} \qquad\qquad \underset{\text{(II)}}{H_3C{-}CO{-}CH_2{-}CH_2{-}CHO}$$

$$\underset{\text{(III)}}{H_3C{-}CO{-}CH_2{-}CH_2{-}COOH}$$

$$\underset{\text{(IV)}}{\sim CH_2{-}\overset{\overset{\displaystyle CH_3}{|}}{C}{=}CH{-}CH_2{-}CH_2{-}\overset{\overset{\displaystyle CH_3}{|}}{C}{=}CH{-}CH_2\sim} \xrightarrow{O_3,\,H_2O}$$

$$\sim CH_2{-}\overset{\overset{\displaystyle CH_3}{|}}{C}{=}O \quad O{=}CH{-}CH_2{-}CH_2{-}\overset{\overset{\displaystyle CH_3}{|}}{C}{=}O \quad O{=}CH{-}CH_2\sim$$

If the isoprene units were joined in some other way, ozonolysis would not yield levulinic aldehyde (and acid). It may be noted that Harries took his results to indicate that rubber hydrocarbon molecules were cyclic structures of low molecular weight which formed aggregates by means of ill-defined forces. This view was in accordance with the then current ideas concerning the nature of polymers. It was not until after 1920, when the concept of the macromolecule was introduced, that rubber hydrocarbon was accepted as consisting of linear polymers composed of isoprene units linked together by normal carbon-carbon bonds in a 1,4-, head-to-tail manner (see also section 1.3). Modern techniques of infrared and NMR spectroscopy have since verified the essential correctness of these conclusions; it has been shown that in rubber hydrocarbon at least 99% of the links between isoprene units are 1,4-bonds.

Examination of the structure of rubber hydrocarbon (IV) indicates the possibility of geometrical isomerism and Staudinger suggested that rubber contains mainly cis-1,4-bonds whilst gutta percha (section 20.2.7) has mainly trans-1,4-bonds. This assignment was made mainly on the basis of the lower density of rubber; it was well known that for simple molecules the cis-isomer generally has a lower density than the trans-isomer and the same relationship was assumed to apply to polymeric materials. That rubber and gutta percha are respectively *cis*- and *trans*-geometrical isomers has subsequently been confirmed by X-ray analysis. It will be appreciated, of course, that ozonolysis of the two materials leads to identical products.

The molecular weight of rubber hydrocarbon is peculiarly difficult to measure and the results of various workers are not always in agreement. It is certain that the range of molecular weights present in the original latex is wide and thus the molecular weight distribution of a test sample is very dependent on the method of isolation. Furthermore, rubber hydrocarbon is rapidly degraded by oxygen and even on standing *in vacuo* in the dark some cross-linking appears to occur. The previous history of a sample is therefore likely to have substantial effects on molecular weight. In one investigation, in which pains were taken to eliminate degradation, an average molecular weight (M_w) of 1.3×10^6 was obtained [1].

Thus by the means outlined above, it has been established that natural rubber is essentially *cis*-1,4-polyisoprene. The final proof in this long endeavour may be considered to have been given in 1954 when the stereospecific polymerization of isoprene was achieved and 'synthetic natural rubber' was obtained (see section 20.3).

20.2.4 Vulcanization

As indicated in section 20.2.1, the term 'vulcanization' originally referred to the treatment of natural rubber with sulphur whereby the plasticity of the rubber was reduced whilst the elasticity was increased. The term now also includes other reagents which have similar effects, although sulphur has remained by far the most important vulcanizing agent. Further, the term now encompasses other elastomers besides natural rubber. From a chemical point of view, vulcanization is principally a process of cross-linking whereby the discreet linear rubber chains are converted into a three-dimensional network. In this way, the chains are prevented from slipping past one another when the sample is stressed and they return to their original positions when the stress is removed.

Technologically, the most important methods of vulcanizing natural rubber are by the use of accelerated sulphur systems and organic peroxides and only these methods are considered in this section.

(a) Accelerated sulphur vulcanization

The interaction between diene rubbers and sulphur is exceedingly complex and has been the subject of extensive investigation. In this book it is possible to give only an extremely brief account of the main conclusions of this work; more detailed reviews may be found elsewhere [2–5].

The reaction between natural rubber and elemental sulphur is relatively slow at the temperatures normally used for vulcanization, namely about 150°C. The process is also inefficient in that between about 40 and 55 atoms of sulphur become chemically combined with the rubber for each cross-link formed. Only 6 to 10 of these sulphur atoms lie in the actual cross-link; the remainder are distributed along the main chains as cyclic sulphide units. This extensive chemical modification of the main polymer chains greatly affects the physical properties of the rubber by inhibiting strain-induced and low-temperature crystallization and by decreasing resilience.

The rate of reaction between natural rubber and sulphur may be substantially increased by the addition of one or more 'accelerators'. Accelerating effects are shown by a wide range of organic compounds, the more important of which are the following:

(i) *Thiazoles*, e.g. mercaptobenzothiazole (MBT)(V), benzothiazyl disulphide (mercaptobenzothiazole disulphide, MBTS)(VI) and sulphenamides (VII).

(ii) *Thiurams*, e.g. tetramethylthiuram disulphide (TMTD)(VIII).

(iii) *Dithiocarbamates*, e.g. zinc dimethyldithiocarbamate $(ZD_MC)(IX)$.

(iv) *Guanidines*, e.g. diphenylguanidine (DPG)(X).

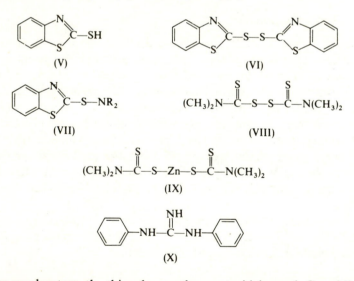

Of these accelerators, the thiazoles are the most widely used. Guanidines are chiefly used as secondary accelerators for thiazoles.

The action of accelerators is improved by the presence of 'activators'; in practice, a combination of zinc oxide and a fatty acid is nearly always used. Specifically, the activator increases the efficiency of the vulcanizing system so that, typically, only about 5 sulphur atoms are combined for each cross-link formed. As a consequence, the vulcanizate shows improved physical pro-

perties, ageing characteristics and appearance (less sulphur bloom). The use
of accelerator-activator systems is standard practice in rubber compounding
and a typical basic tread formulation might be as follows:

Natural rubber (smoked sheet)   100   parts by weight
Carbon black (HAF)   47.5
Sulphur   3.0
Mercaptobenzothiazole   0.85
Zinc oxide   5.0
Stearic acid   3.0
(Cure time: 30 min at 140°C)

It is now supposed that accelerated sulphur vulcanization follows the
general pathway shown in Fig. 20.1. The various steps involved in this scheme
are considered below.

(i) Formation of the active sulphurating agent

In this step, the accelerators and activators interact to give a species which
then reacts with sulphur to form the sulphurating agent. In the case of the
thiazole-zinc oxide-fatty acid system, the initial step may be represented as
follows:

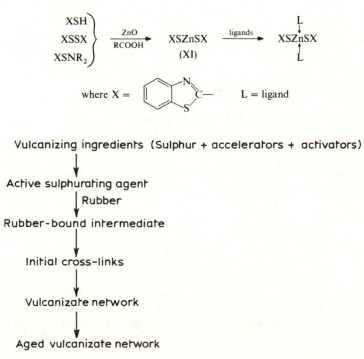

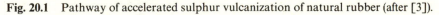

Fig. 20.1   Pathway of accelerated sulphur vulcanization of natural rubber (after [3]).

The initial zinc mercaptide (XI) is sparingly soluble in rubber but is rendered very soluble through co-ordination with nitrogen bases (either present in the raw natural rubber or added as accelerator) or zinc carboxylates; examples of complexes involving co-ordination of amine and carboxylate ligands are as follows:

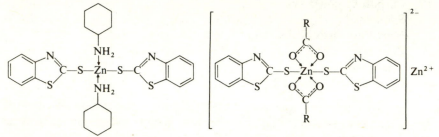

In the next stage, the zinc mercaptide complexes (represented below, for simplicity, as XSZnSX) are thought to react with sulphur to form zinc perthiomercaptides (XII) by means of a series of equilibria involving sulphur insertion and interchange:

$$\underset{XS}{\overset{\delta-}{}}\underset{Zn}{\overset{\delta++}{}}\underset{SX}{\overset{\delta-}{}} \rightleftharpoons XS-S_8-Zn-SX \xrightarrow{\ \ XSZnSX\ \ } XS-S_x-Zn-S_y-SX$$

$$\begin{array}{c} S-\hspace{-0.3em}-S \\ \diagdown\diagup \\ S_6 \end{array}$$

(XII)

The average values of x and y in the perthiomercaptides will be determined by the relative concentrations of sulphur and zinc mercaptide complex. Zinc perthiomercaptide complexes are believed to be the actual sulphurating agents in the vulcanization process.

(ii) Formation of the rubber-bound intermediate

The mechanism by which a zinc perthiomercaptide complex reacts with rubber is not known with certainty. It is possible that reaction involves nucleophilic attack of a terminal perthiomercaptide sulphur atom on an α-methylenic or α-methylic carbon atom in the rubber hydrocarbon (RH):

$$\begin{array}{c} Zn-S \\ \diagup\ \ \diagdown \\ XS-S_x\ \ \ \ S_y-X \\ \diagdown\ \ \ \diagup \\ R-H \end{array} \longrightarrow XSS_xR\ +\ ZnS\ +\ HS_yX$$

(XIII)

$$2HS_yX \xrightarrow{\ \ ZnO\ \ } XS_yZnS_yX\ +\ H_2O$$

As a result of this reaction, pendant polysulphide groups terminated by accelerator moieties are attached to the polyisoprene chain (as represented by XIII) and zinc perthiomercaptide is regenerated and is available for further reaction.

(iii) Formation of initial cross-links

The third step in the vulcanization sequence is considered to be the formation of polysulphide cross-links. The most probable route for the conversion of the rubber-bound intermediate into cross-linked polysulphides is by disproportionation reactions, involving cleavage of S–S bonds; these reactions may be initiated by mercaptide ions (XS^-) derived from the zinc mercaptide (cf. section 19.4):

$$RS_xSX \ + \ XS^- \rightleftharpoons RS_a^- \ + \ XS_bX$$

$$RS_a^- \ + \ RS_xSX \rightleftharpoons RS_zR \ + \ XS_c^- \quad etc.$$

A further possibility involves interchange between the rubber-bound intermediate and zinc perthiomercaptide followed by sulphuration of another polyisoprene chain:

$$RS_xSX \ + \ XSS_xZnS_ySX \rightleftharpoons RS_xZnS_xSX \ + \ XSS_ySX$$

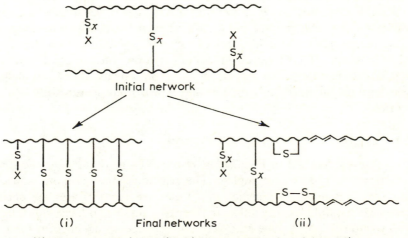

$$R{-}S_x \overset{\displaystyle Zn{-}S}{\diagdown}S_x{-}X \ \longrightarrow \ RS_xR \ + \ ZnS \ + \ HS_xX$$
$$R{-}H$$

The structure of the initial network so obtained is illustrated in Fig. 20.2.

(iv) Formation of vulcanizate network

In this step, the initial polysulphide cross-links undergo further transformations (so called 'maturing' processes). Two competing reactions are

Initial network

(i)   Final networks   (ii)

(i) Network obtained with high accelerator to sulphur ratio
(ii) Network obtained with low accelerator to sulphur ratio

Fig. 20.2 Structural features of networks formed in accelerated sulphur vulcanization (after [3], [4]).

thought to be involved, namely desulphuration and decomposition. The desulphuration process results in progressive shortening of the cross-links, leading in the limit to monosulphide cross-links. Desulphuration is effected by the zinc mercaptide complexes:

$$RS_xR \quad + \quad XSZnSX \longrightarrow RSR \quad + \quad XSS_mZnS_nSX$$

The zinc perthiomercaptide produced is able to form further cross-links.

Decomposition of polysulphide cross-links appears to be an uncatalysed thermal process and leads to cyclic mono- and disulphides, conjugated dienes and trienes, and zinc sulphide. The mechanisms of these reactions are not known, but intramolecular hydrogen transfer may be involved:

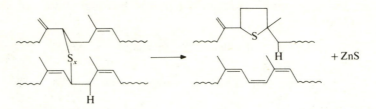

The relative extent of desulphuration and decomposition is determined by the vulcanization conditions. A high concentration of zinc mercaptide complexes will lead to rapid desulphuration of polysulphide cross-links, effectively preventing their alternative thermal decomposition. Consequently, a high concentration of accelerators and activators relative to sulphur will favour a monosulphide cross-linked network with a high degree of cross-linking. Conversely, a low concentration of accelerators and/or activators will favour a mainly polysulphide cross-linked network with a low degree of cross-linking and with modification of the main chains by cyclic sulphide and olefinic groups. The two extreme types of network structure are illustrated in Fig. 20.2. In practice, of course, an actual vulcanizate as taken from the press will have an intermediate structure and will contain features of both networks.

(v) Formation of aged vulcanizate network

The changes depicted in Fig. 20.2 may continue to occur after the formal vulcanization process is complete. The reactions may continue whilst the vulcanizate is in service, particularly if elevated temperatures are encountered. Monosulphide cross-links are thermally stable and hence vulcanizates of Type (i) (Fig. 20.2) show relatively little change on ageing. On the other hand, polysulphide cross-links are thermally unstable and vulcanizates of Type (ii) undergo reversion (loss of cross-links) and main chain modification with corresponding changes in physical properties. Polysulphide cross-links are also susceptible to interchange reactions, initiated by nucleophilic species such as mercaptide ions derived from the zinc mercaptide:

$$RS_aS_bR \quad + \quad XS^- \rightleftharpoons RS_aSX \quad + \quad RS_b^-$$

As a result of these reactions, the vulcanizate will exhibit reduced resistance to compression set and creep. It may be noted that although vulcanizates with predominantly monosulphide cross-links show better heat resistance, they may not have the excellent overall physical properties of predominantly polysulphide cross-linked networks. For this reason, a compromise is usually aimed for in practice and so-called 'semi-efficient' vulcanizing systems are used.

The effects described above which occur when the vulcanizate is aged are, of course, superimposed on any oxidative reactions which may take place (see section 20.2.5).

(b) Peroxide vulcanization

Di-*tert*-alkyl and di-*tert*-aryl peroxides (e.g. di-*tert*-butyl peroxide and di-cumyl peroxide) find some use in the production of heat-stable vulcanizates of natural rubber. However, the particular significance of these reagents is that their mode of reaction is rather straightforward and it has been possible to establish a direct quantitative relationship between the number of cross-links in a vulcanizate and its elastic modulus. This correlation is important to the theory of rubber-like elasticity.

The first step in peroxide vulcanization is thermal decomposition of the peroxide into free radicals e.g. for di-*tert*-butyl peroxide:

$$(CH_3)_3C\!-\!O\!-\!O\!-\!C(CH_3)_3 \longrightarrow 2(CH_3)_3C\!-\!O\cdot$$

$$(CH_3)_3C\!-\!O\cdot \longrightarrow (CH_3)_2C\!=\!O \quad + \quad \dot{C}H_3$$

These primary radicals ($R\cdot$) abstract α-methylenic or α-methylic hydrogen from the polyisoprene chain to give polymeric radicals which then undergo combination, e.g.

$$\sim CH_2\!-\!\underset{\underset{CH_3}{|}}{C}\!=\!CH\!-\!CH_2\sim \; + \; R\cdot \; \longrightarrow \; \sim \dot{C}H\!-\!\underset{\underset{\cdot CH_3}{|}}{C}\!=\!CH\!-\!CH_2\sim \; + \; RH$$

$$2\sim\dot{C}H\!-\!\underset{\underset{CH_3}{|}}{C}\!=\!CH\!-\!CH_2\sim \; \longrightarrow \; \begin{array}{l}\sim CH\!-\!\underset{\underset{CH_3}{|}}{C}\!=\!CH\!-\!CH_2\sim \\ \sim CH\!-\!\underset{\underset{CH_3}{|}}{C}\!=\!CH\!-\!CH_2\sim\end{array}$$

There is also the possibility of cyclization along the main chain:

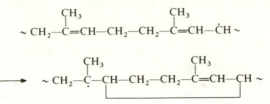

A small amount of chain scission also occurs, possibly as follows:

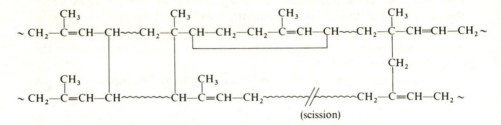

The various structural features of a typical network formed by peroxide vulcanization may be represented schematically as follows [2]:

20.2.5   Properties

As indicated earlier, the essential structural feature of a rubber vulcanizate is its flexible three-dimensional network. It is this arrangement which leads to the characteristic elastic behaviour. Comparative values for some properties of typical vulcanizates of common elastomers are given in Table 20.1. When natural rubber is stretched, crystallization of the highly regular chains occurs and the material shows a high tensile strength. The addition of fillers such as carbon black results in some increase in strength but the effect is not so marked as with elastomers for which stress-induced crystallization is not possible.

Natural rubber vulcanizates are, of course, insoluble unless previously degraded but they are appreciably swollen by a variety of organic solvents including aliphatic and aromatic hydrocarbons, chlorinated hydrocarbons,

Table 20.1 Typical values for various properties of vulcanizates of common diene elastomers [6]

	NR and IR	BR	SBR	SBS	NBR	CR
Gum vulcanizate (unfilled)						
Specific gravity	0.93	0.93	0.94	0.94–1.03	1.0	1.2
Tensile strength (MPa)	17–21	1.4–6.9	1.4–2.8	11–26	3.4–6.9	21–28
(1bf/in²)	2500–3000	200–1000	200–400	1700–3700	500–1000	3000–4000
Reinforced vulcanizate (filled)						
Tensile strength (MPa)	21–28	14–24	14–24	6.9–21	21–28	21–28
(1bf/in²)	3000–4000	2000–3500	2000–3500	1000–3000	3000–4000	3000–4000
Elongation at break (%)	300–700	300–700	300–700	500–1000	300–700	300–700
Hardness (Shore A)	20–100	30–100	40–100	40–85	30–100	20–100
Stiffening temperature (°C)	−30 to −45	−35 to −50	−20 to −45	−50 to −60	−15 to −55	−10 to −30

NR Natural rubber
IR High cis-1,4-polyisoprene (synthetic)
BR Polybutadiene
SBR Styrene-butadiene rubber
SBS Styrene-butadiene-styrene triblock copolymer
NBR Acrylonitrile-butadiene rubber
CR Polychloroprene

esters and ketones. Vulcanizates are generally unaffected by most aqueous reagents but strong oxidizing agents such as nitric acid cause degradation.

Natural rubber also undergoes reaction with atmospheric oxygen. This interaction is a major cause of deterioration in physical properties of the rubber on ageing and has been extensively investigated. In this book it is possible to give only a brief account of these investigations into an extremely complex process; more detailed reviews may be found elsewhere [2, 4, 7, 8].

As described in section 2.3.3, the oxidation of simple hydrocarbon polymers involves the formation of hydroperoxides. The oxidation of natural rubber (and other elastomers containing appreciable amounts of olefinic unsaturation) probably proceeds in a basically similar manner, although there are two important differences. Firstly, only about half of the oxygen absorbed by natural rubber is found as hydroperoxide; this observation is explained by the occurrence of intramolecular radical-addition reactions along the polymer chain. Secondly, natural rubber loses practically all of its useful elastomeric properties by the time only 1% by weight of oxygen has been absorbed; there is also a decrease in the molecular weight of the polymer. Thus scission reactions must be part of the degradation process. These features are illustrated in the following proposal for the oxidative degradation of natural rubber [9]:

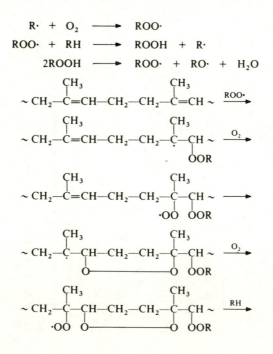

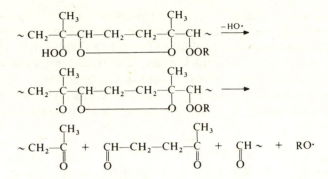

Evidence for this scheme is that levulinic aldehyde has been isolated from aged rubber. Further, it is known that β-peroxyalkoxy radicals are explosively unstable.

The reactions shown above involve the main chain, but it is possible that sulphur-containing cross-links present in a vulcanizate are also sites for oxidative attack. Little is known about such processes except by analogy with simpler sulphides. The general thermal instability of allylic sulphoxides suggests that monosulphide cross-links may undergo the following reactions:

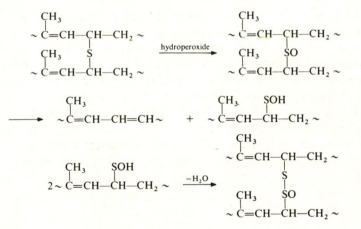

The net result of this sequence is loss of half the original cross-links concerned. Simple disulphides are readily oxidized by hydroperoxides to thiolsulphinates which undergo free radical disproportionation to disulphides and thiolsulphonates:

$$RS-SR \xrightarrow{\text{hydroperoxide}} RSO-SR$$

$$2RSO-SR \longrightarrow RS-SR + RSO_2-SR$$

Oxidation of disulphide cross-links in this way would not lead to a net loss of cross-links but the interchange would be manifested in creep and permanent set if the network were under strain.

As mentioned in section 20.2.3, natural rubber (in common with other diene elastomers) is readily attacked by ozone. The mechanism of the reaction is probably the same as that established for simple olefins [10]. In this case, the initial product is a π-complex which cleaves to form an aldehyde or ketone and a zwitterion. Several subsequent reactions may then occur, depending on the nature of the reactants and conditions. The zwitterion may dimerize or polymerize or react with the carbonyl compound to form an ozonide:

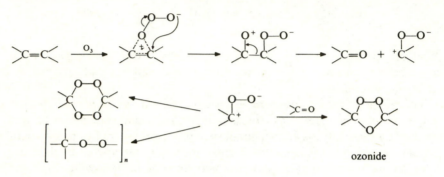

The immediate reduction in viscosity of rubber solutions on exposure to ozone and the immediate appearance of carbonyl groups in the solution favour the suggestion that double bond scission by ozone is essentially complete immediately after the primary attack of the reagent and is not, as was earlier thought, the result of subsequent breakdown of an ozonide. As noted previously, the addition of water to the products of ozonolysis of natural rubber gives rise to high yields of levulinic aldehyde. An important technological consequence of ozone attack is that even the small amounts of ozone in the atmosphere are sufficient to cause severe cracking in stretched natural rubber (and other diene elastomers) within a few months or even weeks.

20.2.6 Derivatives

Various chemical derivatives of natural rubber are of some commercial importance and these are considered in this section.

(a) Chlorinated rubber

Chlorinated rubber is typically prepared by treating a solution of masticated natural rubber in chloroform or carbon tetrachloride with chlorine at 80–100°C until sampling indicates the product has a chlorine content of about 65%. During this time, hydrogen chloride is evolved. After the passage of chlorine has been stopped, the solution is refluxed until the evolution of hydrogen chloride ceases; this results in a product of good stability. The chlorinated rubber is then isolated by precipitation with methanol.

The reaction of natural rubber and chlorine is complex and possibly involves the following sequence [11]:

(i) Cyclization

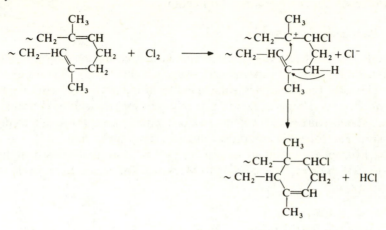

(ii) Substitution and double bond shift

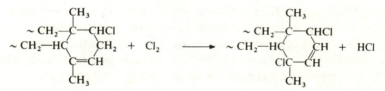

(iii) Addition

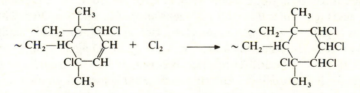

(iv) Substitution

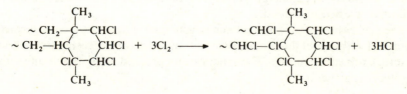

The above scheme has been proposed mainly on the basis of determinations of chlorine content, hydrogen chloride evolution and unsaturation content and the structures shown are somewhat speculative.

Chlorinated rubber films are impermeable to water and highly resistant to most aqueous reagents, including mineral acids and alkalis. The main use of the material is in corrosion resistant coatings.

(b)   Rubber hydrochloride

Typically, rubber hydrochloride is prepared by treating a solution of masticated natural rubber in benzene with hydrogen chloride at 10°C for about 6 hours. Usually, the material is isolated in film form, in which case the solution is neutralized and a plasticizer (e.g. tritolyl phosphate or dibutyl phthalate) added. The mixture is then cast on to a belt which passes through a chamber at 100°C. The solvent evaporates to leave a continuous film.

The hydrochlorination of natural rubber is a straightforward addition reaction which proceeds according to Markownikoff's rule. Some cyclization may also occur.

$$\sim CH_2-\underset{\underset{\displaystyle }{\overset{\displaystyle CH_3}{|}}}{C}=CH-CH_2 \sim \;+\; HCl \;\longrightarrow\; \sim CH_2-\underset{\underset{\displaystyle Cl}{|}}{\overset{\overset{\displaystyle CH_3}{|}}{C}}-CH_2-CH_2 \sim$$

Rubber hydrochloride film has low permeability to water vapour and is resistant to most aqueous reagents. It is, however, degraded by strong oxidizing acids whilst organic bases bring about dehydrohalogenation. The main use of rubber hydrochloride has been as packaging film but it has now been largely replaced by cheaper materials such as polyethylene.

(c)   Cyclized rubber

The cyclization of natural rubber may be brought about by treatment with strong acids (e.g. sulphuric and p-toluenesulphonic acids) or Friedel-Crafts catalysts (e.g. ferric chloride, stannic chloride and chlorostannic acid). In a typical process, the rubber is milled with the cyclizing agent (about 10%) at about 130° for 1–4 hours. As reaction proceeds there is a progressive loss in elasticity until eventually a hard, brittle material is formed. In an alternative procedure, natural rubber latex is stabilized by the addition of a non-ionic stabilizer and then heated with concentrated sulphuric acid at 100°C for 2 hours. The latex is coagulated by pouring it into boiling water and the solid is collected, washed and dried.

The cyclization of natural rubber is generally agreed to involve protonation and addition of the resulting carbenium ion to an adjacent double bond to form a 6-membered ring. This reaction can proceed along the chain to give a polycyclic structure:

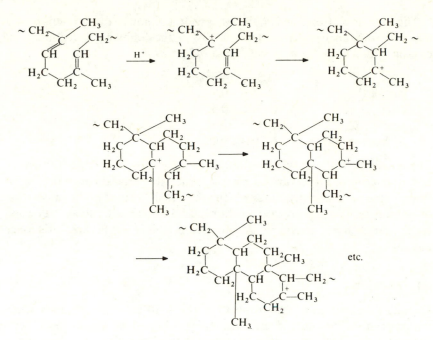

The reaction is eventually terminated by loss of a proton:

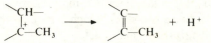

The average number of fused rings per sequence depends on the preparative conditions but is generally in the range 1.5–6.

Cyclized natural rubber is extremely resistant to aqueous acids and alkalis and finds use in surface coatings and adhesives.

(d)   Oxidized rubber

The oxidation processes which occur during normal ageing of natural rubber (section 20.2.5) involve the up-take of less than 1% by weight of oxygen and are relatively slow. In the presence of catalysts, materials with oxygen contents of about 5–15% can be obtained; such products are known as 'oxidized rubber'.

In one process, natural rubber is milled with a metal soap (e.g. cobalt linoleate) and then masticated in an internal mixer at 80°C in the presence of wood flour (which acts as a surface catalyst for oxidation). The mixture is then extracted with benzene and the solvent is removed by distillation to yield the oxidized rubber as a brittle brown solid.

The structure of oxidized rubber is largely unknown. In an analysis of a material with an oxygen content of 13.3% the following distribution of oxygen in functional groups was found [12]:

hydroxyl	$-OH$	4.0%
carbonyl	$C=O$	0.9
ester	$-CO_2R$	0.6
acid	$-CO_2H$	0.3
hydroperoxide	$-OOH$	0.06

The remaining 7.4% oxygen was not assigned but may be present in epoxy and ether groups. These various groups presumably arise through decomposition of the peroxidic structures which are initially formed in the interaction of natural rubber and oxygen (section 20.2.5). Oxidized rubber has good water and chemical resistance and finds use mainly in the field of surface coatings.

(e) Ebonite

Ebonite ('vulcanite' or 'hard rubber') is the hard product obtained when natural rubber (or other diene elastomer) is heated at about 150°C with approximately half its weight of sulphur until reaction is substantially complete.

Typically, the rubber (often cheap reclaimed rubber) is mixed on a mill with sulphur, diphenylguanidine (accelerator), magnesium oxide (activator) and ebonite dust (filler) to give a soft stock which is then compression moulded at 170–180°C until vulcanization is complete.

The structure of ebonite is basically similar to that of a 'soft' vulcanizate prepared by the use of a low accelerator to sulphur ratio (see Fig. 20.2) although, of course, there are far more sulphur-containing structures per unit volume and the material is essentially saturated. When ebonite is pyrolysed, various 2-methylthiophenes are obtained; this suggests that a substantial proportion of the sulphur is present in cyclic monosulphide groups attached to the tertiary carbon atom in the main chain:

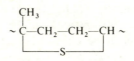

Ebonite shows very low water absorption and has good resistance to most aqueous reagents, except nitric acid. It is swollen by aromatic and chlorinated hydrocarbons. Ebonite has good electrical insulation characteristics and the major application is for car battery cases.

(ƒ) Graft copolymers

Graft copolymers of natural rubber and poly(methyl methacrylate) in latex form have been produced on a small scale. The rubber latex is treated with methyl methacrylate in the presence of an initiator (e.g. hydroperoxide activated by a polyamine). Generally, equal weights of rubber (dry weight) and monomer are used and the product consists of about 80% graft co-polymer, 10% polyisoprene and 10% poly(methyl methacrylate). The material gives rise to vulcanizates with excellent physical properties at high hardness levels.

20.2.7   Gutta percha and balata

Gutta percha is obtained from the latex of trees grown mainly in Malaysia and Indonesia. Commercial material contains about 10% of a resinous substance which can be extracted with petroleum ether to leave a hydrocarbon. This hydrocarbon is high *trans*-1,4-polyisoprene, the molecular weight of which has been determined to be 30 000.

Gutta percha is a tough horny substance which is essentially non-elastic. It is a crystalline material which can exist in two forms. As obtained from the tree it is in the α-form (m.p. 65°C); rapid cooling of the melt gives the β-form (m.p. 56°C). The β-form slowly reverts to the α-form. Gutta percha may be processed as a conventional thermoplastic material and is not normally subjected to vulcanization. Chemically, gutta percha closely resembles its geometrical isomer, natural rubber. Gutta percha has a lower water absorption than natural rubber and is a good dielectric. On account of these properties, gutta percha was once used for underwater cable insulation but it has now been displaced by newer materials.

Balata is obtained from the latex of trees grown mainly in Central America. The coagulated material consists of approximately equal amounts of high *trans*-1,4-polyisoprene and resins. De-resinated balata has been used as an alternative to gutta percha but it is no longer of any importance.

20.3   POLYISOPRENES

20.3.1   Development

As mentioned in section 20.2.3, Bouchardat in 1879 obtained a rubbery material by treating isoprene with hydrochloric acid. Polymerization was presumably due to the presence of impurities (such as peroxides) in the isoprene and several workers were unable to repeat this work. Although Bouchardat's product is often described as 'the first synthetic rubber', it is clear that the hydrogen chloride treatment of isoprene cannot be considered a method of preparing synthetic rubber. A more satisfactory procedure was developed in 1909 by Hofmann; the method simply involved heating isoprene in an autoclave at 90–200°C for 10–150 hours. Although yields were poor, sufficient quantities of material were prepared to ascertain that the polymer had useful technological properties resembling, but inferior to, those of natural rubber. In 1910 Matthews and Harries independently discovered that isoprene could be polymerized to a rubber-like product in high yield by sodium metal at moderate temperature. The product from this process was different from and inferior to that obtained by the simple heating process. However, the high cost of isoprene at the time precluded the development of an economical process for the production of synthetic rubber from this monomer. Accordingly, interest shifted to polymers from more accessible monomers, particularly butadiene and 2,3-dimethylbutadiene (section 20.4).

The polymerization of isoprene appears to have received no further attention until about 1927 when Ziegler began investigations into alkali-metal alkyls as polymerization catalysts. This work was concerned principally with butadiene rather than isoprene and was suspended during the Second World War. When the work was resumed after the war, interest centered mainly on the polymerization of ethylene by the so-called Ziegler-Natta catalysts (section 1.4.2.1). However, Ziegler informed the Goodrich-Gulf Chemicals Co. (USA) of his catalysts, and using such a catalyst Horne succeeded in preparing high cis-1,4-polyisoprene in 1954. Whilst this work was in progress, the Firestone Tire and Rubber Co. (USA) took up the study of diene polymerization with lithium catalysts which Ziegler had discontinued at the beginning of the Second World War and, also in 1954, this company was successful in obtaining high cis-1,4-polyisoprene. Thus two different methods of preparing 'synthetic natural rubber' were discovered almost simultaneously in two different laboratories. Both of these methods are used for the current large scale production of cis-1,4-polyisoprene, the total output of which is running at about 20% that of natural rubber. (Table 20.2.) It will be appreciated that a major contribution to the successful commercial development of cis-1,4-polyisoprene has been, of course, the availability of isoprene as a bulk chemical.

Synthetic trans-1,4-polyisoprene which is very similar to gutta percha is also now manufactured using stereospecific catalysts.

20.3.2 Raw materials

The successful synthesis of high cis-1,4-polyisoprene in 1954 aroused interest in the manufacture of large quantities of isoprene. Prior to this date only relatively small amounts of isoprene, which were used almost exclusively for the manufacture of butyl rubber (section 2.11), were available from naphtha

Table 20.2 Estimated 1986 world production of major rubbers [13]

Type of rubber	Production (million tonnes)
Natural rubber	4.43
Styrene-butadiene	5.02
Polybutadiene	1.28
Polyisoprene	0.92
Ethylene-propylene	0.46
Butyl	0.44
Polychloroprene	0.42
Acrylonitrile-butadiene	0.27
Others	0.44
Total	13.68

cracking operations. Several isoprene processes have been developed; the more important are described in this section.

(a) Thermal dehydrogenation

In this method, isopentane, isopentenes or isopentane/isopentene mixtures are dehydrogenated to give isoprene, e.g.

$$H_3C{-}\overset{\overset{\displaystyle CH_3}{|}}{C}H{-}CH_2{-}CH_3 \xrightarrow{-H_2} \begin{cases} H_3C{-}\overset{\overset{\displaystyle CH_3}{|}}{C}{=}CH{-}CH_3 \\[1em] H_2C{=}\overset{\overset{\displaystyle CH_3}{|}}{C}{-}CH_2{-}CH_3 \end{cases} \xrightarrow{-H_2} H_2C{=}\overset{\overset{\displaystyle CH_3}{|}}{C}{-}CH{=}CH_2$$

The processes are very similar in operation to those developed earlier for the production of butadiene from n-butane and n-butenes (section 20.4.2).

(b) Propylene process

Isoprene is obtained from propylene by the following route:

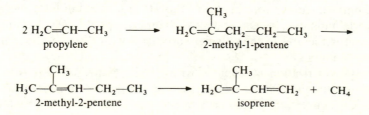

$$2\ H_2C{=}CH{-}CH_3 \longrightarrow H_2C{=}\overset{\overset{\displaystyle CH_3}{|}}{C}{-}CH_2{-}CH_2{-}CH_3 \longrightarrow$$
$$\text{propylene} \qquad\qquad \text{2-methyl-1-pentene}$$

$$H_3C{-}\overset{\overset{\displaystyle CH_3}{|}}{C}{=}CH{-}CH_2{-}CH_3 \longrightarrow H_2C{=}\overset{\overset{\displaystyle CH_3}{|}}{C}{-}CH{=}CH_2 + CH_4$$
$$\text{2-methyl-2-pentene} \qquad\qquad \text{isoprene}$$

In the first step, propylene is dimerized to 2-methyl-1-pentene by passage over a catalyst of tri-n-propylaluminium at about 200°C and 20 MPa (200 atmospheres). This product is then isomerized to 2-methyl-2-pentene by heating at 150–300°C in the presence of a silica-alumina catalyst. The final step in the process is the pyrolysis of the olefin to isoprene at 650–800°C in the presence of a free radical initiator such as hydrogen bromide. The isomerization step is necessary because pyrolysis of 2-methyl-1-pentene gives much poorer yields of isoprene than pyrolysis of 2-methyl-2-pentene.

(c) Isobutene process

Isoprene is prepared from isobutene by the following route:

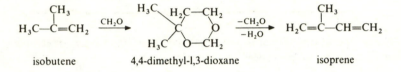

isobutene   4,4-dimethyl-1,3-dioxane   isoprene

Formalin is treated with isobutene in the presence of a strong mineral acid catalyst under mild conditions (about 95°C and a few atmospheres pressure) to give dimethyldioxane, which is separated by distillation under reduced pressure. The dimethyldioxane is then cracked to isoprene at a temperature below 750°C and at atmospheric pressure in the presence of a supported phosphoric acid catalyst.

Isoprene is a colourless liquid, b.p. 34°C.

20.3.3 Preparation

As mentioned previously (section 1.4.2(b)), polymerization of isoprene may be accomplished by a variety of methods. The choice of method has a profound effect on the microstructure of the resulting polymer (Table 1.2) and, consequently, on the technological properties of the product. The only polyisoprenes which have acceptable properties and which are of commercial importance are those with a high cis-1,4-content (although a small amount of high trans-1,4-polyisoprene is produced as a replacement for gutta percha). Three catalyst systems have been used for the production of cis-1,4-polyisoprene, namely coordination catalysts (e.g. titanium tetrachloride-tri-isobutylaluminium), lithium and alkyllithium compounds (e.g. butyllithium). The microstructures of the polymers produced with these catalysts are very similar (Table 1.2).

The polymerization of isoprene with Ziegler-Natta catalysts is typically carried out in an aliphatic solvent such as n-butane or n-pentane at about 50°C. The conversion of monomer to polymer is restricted to 80–90% since chain scission appears to occur at higher conversions, resulting in material of low molecular weight. The product is a slurry of polymer in the aliphatic solvent. This is agitated with aqueous methanol and then the mixture is allowed to separate into two layers. The lower layer, which contains methanol, water and catalyst residues is withdrawn. The remaining polymer slurry is washed with water and then stripped of solvent by treatment with steam to give an aqueous suspension of fine rubber crumbs. The polymer is collected, washed and dried.

The general mechanism of polymerization of vinyl compounds by co-ordination catalysts is described in section 1.4.2(a) and the same principles probably apply to the polymerization of conjugated dienes.

Details of production processes for the polymerization of isoprene by lithium and alkyllithium compounds do not appear to be available. However, the reaction proceeds under mild conditions (i.e. at room temperature or slightly above and at atmospheric pressure) and a typical process is probably not unlike that described above for co-ordination catalysts.

The general nature of polymerization of conjugated dienes by alkali metals and alkali metal alkyls is dealt with in section 1.4.2(b). As discussed, propagation involves an anionic chain-end with its associated metal cation. In the case of lithium, which has the lowest inherent ionic character of any of the alkali metals, there is a strong tendency to form co-ordination complexes. Thus it has been suggested that the diene co-ordinates with hybridized sp^3 orbitals of the lithium ion to give a complex which rearranges to a 6-membered ring transition complex from which a new C–C bond is produced. The process is repetitive and a polymer is formed:

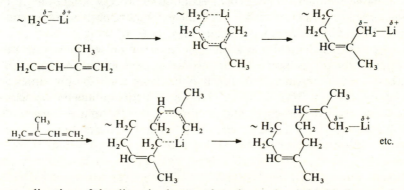

If co-ordination of the diene in the *cisoid* conformation is highly favoured, the resulting polymer will have a high *cis*-configuration; such is the case with isoprene. Also, in isoprene the highest electron density is on the 1,2-bond and it will be this bond in each reacting monomer molecule which is directed toward the positively charged metal ion on the growing chain. Thus a head-to-tail polymer is produced.

Synthetic cis-1,4-polyisoprene latices are also available. These are made by emulsification of polymer solutions followed by solvent stripping.

20.3.4 Properties

The properties of synthetic cis-1,4-polyisoprene are very similar to those of natural rubber and the two materials are essentially interchangeable. The raw synthetic product is softer than raw natural rubber (due to a reduced tendency for stress-induced crystallization) and is therefore more difficult to mill; the band tends to fall apart and compounding ingredients tend to remain in aggregates. On the other hand, the unvulcanized synthetic material flows more readily; this feature is advantageous in such processes as injection moulding. Synthetic cis-1,4-polyisoprene also differs from natural rubber in that it does not contain non-rubber ingredients which act as vulcanization activators; a slightly higher level of added activator is therefore required. At the present time, the synthetic product is somewhat more expensive than natural rubber.

20.4 POLYBUTADIENES

20.4.1 Development

As indicated in section 20.3.1, early attempts to prepare synthetic rubber understandably centred on the polymerization of isoprene but the high cost of the monomer quashed hopes of an economical process. Attention was thus soon turned toward more readily available monomers, particularly butadiene and 2,3-dimethylbutadiene. Polymers from both of these monomers were prepared by the thermal and sodium polymerization techniques described previously for polyisoprene. Of the two monomers, butadiene appeared to yield the better polymers but dimethylbutadiene could be prepared more easily and cheaply (from acetone). Thus polydimethylbutadiene was selected for commercial development and production (by thermal polymerization) was begun in 1911 by F. Bayer and Co. (Germany). This material, known as methyl rubber (because of the presence of one methyl group more per monomer unit than in natural rubber), was produced in relatively large amounts during the First World War. After the war when natural rubber again became available, methyl rubber production was terminated. Synthetic rubber research was restarted, again mainly in Germany, in 1926 and by this time butadiene was available at reasonable cost. I. G. Farbenindustrie A.G. developed a commercial process for sodium-polymerized butadiene and small amounts of this polymer were produced up until the end of the Second World War.

Meanwhile, however, a discovery was made which radically altered the approach to synthetic rubber production. In 1927, it was found that butadiene could be polymerized to a true latex in the presence of suitable emulsifiers and peroxide initiators. This discovery made possible the development of fast, efficient polymerization processes and eliminated many of the handling problems associated with bulk and solution methods. The polybutadienes obtained by emulsion polymerization, however, had disappointing properties and were difficult to process. Thus by about 1929 interest had shifted toward butadiene copolymers which had been found to have very promising properties. Butadiene copolymers, of course, formed the basis of the gigantic output of synthetic rubber which occurred in Germany and the USA during the Second World War but these developments are described in subsequent sections of this chapter.

Interest in straight polybutadienes was revived with the discovery of the stereospecific catalysts. Commercial processes using both co-ordination and alkyllithium catalysts were established by 1960 and polybutadiene is now a significant general purpose rubber (see Table 20.2).

20.4.2   Raw materials

Several processes have been used for the manufacture of butadiene. The more important processes in current use are outlined below.

(a)   Thermal cracking of petroleum fractions

Butadiene is obtained directly by the thermal cracking of petroleum fractions in the naphtha to gas oil range (section 2.2). This process is directed mostly towards the production of ethylene and propylene but butadiene is always a co-product. The butadiene is separated from the C_4 stream by extractive distillation, in which distillation is carried out in the presence of a solvent (such as acetonitrile, furfural or N-methylpyrrolidone) which decreases the volatility of the butadiene relative to the other C_4 compounds. Thermal cracking represents the major source of butadiene in Europe and Japan and is increasing in importance in the USA.

(b)   Dehydrogenation of butenes

Butadiene is obtained when n-butenes are dehydrogenated:

$$H_2C=CH-CH_2-CH_3 \longrightarrow H_2C=CH-CH=CH_2 \ + \ H_2$$
1-butene

$$H_3C-CH=CH-CH_3 \longrightarrow H_2C=CH-CH=CH_2 \ + \ H_2$$
2-butene

n-Butenes are mostly obtained from the catalytic cracking operations carried out on various petroleum fractions; thermal cracking processes usually give low yields of butenes. The dehydrogenation of n-butenes is carried out by mixing the feed with steam (which lowers the partial pressures of the reactants) and passing over a catalyst such as mixed calcium/nickel phosphate stabilized with chromium oxide at about 650°C. The butadiene is recovered by extractive distillation as described above.

(c)   Dehydrogenation of butane

Butadiene is obtained when n-butane is dehydrogenated:

$$H_3C—CH_2—CH_2—CH_3 \longrightarrow H_2C=CH—CH=CH_2 \; + \; 2H_2$$

Various dehydrogenation processes have been developed but the Houdry process is the most extensively used. In this process, butane is passed over a catalyst of 15–20% chromia on activated alumina at an initial temperature of 620–650°C and a pressure of 20 kPa (0.2 atmosphere). After 7–15 minutes the temperature falls and the catalyst becomes coked; the feed is then switched to a fresh reactor. The coke is burnt off the spent catalyst which is thus reactivated and re-heated ready for the next cycle. Butadiene is recovered from the product gas by the methods described above. Butane is the major source of butadiene in the USA at present. However, this route is becoming less important as the amount of butadiene available from naphtha and gas oil cracking increases.

Butadiene is a gas, b.p. −4°C.

20.4.3   Preparation

The polymerization of butadiene is exactly comparable to that of isoprene (section 20.3.3). Similarly, butadiene may be polymerized by a variety of methods, the choice of which determines the microstructure of the resulting polymer (Table 20.3). In the case of polybutadiene, of course, 1,2- and 3,4-addition cannot be distinguished. 1,4-, 1,2- and 1,2-/1,4-Polybutadienes are of commercial interest.

1,4-Polybutadienes are the most important polybutadienes, making up the butadiene rubbers. Three general types are produced:

(i) High cis-1,4-polybutadiene (with approximately 97% cis-1,4-content) using Ziegler-Natta catalysts such as dialkylaluminium chloride-cobalt (II) halide combinations.

(ii) Medium cis-1,4-polybutadiene (with approximately 92% cis-1,4-content) using Ziegler-Natta catalysts such as dialkylaluminium chloride-titanium tetraiodide combinations.

(iii) Low cis-1,4-polybutadiene (with approximately 40% cis-1,4-content) using alkyllithium catalysts.

Table 20.3 Effect of initiator on microstructure of polybutadiene [14]

Initiator	Structural units (%)		
	cis-1,4-	trans-1,4-	1,2-
Free radical	19	60	21
Na	10	25	65
Li	35	52	13
Li n-Bu	33	55	12
TiI_4–Al t-Bu_3	95	2	3
$CoCl_2$–$AlEt_2Cl$–pyridine	98	1	1
VCl_3–$AlEt_3$	0	99	1

Syndiotactic 1,2-polybutadienes (with 90–93% 1,2-content) are prepared using Ziegler-Natta catalysts such as trialkyl aluminium-cobalt(II) halide-ligand combinations [15].

1,2-/1,4-Polybutadienes (with 1–65% 1,2-content) are prepared using alkyllithium-Lewis base combinations [16].

20.4.4 Properties

Butadiene rubbers may be processed and vulcanized in much the same way as natural rubber, but compounding operations are rather more difficult on standard equipment since butadiene rubbers are very resistant to breakdown during mastication. Furthermore, high cis-1,4-polybutadiene bands on a mill only at temperatures below about 40°C whilst low cis-1,4-polybutadiene bands only above about 65°C. This deficiency is overcome by blending with natural or styrene-butadiene rubbers.

Polybutadiene vulcanizates (see Table 20.1 for typical properties) are superior to those of natural rubber with respect to resilience, heat build-up and abrasion resistance. These properties are particularly significant in tyres. On the other hand, polybutadiene vulcanizates have lower tensile strength and tear resistance and polybutadiene tyres have relatively poor road-adhesion in wet conditions. For these reasons and to aid processing, butadiene rubbers are generally used in blends with natural or styrene-butadiene rubbers; such blends usually contain less than 50% polybutadiene. Because of their use in tyre production, butadiene rubbers have become significant tonnage rubbers (Table 20.2). Butadiene rubbers are also widely used in the production of impact polystyrene (section 3.2.6).

Syndiotactic 1,2-polybutadienes are generally processed as thermoplastics rather than elastomers. They are mainly of interest as transparent packaging films with outstanding tear resistance and high permeability to gases and water vapour.

1,2-/1,4-Polybutadienes (or vinyl butadienes) show a structural resemblance to styrene-butadiene copolymers, with 1,2-units replacing styrene, and are of interest as rubbers. As the 1,2-content increases there is a reduction in chain segment mobility leading to poorer low-temperature properties; however, there are lower heat build-up characteristics. Blends of 1,2-/1,4-polybutadienes with other rubbers have been suggested for use in the production of tyres.

20.5 STYRENE-BUTADIENE COPOLYMERS

20.5.1 Development

As indicated in section 20.4.1, the investigation of butadiene copolymers was begun in about 1929 by I.G. Farbenindustrie A.G. (Germany) when it became apparent that emulsion-polymerized butadiene did not represent an entirely acceptable rubber. It was soon found that a copolymer with styrene gave a very satisfactory general purpose rubber. However, the then current low price of natural rubber did not make this type of synthetic rubber an attractive economic proposition. A few years later a political decision to make Germany self-sufficient in rubber resulted in revived interest in styrene-butadiene rubber; production was started in 1937 and reached about 100 000 tons per year by the end of the Second World War.

In the USA, the development of styrene-butadiene rubber followed a not dissimilar course. The early German work was available to the Standard Oil Co. (USA) and formed the basis of subsequent investigations. At first there was no urgency in this work, but the Japanese occupation in 1941 of the rubber plantations in the Far East made imperative the large-scale production of a general purpose rubber. As a result, a cooperative group of several chemical companies was formed under governmental direction. The first production of styrene-butadiene rubber (Government Rubber-Styrene or GR-S) was in 1942 and by 1945 output had risen to over 800 000 tons per year. The design and construction of the synthetic rubber plants in these three years represents an enormous chemical engineering achievement.

After the war when natural rubber became available again the consumption of styrene-butadiene rubber began to fall; however, the trend was reversed in 1949 with the advent of a copolymer made at low temperature. This product gives a passenger-tyre rubber superior to natural rubber and styrene-butadiene rubbers have remained the most important of the large-tonnage rubbers (Table 20.2). In the early 1960s, solution polymerized styrene-butadiene rubbers became available. These rubbers show further improvements in tyre performance. In 1965, styrene-butadiene thermoplastic elastomers were introduced.

20.5.2 Raw materials

The manufacture of styrene is described in section 3.2.2 and that of butadiene is dealt with in section 20.4.2.

20.5.3 Styrene-butadiene rubbers

(a) Preparation

The commercial preparation of styrene-butadiene rubbers is mainly by continuous emulsion polymerization. The following 'Mutual Recipe' was used in the USA during the Second World War and is still the basis of current processes:

Butadiene	75	parts by weight
Styrene	25	
Water	180	
Fatty acid soap	5.0	(emulsifier)
n-Dodecyl mercaptan	0.50	(modifier)
Potassium persulphate	0.30	(initiator)

Polymerization is carried out at 50°C and is allowed to continue for about 12 hours until conversion reaches 72%. Reaction is terminated at this point by the addition of hydroquinone in order to minimize the formation of cross-linked material. Unreacted monomers are removed, butadiene by flash-stripping under reduced pressure and styrene by steam-stripping. An antioxidant (e.g. phenyl-β-naphthylamine) is added to protect the rubber during drying and subsequent storage. The latex is then coagulated by the addition of a sodium chloride-sulphuric acid solution. The coarse crumb is washed with hot water and finally dried for about 2 hours at 80°C.

Processes of the above type are known as 'hot' processes to distinguish them from the 'cold' processes described below. Polymerization proceeds through a free radical mechanism, being initiated by thermal dissociation of the persulphate:

$$S_2O_8^{2-} \longrightarrow 2SO_4^- \cdot$$

The mercaptan present in the system has been found to affect the rate of polymerization and initiation may also stem from the following reaction:

$$S_2O_8^{2-} + C_{12}H_{25}SH \longrightarrow C_{12}H_{25}S \cdot + HSO_4^- + SO_4^- \cdot$$

However, the principal function of the mercaptan is to act as a chain-transfer agent:

$$R \cdot + C_{12}H_{25}SH \longrightarrow RH + C_{12}H_{25}S \cdot \quad (R \cdot = \text{polymeric radical})$$

In this way, the molecular weight of the final polymer is restricted to a level which permits processing without undue difficulty. The average molecular weight (M_n) of a typical styrene-butadiene rubber is about 100 000. The monomer units are randomly distributed in the copolymer; the contributions of the *cis*-1,4- *trans*-1,4- and 1,2-butadiene units are in the approximate ratio 18:65:17.

Most styrene-butadiene rubber is now made by 'cold' processes using redox initiator systems. A typical recipe is as follows:

Butadiene	72 parts by weight	
Styrene	28	
Water	180	
Fatty acid soap	4.5	(emulsifier)
Sodium naphthalene sulphonate	0.3	(stabilizer)
Potassium chloride	0.3	(stabilizer)
tert-Dodecyl mercaptan	0.20	(modifier)
p-Menthane hydroperoxide	0.063	
Ferrous sulphate (FeSO$_4$·7H$_2$O)	0.010	
Ethylenediamine tetraacetic		(initiator
acid sodium salt	0.050	system)
Sodium formaldehyde		
sulphoxylate	0.050	

Polymerization is carried out at 5°C and is allowed to continue for about 12 hours until conversion reaches 60%. Reaction is then stopped by the addition of N,N-dimethyldithiocarbamate. Stripping and isolation operations are carried out as in the 'hot' process.

In the above process, initiation is by the p-menthyloxy radical formed by a redox reaction of a complexed ferrous ion with the hydroperoxide; the sulphoxylate regenerates the ferrous ion from the ferric state. This sequence may be represented as follows:

$$ROOH \ + \ FeY^{2-} \longrightarrow RO· \ + \ FeY^- \ + \ OH^-$$

$$2FeY^- \ + \ HSO_2^- ·HCHO \ + \ 3OH^- \rightleftharpoons$$

$$2FeY^{2-} \ + \ SO_3^{2-}·HCHO \ + \ 2H_2O$$

$$(Y = EDTA \ ligand)$$

The mercaptan level in 'cold' processes is normally designed to yield a product with an average molecular weight (M_n) of about 100 000 but material destined to be oil-extended is allowed to attain a rather higher molecular weight. Compared to the 'hot' process, the 'cold' process leads to a more regular polymer with less branching and cross-linking and slightly higher trans to cis ratio. Also, a narrower molecular weight distribution is obtained. As a result, 'cold' rubbers give tyres with improved abrasion and cut-growth resistance. On the other hand, 'hot' rubbers are somewhat easier to process.

An increasing amount of styrene-butadiene rubber is being manufactured by solution processes using alkyllithium catalysts. Production techniques resemble those used for the polymerization of isoprene (section 20.3.3) and butadiene (section 20.4.3). There is a tendency for alkyllithium initiation to lead to block copolymers since the butadiene in the mixture polymerizes first to the virtual exclusion of the styrene. In order to obtain random copolymers it is necessary to add the butadiene incrementally so that the molar ratio of unreacted styrene to unreacted butadiene is always high. Solution styrene-butadiene rubbers have microstructures similar to those of the emulsion copolymers but show narrower molecular weight distribution and less long chain branching. These features cause the raw rubbers to be liable to cold flow on storage and more difficult to process. These defects may be overcome by adding a stannic compound to the polymerization system. Four 'living' polymer chains become attached to a tin atom to give a star-shaped molecule which shows reduced cold flow and improved processability. The weak C–Sn bonds eventually rupture during processing to regenerate the linear chains.

(b) Properties

Styrene-butadiene rubbers have characteristics very similar to those of natural rubber. They are compounded and processed in much the same way and may be vulcanized with either sulphur systems or peroxides. The styrene-butadiene chain is irregular and there is little tendency to crystallize on stretching (in contrast to natural rubber); thus gum vulcanizates have low tensile strength. However, reinforcement with carbon black leads to vulcanizates which resemble those of natural rubber (see Table 20.1.) and the two products are interchangeable in most applications. The styrene-butadiene vulcanizates show a high heat build-up on flexing which makes them unsatisfactory for the carcasses of large tyres. Solution styrene-butadiene rubbers are generally superior to emulsion styrene-butadiene rubbers in those mechanical properties which are important in tyres; in particular they have reduced rolling resistance.

The chemical resistance of vulcanizates of styrene-butadiene and natural rubbers is similar.Although the copolymers are less unsaturated, their rate of oxidation is slightly greater; however, they are more effectively stabilized by antioxidants and properly compounded styrene-butadiene vulcanizates are actually better than those of natural rubber.

The most important application of styrene-butadiene rubbers is in car tyres but there is also widespread use in mechanical and industrial goods. Latices find use in the carpet, paper-coating and adhesives industries. In these applications the styrene-butadiene copolymer is often carboxylated by the inclusion of up to 5% of a carboxyl-containing monomer (e.g. acrylic, methacrylic or itaconic acid) in the polymerization system. The presence of carboxyl groups permits vulcanization with sulphurless agents (e.g. metal oxides such as zinc oxide) to give vulcanizates with increased strength and resistance to swelling in hydrocarbon oils. Also, the increased polarity of the polymer increases affinity for polar substrates leading to higher adhesive strength.

20.5.4   High-styrene resins

Styrene-butadiene copolymers containing 50–85% styrene are also available
and are commonly known as 'high-styrene resins'. These products can be
vulcanized to give hard products but their chief use is in the reinforcement of
natural and styrene-butadiene rubbers. A typical use of these compositions is
in the production of shoe soles with good abrasion and flex resistance
although there is a trend away from such compositions towards thermoplastic elastomers.

20.5.5   Thermoplastic elastomers

As discussed previously, thermoplastic elastomers are materials which have
the functional properties of conventional vulcanized rubbers but which may
be processed as normal thermoplastics (see section 2.9). Effects of this kind
are shown by styrene-butadiene block copolymers. Two types of styrene-
butadiene block copolymers are produced commercially, namely triblock
and radial block copolymers. The triblock copolymers (denoted by SBS)
consist of a centre block of butadiene units with two terminal blocks of
styrene units. The radial block copolymers (denoted by $(SB)_nX$) consist of
three or more styrene-butadiene diblock copolymers radiating from a central

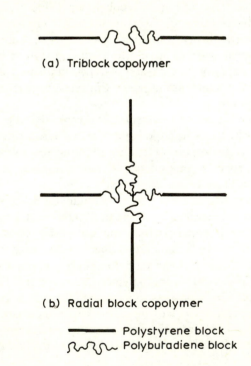

(a) Triblock copolymer

(b) Radial block copolymer

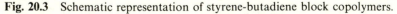

Polystyrene block
Polybutadiene block

Fig. 20.3   Schematic representation of styrene-butadiene block copolymers.

point; in each diblock the butadiene segment is innermost. The two types of structures are represented schematically in Fig. 20.3. In order to achieve an optimum balance between mechanical properties and processability, the overall styrene content of commercial products is within the range 20–40% by weight.

The triblock copolymers are prepared by anionic polymerization (using, for example, ethyllithium as initiator), utilizing the indefinite activity of the growing chains (see section 1.4.2(a)). Essentially, the initiator is added to a solution of styrene in an inert solvent (e.g. benzene); butadiene is then added, followed by more styrene. The copolymer is then precipitated with ethanol. The process may be represented as follows:

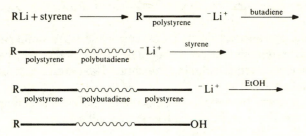

The radial block copolymers are also prepared by anionic polymerization. Firstly, a styrene-butadiene diblock which is active at the butadiene end is formed. A polyfunctional coupling agent such as silicon tetrachloride is then added to produce the radial structure. The process may be represented as follows:

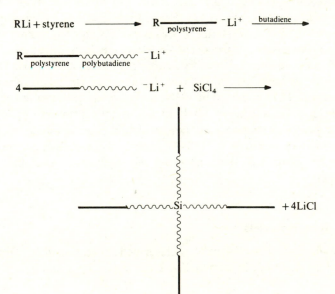

The characteristic properties of the styrene-butadiene thermoplastic elastomers stem from the inherent incompatibility of the polystyrene and polybutadiene blocks. In the bulk material the two different blocks aggregate into separate domains leading to a dispersion of polystyrene domains (as the lesser component) in a continuous matrix of polybutadiene. At ordinary temperatures the polystyrene domains are rigid and immobilize the ends of the polybutadiene segments, in effect serving both as cross-links and filler particles. Thus the system has the strength of a filler-reinforced vulcanized amorphous elastomer (see Table 20.1). At temperatures above the glass-transition temperature of polystyrene (100°C), the polystyrene domains are readily disrupted and the material may be melt-processed. The maximum service temperature of components fabricated from the copolymers is about 60°C.

The chemical properties of the block copolymers are similar to those of unvulcanized styrene-butadiene rubber. Thus the block copolymers are soluble in a range of solvents, including chlorinated and aromatic hydrocarbons, esters and ketones.

Styrene-butadiene thermoplastic elastomers find use in such applications as footwear soles and heels, elastic bands and adhesives.

20.6 ACRYLONITRILE-BUTADIENE COPOLYMERS

20.6.1 Development

As indicated in sections 20.4.1 and 20.5.1, an investigation of butadiene copolymers was undertaken by I.G. Farbenindustrie A.G. (Germany) and in 1930 it was found that acrylonitrile was a promising comonomer in that it gave rise to oil-resistant rubbers. The first small-scale manufacturing unit for acrylonitrile-butadiene rubber (nitrile rubber) was started in 1934. This work was available to the Standard Oil Co. (USA) under the terms of a research agreement with I.G. drawn up in 1930 and production began in the USA in 1939. Substantial quantities of acrylonitrile-butadiene rubber were produced under the government sponsored synthetic rubber programme (as Government Rubber-Acrylonitrile or GR-A). Acrylonitrile-butadiene copolymers have since maintained their position as significant speciality rubbers (Table 20.2).

20.6.2 Raw materials

The manufacture of acrylonitrile is described in section 6.2.5 and that of butadiene is dealt with in section 20.4.2.

20.6.3 Preparation

The commercial preparation of acrylonitrile-butadiene copolymers is generally by emulsion polymerization and the techniques used are entirely similar to those described for styrene-butadiene rubbers (section 20.5.3(a)). Both 'hot' and 'cold' processes may be employed but now only 'cold' processes (using temperatures in the range 5–30°C) are operated commercially. 'Low', 'medium' and 'high' solvent-resistant grades of copolymer, respectively containing approximately 25, 30 and 40% by weight of acrylonitrile, are commonly produced. The butadiene content of a copolymer is made up of *trans*-1,4-, *cis*-1,4- and 1,2-units; in, for example, a copolymer containing 72% butadiene and prepared at 28°C, the respective contributions are 77.5, 12.5 and 10%. Copolymers prepared by 'cold' and 'hot' processes differ in that the former are less branched and have a narrower molecular weight distribution.

Acrylonitrile-butadiene copolymers may also be prepared by the use of Ziegler-Natta catalysts. Catalysts such as a triethylaluminium, aluminium chloride and vanadium chloride combination give copolymers with a high degree of alternation. Such copolymers appear to have improved mechanical properties but have not attracted much commercial interest.

20.6.4 Properties

Acrylonitrile-butadiene rubbers are compounded and processed in much the same manner as natural rubber; they may be vulcanized with either conventional sulphur systems or peroxides. As with styrene-butadiene rubbers, gum vulcanizates have low strength but reinforcement with carbon black leads to vulcanizates with tensile strength approaching that obtained with natural rubber (see Table 20.1).

Acrylonitrile-butadiene rubbers are used almost invariably on account of their resistance toward non-polar solvents such as petrol and oils. The vulcanizates are, however, swollen by aromatic hydrocarbons and polar solvents such as chlorinated hydrocarbons, esters and ketones.

20.7 VINYLPYRIDINE RUBBERS

As indicated in section 20.4.1, butadiene copolymers had become of interest as potential synthetic rubbers in Germany by about 1929. Attention centred largely on styrene and acrylonitrile copolymers; it was also noted that

2-vinylpyridine was a promising comonomer but there were no commercial developments. 2-Vinylpyridine-styrene-butadiene terpolymers were investigated in the USA in the early 1940s in connection with the government sponsored synthetic rubber programme. These materials showed some improved properties, but were not of sufficient merit to warrant large-scale production.

Since this time, latices of vinylpyridine-styrene-butadiene terpolymers have found limited use in the treatment of textile fibres (such as tyre cords) to give improved adhesion to rubber. In this application, various vinylpyridines have been utilized but 2-vinylpyridine is the most commonly used. The vinylpyridine-styrene-butadiene weight ratio is typically 15:15:70.

An interesting characteristic of solid vinylpyridine rubbers is that much improved oil resistance may be obtained by including active organic halides (e.g. α-trichlorotoluene or tetrachloro-p-benzoquinone) in the formulation. These reagents lead to cross-linking by quaternization of the pyridine nitrogen, possibly as follows:

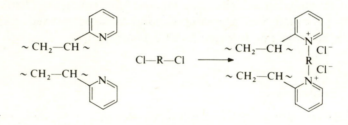

20.8   POLYCHLOROPRENES

20.8.1   Development

Chloroprene (2-chloro-1,3-butadiene) was first synthesized by Carothers and Collins (E.I. du Pont de Nemours and Co.) in 1930 during classical studies on acetylene. The compound was found to undergo spontaneous polymerization with the formation of a rubbery material. This product was shown to have good resistance to oil, heat and weathering and commercial manufacture of chloroprene rubber was started in 1932. Many types and grades of polymers and copolymers of chloroprene have since been introduced and the trivial generic name, *neoprene*, has been coined for these materials. Polychloroprene was produced during World War II under the USA government sponsored synthetic rubber programme (as Government Rubber-Monovinylacetylene or GR-M). Chloroprene polymers have since maintained their position as substantial speciality rubbers (Table 20.2).

20.8.2   Raw materials

The original method for the manufacture of chloroprene utilized acetylene but current processes largely are based on butadiene (section 20.4.2). These processes are more economical and avoid the use of monovinylacetylene, which is liable to undergo explosive decomposition.

(a)   Preparation of chloroprene from butadiene

In this method, the following route is used:

$$H_2C{=}CH{-}CH{=}CH_2 \xrightarrow{\text{Cl}_2} \underset{\text{3,4-dichloro-1-butene}}{\overset{\text{Cl}\quad\text{Cl}}{CH_2{-}CH{-}CH{=}CH_2}} \xrightarrow{-\text{HCl}} \underset{\text{chloroprene}}{\overset{\text{Cl}}{H_2C{=}C{-}CH{=}CH_2}}$$

butadiene

In the first step, butadiene is chlorinated in the vapour phase at 330–420°C and atmospheric pressure. The main products are 3,4-dichloro-1-butene and 1,4-dichloro-2-butene in approximately equal amounts. The latter material is then isomerized to the former by heating with a copper catalyst such as cuprous chloride. The 3,4-dichloro-1-butene is dehydrochlorinated by treatment with 10% aqueous sodium hydroxide at 85°C. Chloroprene is isolated by distillation under reduced pressure in the presence of polymerization inhibitors.

(b)   Preparation of chloroprene from acetylene

In this method, the following route is used:

$$\underset{\text{acetylene}}{HC{\equiv}CH} \xrightarrow{\text{CuCl,NH}_4\text{Cl}} \underset{\text{monovinylacetylene}}{H_2C{=}CH{-}C{\equiv}CH} \xrightarrow{\text{HCl}} \underset{\text{chloroprene}}{\overset{\text{Cl}}{H_2C{=}CH{-}C{=}CH_2}}$$

In the first step, acetylene is dimerized to monovinylacetylene by passage into an aqueous solution of cuprous chloride and ammonium chloride at 80°C. The purified monovinylacetylene is then hydrochlorinated by concentrated hydrochloric acid containing cuprous chloride at 30–60°C.

Chloroprene is a colourless liquid, b.p. 59°C.

20.8.3   Preparation

The commercial preparation of chloroprene polymers is invariably by emulsion polymerization. A typical recipe might be as follows [17]:

Chloroprene	100 parts by weight
Water	150
Rosin	4
Sodium hydroxide	0.8 (stabilizer)
Methylenebis(naphthalenesulphonic acid) sodium salt	0.7 (emulsifier)
Sulphur	0.6 (modifier)
Potassium persulphate	0.2–1.0 (initiator)

Polymerization is carried out at 40°C and is allowed to continue until a 90% conversion is reached. The latex is cooled to 20°C and tetraethylthiuram disulphide is added; the product is then allowed to stand for about 8 hours until the desired molecular weight is reached. The latex is then acidified with acetic acid just short of coagulation, which is ultimately accomplished by freezing on the surface of a cooled (about -15°C) rotating drum partially immersed in the latex. The resulting film of polymer is stripped from the roll, washed, dried in air at 120°C and cut into small pieces. This method of isolation is used because polychloroprene is tacky and it is difficult to coagulate the latex directly to crumb.

Polychloroprene obtained by free radical processes such as that described above consists essentially (about 85%) of linear sequences of *trans*-1,4-units with minor amounts of *cis*-1,4-(10%), 1,2-(1.5%) and 3,4-units (1.0%). The contribution of the *trans*-1,4-structure increases as the polymerization temperature is lowered; for example, it is about 95% when a temperature of -40°C is used.

The pendant unsaturated groups along the polychloroprene chain are potential sites for branching and cross-linking and the probability of such reactions increases as the ratio of polymer to monomer in the system increases. Thus at conversions above about 70%, unmodified polychloroprene is quite highly cross-linked and is difficult to process. In order to obtain polymers of limited molecular weight which are readily processable, a sulphur-tetraethylthiuram disulphide modification is used. In this modification, carried out as described above, the chloroprene is polymerized in the presence of sulphur to give a copolymer of the following type:

$$\sim S_x \left[CH_2 - \overset{\displaystyle Cl}{\underset{\displaystyle |}{C}} = CH - CH_2 \right]_n S_x \left[CH_2 - \overset{\displaystyle Cl}{\underset{\displaystyle |}{C}} = CH - CH_2 \right]_n S_x \sim$$

where x is 2–6 and n is, on average, 80–$100x$. When the copolymer is treated with the disulphide, interchange reactions occur at the polysulphide links to give stable thiuram-terminated polymers of reduced molecular weight. The following type of free radical mechanism has been suggested [18] for this

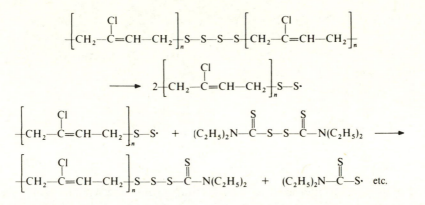

cleavage (cf. section 19.6): However, since the latex is alkaline at this stage (sodium hydroxide is included in the emulsifying system), the thiol anion, $(C_2H_5)_2NC(S)S^-$, may be the active species. The products obtained by this process have average molecular weights of about 100 000. In addition to sulphur-modified polychloroprenes (the so-called G type neoprenes) there are available non-sulphur modified products (W type neoprenes). In the production of these polymers, the average molecular weight is kept to about 200 000 by the use of mercaptan transfer agents. Copolymers of chloroprene with various monomers (e.g. 2,3-dichloro-1,3-butadiene and acrylonitrile) are also manufactured to give products with specific properties such as less tendency to crystallize, improved adhesion and superior oil resistance.

20.8.4   Vulcanization

Polychloroprenes differ from other polydienes in that conventional sulphur vulcanization is not very effective. The double bonds are deactivated by the electronegative chlorine atoms and direct reaction with sulphur is limited. The vulcanization of polychloroprenes is normally achieved by heating at about 150°C with a mixture of zinc and magnesium oxides; W type neoprenes also require an organic accelerator (commonly either a diamine or ethylene thiourea) but G types cure quite rapidly without acceleration. The mode of reaction has not been established with certainty, but it is generally supposed that cross-linking occurs at the tertiary allylic chloride structures generated by 1,2-polymerization (see section 20.8.3) and that a 1,3-allylic shift is the first step. The metal oxides may lead to ether cross-links as follows:

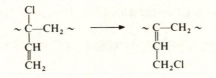

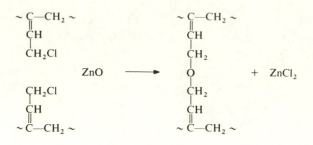

The zinc chloride produced acts as a catalyst and can lead to scorching (premature vulcanization). The tendency to scorch is reduced by the presence of magnesium oxide, which is thought to remove zinc chloride by the following reaction:

$$2ZnCl_2 + MgO + H_2O \longrightarrow 2Zn(OH)Cl + MgCl_2$$

When diamines are included in the vulcanizing system, a similar cross-linking reaction can occur:

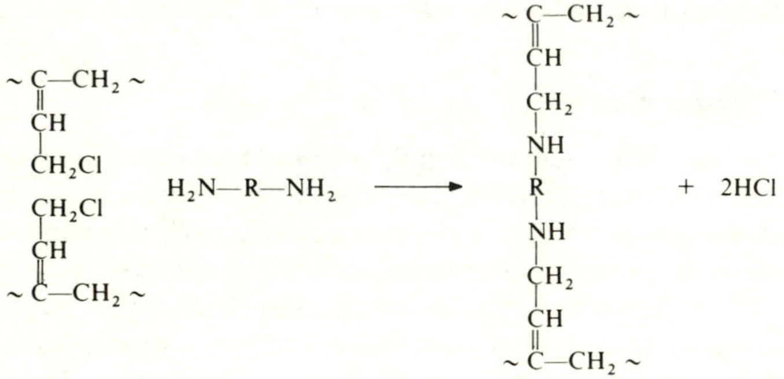

Evidence for cross-links of the above type is that polymers cannot be vulcanized after they have been treated with a monoamine (e.g. aniline), which removes allylic chloride. Further, excess of diamine in the vulcanizing system results in a poor vulcanizate; in this case it may be supposed that some amine molecules react with only one chlorine atom instead of the two required for cross-link formation.

For ethylene thiourea, the following cross-linking process has been proposed [19]:

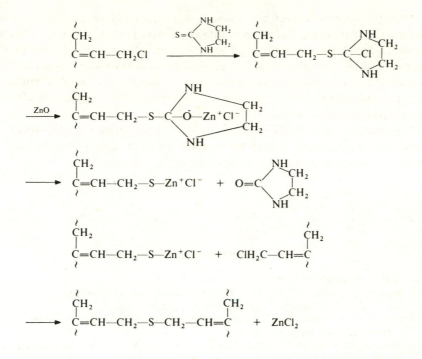

Evidence for this mechanism is that ethylene urea has been isolated from an ethylene thiourea-accelerated vulcanizate. It may be noted that although ethylene thiourea has been widely used in the past to accelerate the vulcanization of polychloroprenes, the more recent recognition of its toxicity has led to a decline in its use.

20.8.5 Properties

Polychloroprenes give vulcanizates which are broadly similar to those of natural rubber in physical strength and elasticity (see Table 20.1). However, the polychloroprenes show much better heat resistance in that these physical properties are reasonably well maintained up to about 150°C in air. The heat resistance and compression set resistance of the G type polymers are inferior to those of the W type polymers due to the presence of labile polysulphide bonds. On the other hand, the G types show less nerve on processing, better building tack and higher resilience, tear strength and resistance to stress cracking. As might be expected from the highly regular structure of polychloroprene, normal grades readily crystallize and become stiff when cooled below − 10°C. Crystallization-resistant grades of polychloroprene are available for low temperature use but fast-crystallizing grades of polymer are used for quick-grab adhesives.

Polychloroprene vulcanizates have a high order of oil and solvent resistance (but are generally less resistant than those of nitrile rubber). Aliphatic solvents have little effect although aromatic and chlorinated solvents cause some swelling. The presence of chlorine-substituted double bonds makes the polymer rather unreactive and leads to good resistance to most chemicals, oxygen and ozone. The high chlorine content of the polymer results in products which are generally self-extinguishing.

Polychloroprene rubbers find use in such applications as cable-sheaths, hose and weather strips. Latices are used in the production of dipped goods, such as gloves, adhesives and corrosion-resistant coatings.

REFERENCES

1. Schulz, K. *et al.* (1956) *Makromol. Chem.*, **21**, 13.
2. Bateman, L. *et al.* (1963) in *The Chemistry and Physics of Rubber-like Substances*, (ed. L. Bateman), Maclaren and Sons Ltd., London, ch. 15.
3. Porter, M. (1965) in *The Chemistry of Sulfides*, (ed. A. V. Tobolsky), Interscience Publishers, New York, p. 165.
4. Barnard, D. *et al.* (1970) in *Encyclopedia of Polymer Science and Technology*, Vol. 12, Interscience Publishers, New York.
5. Brydson, J. A. (1978) *Rubber Chemistry*, Applied Science Publishers Ltd, London, p. 198.
6. Rodriguez, F. (1970) *Principles of Polymer Systems*, McGraw-Hill Book Company, New York, p. 528.
7. Smith, J. F. (1961) in *The Applied Science of Rubber*, (ed. W. J. S. Naunton), Edward Arnold (Publishers) Ltd, London, ch. 13.
8. Brydson, J. A. (1978) *Rubber Chemistry*, Applied Science Publishers Ltd, London, p. 260.
9. Mayo, F. R. (1960) *Ind. Eng. Chem.*, **52**, 614.
10. Criegee, R. (1951) *Abstracts*, 120th Meeting of Amer. Chem. Soc., p. 22M; Bailey, P. S. (1958) *Chem. Rev.*, **58**, 925.
11. Bloomfield, G. F. (1943) *J. Chem. Soc.*, 289.
12. Naylor, R. F. (1945) *I.R.I. Trans.*, **20**, 45.
13. Private communication from International Rubber Study Group, July 1987.
14. Cooper, W. and Vaughan, G. (1967) in *Progress in Polymer Science*, Vol. I, (ed. A. D. Jenkins), Pergamon Press, Oxford, ch. 3.
15. Elias, H-G. and Vohwinkel, F. (1986) *New Commercial Polymers 2*, Gordon and Breach Science Publishers, New York, p. 150.
16. Elias, H-G. and Vohwinkel, F. (1986) *New Commercial Polymers 2*, Gordon and Breach Science Publishers, New York, p. 80.
17. Gintz, F. P. (1968) in *Vinyl and Allied Polymers*, Vol. I, (ed. P. D. Ritchie), Iliffe Books Ltd, London, p. 241.
18. Klebanski, A. L. *et al.* (1958) *J. Polymer Sci.*, **30**, 363; (1959) *Rubber Chem. Technol.*, **32**, 588.
19. Pariser, R. (1960) *Kunstoffe*, **50**, 623.

BIBLIOGRAPHY

Whitby, G. S. (ed.) (1954) *Synthetic Rubber*, John Wiley & Sons, Inc., New York.

Davies, B. L. and Glazer, J. (1955) *Plastics Derived from Natural Rubber*, Plastics Institute, London.

Naunton, W. J. S. (ed.) (1961) *The Applied Science of Rubber*, Edward Arnold (Publishers) Ltd, London.

Bateman, L. (ed.) (1963) *The Chemistry and Physics of Rubber-like Substances*, Maclaren and Sons Ltd, London.

Hofmann, W. (1964) *Rubber Review for 1963: Nitrile Rubber*, Rubber Chem. Technol., **37**.

Murray, R. M. and Thompson, D. C. (1964) *The Neoprenes*, E. I. du Pont de Nemours & Co., (Inc), Wilmington.

Kennedy, J. P. and Törnqvist, E. G. M. (eds) (1968, 1969) *Polymer Chemistry of Synthetic Elastomers*, Part I, Part II, Interscience Publishers, New York.

Ritchie, P. D. (ed.) (1968) *Vinyl and Allied Polymers*, Vol. I, Iliffe Books Ltd, London.

Saltman, W. M. (ed.) (1977) *The Stereo Rubbers*, John Wiley and Sons, New York.

Brydson, J. A. (1978) *Rubber Chemistry*, Applied Science Publishers Ltd, London.

Blackley, D. C. (1983) *Synthetic Rubbers: Their Chemistry and Technology*, Applied Science Publishers, London.

INDEX